Translations of
MATHEMATICAL MONOGRAPHS

Volume 20

Statistical Problems with Nuisance Parameters

Ju. V. Linnik

American Mathematical Society
Providence, Rhode Island

СТАТИСТИЧЕСКИЕ ЗАДАЧИ
С МЕШАЮЩИМИ ПАРАМЕТРАМИ

Ю. В. ЛИННИК

ТЕОРИЯ ВЕРОЯГНОСТЕЙ
И МАТЕМАТИЧЕСКАЯ СТАТИСТИКА

Издательство "Наука"
Главная Редакция
Физико-Математической Литературы
Москва 1966

Translated from the Russian by Scripta Technica

2000 *Mathematics Subject Classification.* Primary 62–XX.

Library of Congress Cataloging-in-Publication Data

Linnik, ĨU. V. (ĨUriĭ Vladimirovich), 1915–1972.

Statisticheskie zadachi s meshaiushchimi parametrami. English

Statistical problems with nuisance parameters, by Ju. V. Linnik. [Translated from the Russian by Scripta Technica].

p. cm. — (Translations of mathematical monographs, v. 20)

Includes bibliographical references.

1. Statistical hypothesis testing. 2. Estimation theory. I. Title. II. Series.

QA277.L513
519′.1 67030101

ISBN 978-0-8218-4687-2

∞ The paper used in this book is acid-free and falls within the guidelines
established to ensure permanence and durability.
Visit the AMS home page at http://www.ams.org/

10 9 8 7 6 5 4 3 2 1 13 12 11 10 09 08

PREFACE

The present book is devoted to the analytic theory of elimination of nuisance parameters in the testing of statistical hypotheses and to the theory of unbiased estimates. Our attention is concentrated on the analytic properties of tests and estimates and on the mathematical foundations for obtaining a test or unbiased estimate that is optimal in some sense or other. It does not, however, include either computational algorithms (which in many cases reduce to certain forms of linear programming) or tables. Thus the book does not contain individual statistical recipes for problems with nuisance parameters, but rather attempts to point out procedured for constructing such recipes.

In the introduction and later, we recall certain standard theorems on σ-algebras, probabilistic measures, and statistics. In Chapter I, we treat multiple Laplace transforms and describe the simpler properties and applications of analytic sheaves along the lines developed by H. Cartan. In later chapters these properties will be applied to the theory of exponential families. Chapter II gives the fundamentals of the theory of sufficient statistics for distributions in Euclidean spaces and exponential families associated with them (for repeated samples). Chapter III presents some of the problems themselves with nuisance parameters. Chapter IV treats the theory of similarity following J. Neyman, E. Lehmann, and H. Scheffé. Chapters V and VII–X discuss the recent researches of statisticians at Leningrad University in the theory of similar tests and unbiased estimates, particularly in connection with the Behrens-Fisher problem. In Chapter VI, an exposition is given of the remarkable method of R. A. Wijsman; however, this method does not yield all desirable tests. The role of the theory of sheaves of ideals of functions as an analytic foundation of the theory of similar tests and unbiased estimates for imcomplete exponential families is clarified in Chapters V and VII. Here exponential families are considered not only for repeated samples but for other cases as well.

In Chapter XI, an exposition is given of the problem of many small samples, and, in particular, of the researches of A. A. Petrov.

At the end of the book are several unsolved problems, which constitute only a small portion of the esthetically pleasing and varied problems that arise in analytical statistics. Our purpose of the present book is to draw the attention of persons interested in mathematical statistics to its analytical aspects.

A. M. Kagan, I. L. Romanovskaja, and V. N. Sudakov had a share in the writing of this book. Sections 2 and 3 of Chapter VII were written by the author in collaboration with A. M. Kagan, and section 4 of Chapter VIII with I. L. Romanovskaja. Section 2 of Chapter X was written by V. N. Sudakov. A considerable amount of help in the writing of Chapter I was provided by N. M. Mitrofanova and V. L. Eĭdlin.

I wish to express my gratitude to O. I. Rumjanceva and S. I. Čirkunova for their great help in the preparation of the manuscript.

Ju. V. Linnik

PREFACE TO THE AMERICAN EDITION

The American translation of this book takes account of several corrections of misprints and author's errors that were noticed by readers or by the author. It also contains a supplement to the book written by A. M. Kagan and V. P. Palamodov, expounding their important contributions published recently in "Teorija Verojatnostei i ee Primenenija". The answers to several questions raised at the end of the book are provided by the supplement. This new material includes a considerable advance in the theory of nonsequentially verifiable functions, the construction of all randomized similar tests for the Behrens-Fisher problem, and important progress in the estimation theory for incomplete exponential families, based on the introduction into statistics of the elements of homological algebra (in particular, flat modules).

The analytical sheaf theorems on which a large part of the book is based are replaced in the supplement by the Hörmander-Malgrange theory of linear differential operators with constant coefficients. This theory enables us to solve problems involving convex supports, rather than merely the polygonal ones discussed in the book. Thus we can now construct all similar tests for a linear hypothesis with unknown variances (least square method with unknown observation weights) and for many other problems of testing hypotheses and unbiased estimation. The optimization problems are thus reduced to purely analytic (variational) ones.

I am very grateful to the American Mathematical Society for publishing a translation of my book with the supplement. It is my pleasant duty to thank S. H. Gould and G. L. Walker for their interest in my book.

Ju. V. Linnik

TABLE OF CONTENTS

INTRODUCTION

§1. PROBABILITY MEASURES. INTEGRATION. STATISTICS

In the present section we shall review the basic material from measure theory, which usually constitutes the foundation of the theory of statistical inferences. For the proofs of the corresponding theorems, we shall refer the reader to certain well-known texts. We shall, however, give the proofs of the less widely used theorems. We employ the usual set-theory notation.

As a rule, we consider a space $\mathfrak{X}$ of elementary events. Along with this space, we also consider the σ-algebra $\mathfrak{A}$ of measurable subsets A of $\mathfrak{X}$. The pair $(\mathfrak{X}, \mathfrak{A})$ is called a *measurable space*. Let us consider a countably additive (though not necessarily finite) nonnegative set function $\mu = \mu(A)$ defined on the sets $A \in \mathfrak{A}$. We take $\mu(\emptyset) = 0$, where $\emptyset$ denotes the empty set. If $\mu(\mathfrak{X}) = 1$, then μ is a probability measure and we shall usually denote it by the letter P. If $\mathfrak{A}$ contains a countable family of disjoint sets $A_1, A_2, \cdots$ such that $\mu(A_i) < \infty$ (for $i = 1, 2, \cdots$) and $\bigcup_i A_i = \mathfrak{X}$ then the measure μ is said to be σ-finite. We shall usually make a given μ *complete*, without changing the notation for it, by supplementing, if necessary, the σ-algebra $\mathfrak{A}$ with all subsets of sets of zero measure and assigning to them the measure 0. The most important particular measures that we shall be using are Lebesgue measure and "counting measure".

Lebesgue measure is defined for $\mathfrak{X} = E_n$ (that is, n-dimensional Euclidean space). Here the σ-algebra is composed of the Borel sets generated by all the parallelepipeds $x_i \in (a_i, b_i]$ (for $i = 1, 2, \cdots, n$), where $x_1, \cdots, x_n$ are the coordinates of a point in E_n. This is the minimal σ-algebra containing all such parallelepipeds. For these sets, the measure is simply their geometric volume.

Counting measure is defined on a countable set $\mathfrak{X}$. The σ-algebra $\mathfrak{A}$ is the family of all subsets of $\mathfrak{X}$. For $A \in \mathfrak{A}$, $\mu(A)$ is defined as the number of elements of A, so that $\mu(A) = \infty$ if A is an infinite set.

Consider a measurable space $(\mathfrak{X}, \mathfrak{A})$. Let $\mathfrak{T}$ denote a space other than $\mathfrak{X}$. Let T denote a mapping defined on $\mathfrak{X}$ into $\mathfrak{T}$:

$$T: \mathfrak{X} \to \mathfrak{T}.$$

Let the space $\mathcal{T}$ also be provided with a σ-algebra $\mathcal{B}$ of measurable subsets B. The mapping T is said to be measurable if the pre-images under T of measurable sets are measurable, that is, if $B \in \mathcal{B}$ implies $T^{-1}(B) \in \mathcal{A}$. In the case of the measurable space $(\mathcal{T}, \mathcal{B})$, we customarily define a measure ν by $\nu(B) = \mu(T^{-1}(B))$ for $B \in \mathcal{B}$.

When we interpret the original measurable space $(\mathcal{X}, \mathcal{A})$ as a sample space, we shall call the measurable mapping $T: \mathcal{X} \to \mathcal{T}$ a *statistic*. In the majority of the particular cases that we shall consider, $\mathcal{X}$ and $\mathcal{T}$ will be Euclidean spaces, while the sample element in $\mathcal{X}$ and the value of the statistic T will be random vectors. In particular, if $\mathcal{T}$ is the real axis, the mapping T is a measurable function. Here we may consider ordinary integration with respect to the measure μ. If μ is the measure corresponding to the measurable space $(\mathcal{X}, \mathcal{A})$ and ϕ is a nonnegative measurable function, then the expression

$$\nu(A) = \int_A \varphi \, d\mu$$

for $A \in \mathcal{A}$, defines a new measure ν on $(\mathcal{X}, \mathcal{A})$. We write $\phi = d\nu/d\mu$ and we call ϕ the *Radon-Nikodym derivative* of the measure ν with respect to μ. Here, if $\nu(\mathcal{X}) = 1$ and ν is a probability measure, then ϕ is called the *probability density* for ν with respect to μ. The function ϕ is uniquely determined up to its values on sets of measure zero.

Conditions for existence of the function ϕ are given by the well-known Radon-Nikodym theorem:

Theorem 0.1.1. *Suppose that μ and ν are σ-finite measures over $(\mathcal{X}, \mathcal{A})$. A necessary and sufficient condition for existence of the Radon-Nikodym derivative of the measure ν with respect to μ is that the measure ν with respect to μ be absolutely continuous, that is, that $\nu(A) = 0$ for all $A \in \mathcal{A}$ such that $\mu(A) = 0$.*

In what follows the concept of a product measure $\rho = \mu \times \nu$ will be useful. Suppose that $(\mathcal{X}, \mathcal{A}, \mu)$ and $(\mathcal{Y}, \mathcal{B}, \nu)$ are two measurable spaces equipped with measures μ and ν respectively. Consider the Cartesian product $\mathcal{X} \times \mathcal{Y}$ of $\mathcal{X}$ and $\mathcal{Y}$ and the Cartesian product $\mathcal{A} \times \mathcal{B}$ of the σ-algebras $\mathcal{A}$ and $\mathcal{B}$, this last product meaning the minimal σ-algebra containing all sets $A \times B$ for $A \in \mathcal{A}$ and $B \in \mathcal{B}$. Thus we obtain a measurable space $(\mathcal{X} \times \mathcal{Y}, \mathcal{A} \times \mathcal{B})$. The product measure for this space is defined as

$$\rho(A \times B) = \mu(A) \cdot \nu(B).$$

For probability measures this definition corresponds to the concept of independent events.

In connection with this concept, we shall need the well-known theorem of Fubini.

Theorem 0.1.2. *Suppose that* μ *and* ν *are* σ*-finite measures over measurable space* $(\mathfrak{X}, \mathfrak{A})$ *and* $(\mathfrak{Y}, \mathfrak{B})$ *respectively. Suppose that* $\rho = \mu \times \nu$. *If a function* $\phi(x, y)$ *is integrable with respect to the measure* ρ, *it is integrable with respect to* μ *for almost all (in the sense of the measure* ν*) values of* y. *Furthermore, the function* $\int f(x, y)\,d\mu(x)$ *is integrable with respect to* ν, *and*

$$\int_{\mathfrak{X}\times\mathfrak{Y}} f(x, y)\,d\rho(x, y) = \int_{\mathfrak{Y}} d\nu(y) \int_{\mathfrak{X}} f(x, y)\,d\mu(x). \tag{0.1.1}$$

Proofs of these two theorems can be found, for example, in the book by Halmos [69].

§2. SOME PROPERTIES OF STATISTICS

Let us examine some properties of statistics that will be needed below. Our exposition follows to some extent the well-known book by Lehmann [36].

Suppose that a statistic T maps a measurable space $(\mathfrak{X}, \mathfrak{A})$ into a measurable space $(\mathfrak{T}, \mathfrak{B})$. If $B \in \mathfrak{B}$, then $T^{-1}(B) \in \mathfrak{A}$, but the family of measurable sets $\{T^{-1}(B)\} = \mathfrak{A}_0$, although it is a subset of the family $\mathfrak{A}$, does not necessarily coincide with it. Obviously the family $\mathfrak{A}_0$ constitutes a σ-subalgebra, known as the *σ-subalgebra induced by the statistic* T. Let us consider the measurable space $(\mathfrak{X}, \mathfrak{A}_0)$ and assign the measurable real functions in it to the same functions for the measurable space $(\mathfrak{T}, \mathfrak{B})$. This enables us to consider the σ-algebras $\mathfrak{A}_0$ and $\mathfrak{B}$ in a certain sense equivalent with respect to the statistic T.

Theorem 0.2.1. *Suppose that a statistic* $T\colon (\mathfrak{X}, \mathfrak{A}) \to (\mathfrak{T}, \mathfrak{B})$ *induces a* σ*-subalgebra* $\mathfrak{A}_0$. *Let* f *denote a real* $\mathfrak{A}$*-measurable function. The function* f *is* $\mathfrak{A}_0$*-measurable if and only if there exists a* $\mathfrak{B}$*-measurable function* g *such that*

$$f(x) = g(T(x)) \tag{0.2.1}$$

for all x.

For the proof, see the book by Lehmann [36].

Another theorem that we shall find useful is

Theorem 0.2.2. *Let* $T\colon (\mathfrak{X}, \mathfrak{A}) \to (\mathfrak{T}, \mathfrak{B})$ *denote a measurable mapping, let* μ

denote a σ-finite measure over $(\mathfrak{X}, \mathfrak{A})$, let g denote a real measurable function of $t \in \mathfrak{T}$, and let μ^* denote the measure defined over $(\mathfrak{T}, \mathfrak{B})$ by

$$\mu^*(B) = \mu(T^{-1}(B)) \text{ for } B \in \mathfrak{B}. \tag{0.2.2}$$

Then for arbitrary $B \in \mathfrak{B}$,

$$\int_{T^{-1}(B)} g(T(x))\, d\mu(x) = \int_B g(t)\, d\mu^*(t); \tag{0.2.3}$$

that is, if one of these integrals exists, so does the other, in which case they coincide.

For proof see [36].

§3. ON CONDITIONAL PROBABILITIES

In this section we state without proof the information that we shall need regarding conditional mathematical expectations and probabilities. Detailed proofs can be found, for example, in [36] and [8].

Suppose that a measurable space $(\mathfrak{X}, \mathfrak{A})$ is equipped with a probability measure P and that a statistic T maps $(\mathfrak{X}, \mathfrak{A})$ into $(\mathfrak{T}, \mathfrak{B})$, inducing a σ-subalgebra $\mathfrak{A}_0$. Let f denote a nonnegative function. Suppose that $\mathfrak{A}$ is measurable and integrable with respect to the measure P. Then the integrals $\int_A f\,dP$ exist for all $A \in \mathfrak{A}$ and a fortiori for all $A_0 \in \mathfrak{A}_0$. It follows from the Radon-Nikodym Theorem (Theorem 0.1.1) that there exists a function f_0 that is $\mathfrak{A}_0$-measurable with respect to the measure P and that satisfies the equation

$$\int_{A_0} f\, dP = \int_{A_0} f_0\, dP \tag{0.3.1}$$

for all $A_0 \in \mathfrak{A}_0$.

In accordance with Theorem 0.2.1, f_0 is a measurable function of $T(x)$. This function has the following two important properties.

1) Equation (0.3.1) holds for an arbitrary set $A_0 \in \mathfrak{A}_0$.

2) The function f_0 is a measurable function of $T(x)$.

On the basis of Theorem 0.1.1 these properties determine f_0 uniquely up to its values on sets P of measure 0. By definition, the function f_0 is taken as the conditional mathematical expectation $f(x)$ for a given value of the statistic $T(x)$:

$$f_0 = E(f(x)|T(x)) = E(f(x)|T = t) = g(t). \tag{0.3.2}$$

Let $P^*(t)$ denote the measure induced by the statistic T in the measurable space $(\mathcal{T}, \mathfrak{B})$. Then by virtue of Theorem 0.2.2,

$$\int_{T^{-1}(B)} f(x)\,dP(x) = \int_B g(t)\,dP^*(t). \tag{0.3.3}$$

for arbitrary $B \in \mathfrak{B}$.

To extend the definition of conditional mathematical expectation to sign-variable functions $f(x)$, we define

$$f^+(x) = \frac{f(x) + |f(x)|}{2}; \quad f^-(x) = \frac{|f(x)| - f(x)}{2}.$$

Then $f(x) = f^+(x) - f^-(x)$. We now define

$$E(f(x)\,|\,t) = E[f^+(x)\,|\,t] + E[f^-(x)\,|\,t],$$

provided the two expressions on the right are meaningful and these quantities exist. The basic properties of conditional mathematical expectations can be found in the books [36] and [8]. We shall state two theorems that will be needed in our study of sufficient statistics and exponential families.

In view of the definition given above for conditional mathematical expectation, we can, in particular, take for the measurable function $f(x)$ the characteristic function $I_A(x)$ of any set $A \in \mathfrak{A}$, that is, the function that is equal to 1 on the set A and equal to 0 on $\mathfrak{X} \setminus A$.

The conditional probability of A for a fixed value of the statistic $T = t$ is defined as

$$P(A\,|\,t) = E(I_A(x)\,|\,t). \tag{0.3.4}$$

On the basis of (0.3.3), we have

$$P(A \cap T^{-1}(B)) = \int_{T^{-1}(B)} I_A(x)\,dP(x) = \int_B P(A\,|\,t)\,dP^*(t),$$

where $B \in \mathfrak{B}$ is an arbitrary measurable set and $P^*(t)$ is, just as before, the measure induced by the statistic T in the measurable space $(\mathcal{T}, \mathfrak{B})$.

We now turn to the important special case in which $\mathfrak{X} = E_n$ is n-dimensional Euclidean space and $\mathfrak{A}$ is the family of Borel subsets of E_n. We note that the more general case in which $\mathfrak{X}$ is a Borel subset of E_n reduces to this case since we can extend the probability measure to all E_n by taking its value equal to 0 on $E_n \setminus \mathfrak{X}$.

We have the following important theorems, for proof of which see for example [36].

Theorem 0.3.1. *If $\mathfrak{X}$ is n-dimensional Euclidean space, then there exist definitions of conditional probabilities $P(A \mid t)$ such that $P(A \mid t)$ is a probabilistic measure on $(\mathfrak{X}, \mathfrak{A})$ for every t.*

Let x denote a random vector defined on $\mathfrak{X} = E_n$, let $T: (\mathfrak{X}, \mathfrak{A}) \to (\mathfrak{T}, \mathfrak{B})$ denote a statistic, and let $P(x \mid t)$ denote the conditional distribution (determined by Theorem 0.3.1) of the random vector x. From the preceding theorems we easily obtain

Theorem 0.3.2. *If $f(x)$ is a measurable function of the random vector x in $\mathfrak{X} = E_n$ and if $E\,|f(x)| < \infty$, then*

$$E(f(x) \mid t) = \int f(x)\, dP(x \mid t). \tag{0.3.5}$$

CHAPTER I

THE MULTIPLE LAPLACE TRANSFORMATION, FUNCTIONS OF SEVERAL COMPLEX VARIABLES, AND ANALYTIC SHEAVES

§1. THE MULTIPLE LAPLACE TRANSFORMATION

A considerable portion of the present book is devoted to statistical problems associated with exponential families of distributions. A natural analytical tool in the theory of such families is the multiple bilateral Laplace transformation. In the present section we shall state without proof the theorems in the theory of such transformations that we shall need. Proofs either can be found in the books [10] and [16] or can be obtained by obvious modifications in the reasoning in those books.

Consider the s-fold Laplace transform

$$L(m|\theta)=\int_{-\infty}^{\infty} dT_1 \dots \int_{-\infty}^{\infty} dT_s m(T_1, \dots, T_s) \times \exp[-(\theta_1 T_1 + \dots + \theta_s T_s)], \qquad (1.1.1)$$

where $m(T_1, \cdots, T_s)$ is a complex-valued function that is continuous almost everywhere with respect to Lebesgue measure and $\theta_1, \cdots, \theta_s$ are complex parameters such that

$$\theta_j = x_j + iy_j \qquad (j = 1, 2, \dots, s).$$

In many problems that we shall be studying, $m(T_1, \cdots, T_s)$ vanishes for $T_j \leq 0$, where j is one of the numbers $1, 2, \cdots, s_1 \leq s$. If $s_1 = s$, what we have is a unilateral Laplace transformation. In the general case $s_1 < s$. We shall assume that the function $m(T_1, \cdots, T_s)$ is such that $L(m \mid \theta)$ converges absolutely in the Cartesian product P of the s_1 half-planes R_j: $x_j > 0$ (where $j = 1, 2, \cdots, s_1$) and the $s - s_1$ strips S_j: $0 < x_j \leq A_j$ (for $j = s_1+1, \cdots, s$). (We can reduce arbitrary strips to this last type by a linear transformation of the parameters and variables.) Let m_1 and m_2 denote two functions that have

Laplace transforms of the type described. We have the following important convolution theorem.

Theorem 1.1.1.

$$L(m_1 * m_2|\theta) = L(m_1|\theta)\, L(m_2|\theta), \tag{1.1.2}$$

where

$$m_1 * m_2 = \int_0^{T_1} d\xi_1 \ldots \int_0^{T_{s_1}} d\xi_{s_1} \int_{-\infty}^{\infty} d\xi_{s_1+1} \ldots$$

$$\ldots \int_{-\infty}^{\infty} d\xi_s m_1(\xi_1, \ldots, \xi_s)\, m_2(T_1 - \xi_1, T_2 - \xi_2, \ldots, T_s - \xi_s).$$

Here $L(m_1 * m_2|\theta)$ converges absolutely in a region of the type described above.

We have the following theorem on the inverse transformation.

Theorem 1.1.2. *If a point $(c_1, \cdots, c_s)$ belongs to $\mathcal{P}$ and if the integral*

$$\frac{1}{(2\pi i)^s} \int_{c_1 - i\infty}^{c_1 + i\infty} d\theta_1 \ldots \int_{c_s - i\infty}^{c_s + i\infty} d\theta_s L(m|\theta) \exp(\theta_1 T_1 + \ldots + \theta_s T_s), \tag{1.1.3}$$

over the product of the vertical contours converges absolutely, it is equal to the function $m(T_1, \cdots, T_s)$ at all its points of continuity.

Since $L(m \mid \theta)$ is holomorphic in the region $\mathcal{P}$, every contour in the product can be deformed in a rather arbitrary manner by replacing it with rectifiable curves of a type that is convenient in some respect or other. In particular, if $L(m \mid \theta)$ satisfies the inequality

$$|L(m|\theta)| < \frac{C_0}{|(|\theta_1|+1) \ldots (|\theta_s|+1)|^r}, \tag{1.1.4}$$

for $r > 1$, $\theta \in \mathcal{P}$ and $C_0 =$ const., then by translating the contours for the integrations with respect to $\theta_1, \cdots, \theta_s$ to the right, we see that

$$m(T_1, \ldots, T_s) = 0 \qquad \text{for} \qquad T_1 < 0, \ldots, T_{s_1} < 0.$$

We note also that for positive integers $k_1, \cdots, k_s$ the fraction $1/\theta_1^{k_1} \ldots \theta_s^{k_s}$ is the unilateral Laplace transform of

$$\frac{1}{(k_1 - 1)! \ldots (k_s - 1)!} T_1^{k_1} \ldots T_s^{k_s}.$$

Theorem 1.1.3. *Let $E(\theta_1, \cdots, \theta_s)$ denote a function that is holomorphic in a region $\mathcal{P}$. Suppose that for arbitrary $\eta > 0$ and $c_j \geq \eta$ (where $j = 1, 2, \cdots, s$) the inequality*

$$|E(\theta_1, \ldots, \theta_s)| \leqslant \frac{K_1(\eta)}{(|\theta_1|+1)^r \ldots (|\theta_s|+1)^r} \quad (r \geqslant 3) \tag{1.1.5}$$

holds on the vertical contours (mentioned in Theorem 1.1.2) *passing through the region* $\mathcal{P}$. *Then* $E(\theta_1, \cdots, \theta_s)$ *is the Laplace transform* $L(H \mid \theta)$, *where* $H = H(T_1, \cdots, T_s)$ *vanishes for* $T_j < 0$ *(for* $j = 1, 2, \cdots, s_1$) *and has partial derivatives of the first* $r - 2$ *orders. Furthermore,*

$$|H(T_1, \ldots, T_s)| \leqslant K_2 \exp \zeta (|T_1| + \cdots + |T_s|). \tag{1.1.6}$$

where $\zeta > 0$ *may be arbitrarily small. Here and in what follows the* K_i *are positive constants.*

To prove this, we substitute the function $E(\theta_1, \cdots, \theta_s)$ for $L(m \mid \theta)$ in formula (1.1.4). By virtue of inequality (1.1.5), the integral converges absolutely. Denoting it by $H(T_1, \cdots, T_s)$, we obtain a function satisfying inequality (1.1.6). If one of the variables $T_1, T_2, \cdots, T_{s_1}$ is negative, we find, by increasing the abscissa c_j of the corresponding contour, that $H(T_1, \cdots, T_s) = 0$. Specifically, the integrand in (1.1.3) contains the factor $\exp(\theta_1 T_1 + \cdots + \theta_s T_s)$, which converges uniformly to 0. Furthermore, by differentiating formula (1.1.3) formally with respect to $\theta_1, \cdots, \theta_s$, we obtain a partial derivative of order $r - 2$. By virtue of inequality (1.1.5), this integral converges absolutely, so that the corresponding derivatives of $H(T_1, \cdots, T_s)$ exist.

We still need a theorem on the Laplace transforms of functions that vanish outside a finite interval (that is, functions of compact support).

Theorem 1.1.4. *Suppose that the function* $\delta(T_1, \cdots, T_s)$ *has partial derivatives of the first* r (≥ 1) *orders almost everywhere and that* δ *vanishes outside a finite interval* $[0, b]$. *Then for arbitrary* $(\theta_1, \cdots, \theta_s)$ $(\operatorname{Re} \theta > 0)$,

$$L(\delta \mid \theta) = O\left(\frac{1}{(|\theta_1|+1)^r \ldots (|\theta_s|+1)^r}\right). \tag{1.1.7}$$

To prove this we note that here the Laplace transform is unilateral. Integrating by parts, we obtain

$$\begin{aligned} L(\delta \mid \theta) = & \int_0^b dT_1 \ldots \int_0^b dT_s \delta(T_1, \ldots, T_s) \\ & \times \exp[-(\theta_1 T_1 + \ldots + \theta_s T_s)] \\ = & \frac{1}{\theta_1} \int_0^b dT_1 \ldots \int_0^b dT_s \delta(T_1, \ldots, T_s) \exp[-(\theta_1 T_1 + \ldots + \theta_s T_s)]. \end{aligned}$$

Repeating this operation r times with respect to the corresponding variables, we obtain inequality (1.1.7). We note that for θ_1 close to 0 this operation is not suitable. It should be applied to those variables T_j for which $|\theta_j| \geq 1$. If every

$|\theta_j| \leq 1$ then inequality (1.1.7) is trivial.

We shall say that the function $E(\theta_1, \cdots, \theta_s)$ is *holomorphic in a neighborhood of the point at infinity* $(\theta_1 = \infty, \cdots, \theta_s = \infty)$ if it can be represented as a power series in $p_1 = 1/\theta_1, \cdots, p_s = 1/\theta_s$ that converges in a neighborhood of $(0, \cdots, 0)$.

Theorem 1.1.5. *If the function $E(\theta_1, \cdots, \theta_s)$ is holomorphic in a neighborhood of the point at infinity $(\theta_1 = \infty, \cdots, \theta_s = \infty)$ and if it vanishes there, then it is the unilateral Laplace transform $E(\theta_1, \cdots, \theta_s) = L(H|\theta)$, where the function $H = H(T_1, \cdots, T_s)$ is holomorphic for $T_1 > 0, \cdots, T_s > 0$.*

Suppose that

$$E(\theta_1, \ldots, \theta_s) = \sum_{n_1, \ldots, n_k = 1}^{\infty} a_{n_1, \ldots, n_k} \theta_1^{-n_1} \ldots \theta_s^{-n_s}$$

and that the series converges in a neighborhood of $(0, \cdots, 0)$. By virtue of the correspondence mentioned above,

$$\theta_1^{-n_1} \ldots \theta_s^{-n_s} = L\left[\frac{1}{(n_1 - 1)! \ldots (n_s - 1)!} T_1^{n_1} \ldots T_s^{n_s} \,\middle|\, \theta\right].$$

Multiplying these expressions by $a_{n_1, \cdots, n_k}$ and summing, we obtain an everywhere-holomorphic function under the symbol $L(H|\theta)$. We need to consider it only for $T_1 > 0, \cdots, T_s > 0$.

§2. FUNCTIONS OF SEVERAL COMPLEX VARIABLES. THEOREMS YIELDING BOUNDS

We saw above that functions of the form $L(m|\theta)$ are holomorphic functions of several complex variables. We shall use certain concepts and theorems from the theory of such functions. These can be found in the monographs [5], [55], [75], and [76]. In particular, we shall need the concept of a holomorphic function in a region of superposition.

Let us denote by C^s the Cartesian product of s copies of the Euclidean complex plane. A *domain of holomorphy* is defined as an open region $D \subset C^s$ for which there exists a function $f(\theta_1, \cdots, \theta_s)$ that is holomorphic in D but does not have an analytic continuation to any open region of which D is a proper subset. In particular, the polycylinders $Z_1 \times Z_2 \times \cdots \times Z_s$, where the components Z_j (for $j \leq s$) are open simply-connected regions of the complex variables $x_j + iy_j$, are domains of holomorphy. In what follows, we shall be concerned primarily with such regions.

Suppose that the Cartesian product $I_1 \times \cdots \times I_s$ of real intervals $I_1, \cdots, I_s$ of positive length is contained in the interior of the polycylinder $Z_1 \times \cdots \times Z_s$. Then we have

Theorem 1.2.1. *If a function that is holomorphic in* $Z_1 \times \cdots \times Z_s$ *vanishes at all points of the set* $I_1 \times \cdots \times I_s$, *then it also vanishes everywhere in* $Z_1 \times \cdots \times Z_s$.

Let $f_1(\theta_1, \cdots, \theta_s), \cdots, f_r(\theta_1, \cdots, \theta_s)$ denote r holomorphic functions defined in the polycylinder $Z_1 \times \cdots \times Z_s$. Consider the system of equations

$$f_1 = 0, \cdots, f_r = 0. \tag{1.2.1}$$

The set of solutions of these equations $(\theta_1, \cdots, \theta_s)$ inside $Z_1 \times \cdots \times Z_s$ is called the *analytic set* generated by the system (1.2.1). Let us denote this set by $V_{f_1, \cdots, f_r}$. Suppose that the functions $f_j(\theta_1, \cdots, \theta_s)$ for $j = 1, 2, \cdots, s$ have real values on the real axes. Furthermore, suppose that $r < s$ and that the analytic set $V_{f_1, \cdots, f_s}$ inside the polycylinder $Z_1 \times \cdots \times Z_s$ with bounded simply connected components Z_j (for $j = 1, 2, \cdots, s$) can be decomposed into a finite number of components $V^q_{f_1, \cdots, f_s}$, where $q = 1, 2, \cdots, M$, that are connected along strips,[1] each of these being of complex dimension $s - r$, and suppose that these components contain the connected set $R^q_{f_1, \cdots, f_s}$ of real points of real dimension $s - r$.

Theorem 1.2.2. *Suppose that each* $R^q_{f_1, \cdots, f_s}$ *has an interior point at which*

$$\operatorname{rank} \begin{Vmatrix} \frac{\partial f_1}{\partial \theta_1} & \cdots & \frac{\partial f_1}{\partial \theta_s} \\ \cdot & \cdots & \cdot \\ \frac{\partial f_r}{\partial \theta_1} & \cdots & \frac{\partial f_r}{\partial \theta_s} \end{Vmatrix} = r. \tag{1.2.2}$$

Then every function $F(\theta_1, \cdots, \theta_s)$ *that is holomorphic in* $Z_1 \times \cdots \times Z_s$ *and vanishes on* $R^q_{f_1, \cdots, f_s}$ (where $q = 1, \cdots, M$) *also vanishes on the entire analytic set* $V_{f_1, \cdots, f_s}$.

To prove this we note that, on the basis of (1.2.2) and the continuity of the matrix on the left-hand side of that equation, we can express r of the variables

[1] That is, any two points of a component can be connected by an open strip of maximum dimension in it.

$\theta_1, \cdots, \theta_s$ in a neighborhood of the point inside $R^q_{f_1,\cdots,f_s}$ in question in terms of $f_1, \cdots, f_r$. Let us suppose that this can be done for the parameters $\theta_1, \cdots, \theta_r$. Then in that neighborhood

$$F(\theta_1, \cdots, \theta_s) = F_1(f_1, \cdots, f_r, \theta_{r+1}, \cdots, \theta_s),$$

where F_1 is holomorphic in that neighborhood. Furthermore, $F_1(0, \cdots, 0, \theta_{r+1}, \cdots, \theta_s) = 0$ for some set of values $\theta_{r+1}, \cdots, \theta_s$ of real dimension $(s - r)$. Then the same is true of some complex neighborhood of this set and $F(\theta_1, \cdots, \theta_s)$ will vanish in that $(s - r)$-dimensional neighborhood. This last follows from standard theorems in the theory of functions of several complex variables, in particular from Theorem 1.2.1.

We shall also need certain special theorems providing bounds. We present these along with their proofs. They depend on other theorems of the same type, which appear in the literature on the subject. In particular, we shall need the following theorem proved by Lojasiewicz in 1959 in the article [52].

Theorem 1.2.3. *Let $f(x_1, \cdots, x_n)$ denote a holomorphic function of n real variables defined on an open set $U \subset E_n$. Let X denote the set of real zeros of f and suppose that X is not empty. Let A denote a compact subset of U. Then there exist positive constants c and q such that for all $x = (x_1, \cdots, x_n) \in A$,*

$$|f(x_1, \cdots, x_n)| \geq c(\rho(x, X))^q,$$

where $\rho(x, X)$ is the distance in E_n.

From this theorem we can easily derive a corollary on functions of several complex variables. Let $f(\theta_1, \cdots, \theta_s)$ denote a function of s complex variables $\theta_1, \cdots, \theta_s$ that is holomorphic in an open subset U of E_s, where E_s is s-dimensional complex space. If we set $\theta_j = x_j + iy_j$, we obtain $f(\theta_1, \cdots, \theta_s) = \phi_1(x, y) + i\phi_2(x, y)$, where ϕ_1 and ϕ_2 are real functions of the variables $x_1, \cdots, x_n, y_1, \cdots, y_n$, both holomorphic in U (see [5]). If X denotes the set of complex zeros of $f(\theta_1, \cdots, \theta_s)$ in U, then X is also the set of real zeros of the function $\phi = \phi_1^2 + \phi_2^2$, which is a real function in U. If A is a compact subset of U, then Theorem 1.2.3 provides the corollary that we need: *for $\theta = (\theta_1, \cdots, \theta_s)$, there exist constants $c = c(A) > 0$ and $q = q(A) > 0$ such that*

$$|f(\theta_1, \cdots, \theta_s)| \geq c(\rho(\theta, X))^q. \tag{1.2.3}$$

Let $Z = Z_1 \times \cdots \times Z_s$ denote a polycylinder with open bounded simply-connected components and let $\overline{Z}$ denote its closure.

If a function $Q(\theta_1, \cdots, \theta_s)$ is holomorphic on $\overline{Z}$, then it is holomorphic in some open set U containing $\overline{Z}$. Suppose that the set X of complex zeros $Q(\theta_1, \cdots, \theta_s)$ has no points inside $\overline{Z}$. Then we can apply (1.2.3) with $A = \overline{Z}$, so that

$$|Q(\theta_1, \cdots, \theta_s)| > c(\rho(x, X))^q. \tag{1.2.4}$$

We turn now to the theorems on bounds that we shall need. The letters K_0, $K_1, \cdots$ will denote positive constants. Let Z denote an open polycylinder of the type described and let $\overline{Z}$ denote its closure. Let $f_1(\theta_1, \cdots, \theta_s)$ denote a function that is holomorphic on $\overline{Z}$ and let $F(\theta_1, \cdots, \theta_s)$ denote a function that is holomorphic in Z. Suppose also that, throughout Z,

$$F(\theta_1, \ldots, \theta_s) = f_1(\theta_1, \ldots, \theta_s)\, G(\theta_1, \ldots, \theta_s). \tag{1.2.5}$$

Theorem 1.2.4. *Suppose that there exists a constant M such that $|F(\theta_1, \cdots, \theta_s)| \le M$ everywhere in Z, and that* rank $\|\partial f_1/\partial\theta_1, \cdots, \partial f_1/\partial\theta_s\| = 1$. *Then everywhere inside Z,*

$$|G(\theta_1, \ldots, \theta_s)| \leqslant \frac{K_0 M}{\delta^{K_1}}, \tag{1.2.6}$$

where δ is the distance from $(\theta_1, \cdots, \theta_s)$ to the boundary of Z.

Proof. By the hypothesis of the theorem, the polycylinder Z can be covered by a finite number of open circular polycylinders $U_1, \cdots, U_k$ such that in each of them,

$$F(\theta_1, \ldots, \theta_s) = F_1(f_1, \theta_{i_2}, \ldots, \theta_{i_s}) = f_1 G_1(f_1, \theta_{i_2}, \ldots, \theta_{i_s}). \tag{1.2.7}$$

Here $(i_2, \cdots, i_s)$ is a sample of $(s-1)$ numbers of the set $1, 2, \cdots, s$; $G_1 = G_1(\theta_1, \cdots, \theta_s)$ for $f_1 = f_1(\theta_1, \cdots, \theta_s)$; the functions F_1 and G_1 are holomorphic in the images V_j of the polycylinders U_j under the mapping $\phi: (\theta_1, \cdots, \theta_s) \to (f_1, \theta_{i_2}, \cdots, \theta_{i_s})$.

Consider a point $(\theta_1, \cdots, \theta_s) \in U_j \cap Z$ and suppose that the mapping ϕ assigns to it the point $(f_1, \theta_{i_2}, \cdots, \theta_{i_s})$. If the distance from the point $(\theta_1, \cdots, \theta_s)$ to the boundary of Z is equal to δ, then it follows from inequality (1.2.4) and the familiar properties of Jacobians that the distance from $(f_1, \theta_{i_2}, \cdots, \theta_{i_s})$ to the corresponding boundary is $\gamma_j \delta^q$, where q is a positive constant and γ_j is a positive function bounded above and below by constants depending only on j and, since $j \le k$, by certain absolute constants. If the value of f_1 satisfies the inequality $|f_1| > \gamma_j \delta^q/10$, we obtain from (1.2.7)

$$|G_1(f_1, \theta_{i_2}, \ldots, \theta_{i_s})| \leqslant \frac{10M}{\gamma_j \delta^q} \leqslant \frac{K_2 M}{\delta^q}. \tag{1.2.8}$$

On the other hand if $|f_1| \leq \gamma_j \delta^q/10$, then the circle $C_{\gamma_j}: |f'_1 - f_1| = \gamma_j \delta^q/5$ lies inside the image $\phi(U_j \cap Z)$ and from Cauchy's theorem we have

$$G_1(f_1, \theta_{i_2}, \ldots, \theta_{i_s}) = \frac{1}{2\pi i} \oint_{C_{\gamma_j}} \frac{F_1(f'_1, \theta_{i_2}, \ldots, \theta_{i_s})}{(f'_1 - f_1) f'_1} df'_1,$$

so that

$$|G(\theta_1, \ldots, \theta_s)| = (G_1(f_1, \theta_{i_2}, \ldots, \theta_{i_s})| \leqslant \frac{10M}{\gamma_j \delta^q}.$$

Combining this inequality with (1.2.8), we obtain the proof of the theorem.

A modification of Theorem 1.2.4 that will be important in what follows is a theorem dealing with the case in which the polycylinder Z is the Cartesian product of s open vertical strips $S_1 \times \cdots \times S_s$.

Theorem 1.2.5. *Suppose that the polycylinder Z is the product of vertical strips S_j: $0 < \operatorname{Re} \theta_j < \eta_j$, $0 < \eta_j \leq \infty$, where $j = 1, 2, \cdots, s$. Suppose that equation* (1.2.5) *holds and that $|F(\theta_1, \cdots, \theta_s)| \leq M$ inside Z. Suppose that $f_1(\theta_1, \cdots, \theta_s)$ is a holomorphic function of the variables $1/\theta_1 + 1, \cdots, 1/\theta_s + 1$ on the closure of Z. Then*

$$G(\theta_1, \ldots, \theta_s)| \leqslant \frac{K_3 M}{\delta^{K_2}} (|\theta_1| + \ldots + |\theta_s| + 1)^{K_2}, \tag{1.2.9}$$

where δ is the distance from $(\theta_1, \cdots, \theta_s)$ to the boundary of Z.

To prove this we note that the mapping

$$p_1 = \frac{1}{\theta_1 + 1}, \ldots, p_s = \frac{1}{\theta_s + 1}$$

maps Z into the Cartesian product of the "half-moons" tangent to vertical lines, so that we obtain an open polycylinder $\mathfrak{S}$. Here, the points $(\theta_1, \cdots, \theta_s)$, where $\theta_j = \infty$, are mapped into points at which $p_j = 0$. The distances from these points to the new boundary can be expressed in terms of δ and the quantities $|\theta_j + 1|$, for $j = 1, 2, \cdots, s$. Thus the theorem reduces to the preceding one.

A generalization of equation (1.2.5) is the equation

$$F(\theta_1, \ldots, \theta_s) = f_1(\theta_1, \ldots, \theta_s) G_1(\theta_1, \ldots, \theta_s) + \ldots \\ \ldots + f_r(\theta_1, \ldots, \theta_s) G_r(\theta_1, \ldots, \theta_s), \tag{1.2.10}$$

where (in the previous notation) the functions $f_1, \cdots, f_r$ are holomorphic on $\overline{Z}$

and the functions $F, G_1, \cdots, G_r$ are holomorphic in Z. Here Z is a polycylinder with open connected bounded components and $2 \leq r < s$.

The functions $G_1, \cdots, G_r$ are not uniquely defined. To these functions we can add arbitrary functions $A_1, \cdots, A_r$ that are holomorphic in Z and that satisfy the condition

$$A_1 f_1 + \ldots + A_r f_r = 0.$$

To obtain inequalities analogous to (1.2.4) we must make the comparatively stringent requirement that there exist constants $\theta_{\alpha_1}, \theta_{\alpha_2}, \cdots, \theta_{\alpha_r}$, such that

$$\begin{vmatrix} \frac{\partial f_1}{\partial \theta_{\alpha_1}} & \cdots & \frac{\partial f_r}{\partial \theta_{\alpha_1}} \\ \cdot & \cdot \cdot \cdot & \cdot \\ \frac{\partial f_1}{\partial \theta_{\alpha_r}} & \cdots & \frac{\partial f_r}{\partial \theta_{\alpha_r}} \end{vmatrix} \neq 0. \qquad (1.2.11)$$

in the polycylinder $\overline{Z}$.

Theorem 1.2.6. *Suppose that the functions* $f_1, \cdots, f_r$ *have a common zero in the region* Z, *that equation* (1.2.10) *holds, and that* $|F(\theta_1, \cdots, \theta_s)| \leq M$ *in the region* Z. *Then equation* (1.2.10) *remains valid if we replace the functions* $G_1, \cdots, G_r$ *with functions* $G'_1, \cdots, G'_r$ *such that*

$$|G'_i(\theta_1, \ldots, \theta_s)| \leqslant \frac{K_4 M}{\delta^{K_5}} \quad (i = 1, 2, \ldots, r),$$

where δ *is the distance from the point* $(\theta_1, \cdots, \theta_s)$ *to the boundary of* Z.

Proof. We use induction. The theorem is already proved for $r = 1$. Let us suppose it true for $1, 2, \cdots, r-1$. Let F denote the analytic set of common zeros of $f_1, \cdots, f_r$ inside Z. By hypothesis F is nonempty.

Let us number the functions $f_1, \cdots, f_r$ and the variables $\theta_{\alpha_1}, \cdots, \theta_{\alpha_r}$ in such a way that the corner minors of orders $1, 2, \cdots, r$ in the matrix $\|\partial f_i / \partial \theta_j\|$ $(i, j = 1, 2, \cdots, r)$ are nonsingular in Z. The variables $\theta_1, \cdots, \theta_r$ can be expressed as functions $\theta_i = \theta_i(f_1, \cdots, f_r, \theta_{r+1}, \cdots, \theta_s)$ that are holomorphic in the domain of superposition which is obtained by representing Z in terms of the functions $f_1, \cdots, f_r, \theta_{r+1}, \cdots, \theta_s$, and ϕ where

$$\varphi: (\theta_1, \ldots, \theta_s) \to (f_1, \ldots, f_r, \theta_{r+1}, \ldots, \theta_s).$$

The functions $G_j(\theta_1, \cdots, \theta_s)$ can also be represented in this region as holomorphic functions $G'_j(f_1, \cdots, f_r, \theta_{r+1}, \cdots, \theta_s)$, where $j = 1, 2, \cdots, r$. Consider a point $(\theta_1^{(0)}, \cdots, \theta_s^{(0)}) \in F$. In a neighborhood of the point $f_1 = f_1(\theta_1^{(0)}, \cdots, \theta_s^{(0)}), \cdots$ $\cdots, f_r = f_r(\theta_1^{(0)}, \cdots, \theta_s^{(0)}), \theta_{r+1}^{(0)}, \cdots, \theta_s^{(0)}$ we may write

$$G_1(\theta_1, \dots, \theta_s) = G_{11}(f_1, 0, \dots, 0, \theta_{r+1}, \dots, \theta_s) + f_2 G_{12}(f_1, \dots, f_r, \theta_{r+1}, \dots, \theta_s) + \dots \dots + f_r G_{1r}(f_1, \dots, f_r, \theta_{r+1}, \dots, \theta_s),$$

where $G_{12}, \cdots, G_{1r}$ are holomorphic functions of $\theta_1, \cdots, \theta_s$. With the aid of equation (1.2.10) we obtain

$$F(\theta_1, \dots, \theta_s) = f_1 G_{11}(f_1, 0, \dots, 0, \theta_{r+1}, \dots, \theta_s) + f_2 G_2'(\theta_1, \dots, \theta_s) + \dots + f_r G_r'(\theta_1, \dots, \theta_s), \qquad (1.2.12)$$

where $G_j' = G_j - f_1 G_j$ $(j = 2, 3, \cdots, r)$.

The representation (1.2.12) is valid for all values of $(\theta_1, \cdots, \theta_s)$ and the values of $f_1, \cdots, f_r$ corresponding to them. This equation should be understood to mean that the function G_r depends only on f_1 and $\theta_{r+1}, \cdots, \theta_s$. Therefore we need only find a bound for it at the values $f_1 = 0, \cdots, f_r = 0$. Then

$$F(\theta_1, \dots, \theta_s) = F_1(f_1, 0, \dots, 0, \theta_r, \dots, \theta_s) = f_1 G_{11}(f_1, 0, \dots, 0, \theta_{r+1}, \dots, \theta_s),$$

which corresponds to the case considered in the preceding theorem, so that we obtain an inequality of the form (1.2.8) for G_{11}. We can now write the expression (1.2.12) in the form

$$F - f_1 G_{11} = f_r G_2' + \dots + f_r G_r'.$$

The conditions of the theorem are satisfied and $|F - f_1 G_{11}| \le K_6 M/\delta^{K_7}$ in Z. The rest follows by the induction hypothesis. This completes the proof.

Theorem 1.2.7. *Suppose that a polycylinder Z is the Cartesian product of vertical strips S_j: $0 < \mathrm{Re}\,\theta_j < \eta_j$, $\eta_j > 0$, for $j = 1, 2, \cdots, s$. (The possibility $\eta_j = \infty$ is admitted.) Suppose that conditions* (1.2.11) *and* (1.2.10) *are satisfied, where $|F(\theta_1, \cdots, \theta_s)| \le M$ inside Z and the functions $f_1(\theta_1, \cdots, \theta_s), \cdots \cdots, f_r(\theta_1, \cdots, \theta_s)$ are holomorphic functions of the variables $1/\theta_1 + 1, \cdots \cdots, 1/\theta_s + 1$ on the closure of Z. Then equation* (1.2.10) *remains valid if we replace the functions $G_1, \cdots, G_r$ with functions $G_1', \cdots, G_r'$ such that*

$$|G_j'(\theta_1, \dots, \theta_s)| \leqslant \frac{K_7 M}{\delta^{K_8}} (|\theta_1| + \dots + |\theta_s| + 1)^{K_8}.$$

This theorem reduces to the preceding one just as Theorem 1.2.5 reduced to Theorem 1.2.4.

§3. IDEALS IN RINGS OF HOLOMORPHIC FUNCTIONS. ANALYTIC SHEAVES

The study of similar tests and unbiased estimates for exponential families of distributions leads us in a natural way to a study of ideals in rings of certain holomorphic functions. Let O denote any commutative ring. Then a subring I such that $aI \subset I$ for arbitrary $a \in O$ is called an *ideal* of O. Let $\mathfrak{B}$ denote a subset of O. Then the set of all finite sums of the form $\Sigma_{i=1}^{M} a_i b_i$, where $a_i \in O$ and $b_i \in \mathfrak{B}$, constitutes an ideal known as the *ideal generated by* $\mathfrak{B}$ in O. In particular, if $\mathfrak{B}$ consists of a single element b, the ideal $\{ab: a \in 0\}$ generated by it is called a *principle ideal* and is denoted by (b). If an arbitrary element of an ideal I can be represented as a finite sum of the form $\Sigma_{i=1}^{M} a_i b_i$, where $b_1, \cdots, b_M$ are given elements of O and $a_i \in O$, then the ideal I is called an *ideal with finite basis*. (To obtain a basis from the elements $b_1, \cdots, b_M$ one removes all elements that can be expressed as linear combinations of the preceding elements of that set.) We shall consider ideals I in the ring O of functions that are holomorphic in an open polycylinder $Z = Z_1 \times \cdots \times Z_s$, where $Z_1, \cdots, Z_s$ are simply-connected bounded open regions of the variables $\theta_1, \cdots, \theta_s$ respectively. The study of such ideals, in accordance with the ideas of H. Cartan [25]–[28] and K. Oka [59], is based on the local properties of ideals, from which we derive their global properties with the aid of the theory of analytic sheaves.

Of special importance in the study of local properties of ideals of functions is the following theorem, proved by Rückert [65] in 1932.

Theorem 1.3.1. *Every ideal of functions that are holomorphic in a given region of a given point has a finite basis in a sufficiently small neighborhood of that point.*

The following theorem, proved by Cartan [26] in 1940, on the "gluing" of bases is also extremely important.

Theorem 1.3.2. *Let Δ' and Δ'' denote compact polycylinders all but one of the components of which are the same. Suppose that $\Delta = \Delta' \cap \Delta''$ is not empty. Let I, I', I'' denote ideals in rings of holomorphic functions defined on Δ, Δ', and Δ'' respectively. Suppose that I' and I'' generate I in the compact set Δ. Then there exists in the region $\Delta' \cup \Delta''$ an ideal I^0 that has a finite basis and that generates all three of these ideals in their domains.*

By a function that is holomorphic on a compact polycylinder we mean a function that is holomorphic in some neighborhood of that polycylinder.

The general local properties of ideals of functions are formulated with the aid of the concept of the germ of a function. Let f denote a function that is holomorphic in a neighborhood of $\theta = (\theta_1, \cdots, \theta_s) \in Z$. The *germ* of f is defined as the class of functions that are holomorphic in a neighborhood of θ and that coincide with f in a neighborhood of θ. This definition can be made more precise with the aid of the concept of an inductive limit (see Cartan [28]). Consider the open subsets U of Z. In each set U consider the ring S_U of functions $f(\theta)$ that are holomorphic in U. Let V denote an open subset of U and introduce the homomorphism r_V^U that consists of the restriction to V of functions that are holomorphic in U. For $W \subset V \subset U$ we will have $r_W^U = r_W^V \circ r_V^U$. The inductive limit of the groups S_U for $\theta \in U$ constitutes a group S_θ. Every element of this group is a germ of functions f_θ. If $f \in S_U$, $g \in S_V$, and $\theta \in U \cap V$ and if there exists an open set $W \subset U \cap V$ such that $r_W^U f = r_W^V g$, then $f_\theta = g_\theta$.

Addition and multiplication of germs of functions f_θ are defined in a natural way. They constitute a ring O_θ in which we may consider the ideals I_θ. In the space of all germs f_θ we define a topology (of the general type), and for this we need only exhibit a basis of the family of open sets. Let U denote an open set Z such that $f \in S_U$. For every point $\theta \in Uf$ we define a germ $f_\theta \in S_\theta$. We denote the set of such germs by f_U. We form a basis of the family of open sets from the sets f_U for all f and U.

We shall now present briefly the information on sheaves that we shall need. (For a more detailed exposition see [17] or [55].) Although we shall need only sheaves of ideals of germs of holomorphic functions over Z, it will be useful to give the general definition of a sheaf of rings. A sheaf $\mathfrak{F}$ of rings is determined when the following four things are defined:

1) a topological space X (the base of the sheaf),

2) a function $x \longrightarrow F(x)$ that assigns to each $x \in X$ a ring F_x,

3) a (general) topology in the union F of all the sets F_x,

4) a "projection" $p\colon F \longrightarrow X$ that assigns to all elements of F_x the element x and that is a local homeomorphism.

In addition we require continuity of the algebraic operations. More precisely, the mapping $\alpha \longrightarrow -\alpha$ $(\alpha \in F)$ of the space F onto itself must be continuous; also, the mappings $(\alpha, \beta) \longrightarrow \alpha + \beta$ and $(\alpha, \beta) \longrightarrow \alpha\beta$ of those pairs $(\alpha, \beta) \in F \times F$ for which $p(\alpha) = p(\beta)$ (so that the operations are defined) must be continuous (the topology in $F \times F$ is defined as usual).

The rings F_x are called the *stems* of the sheaf.

Let U denote an open subset of X. A continuous mapping $s: U \to F$ for which the mapping $p \circ s$ is the identity mapping is called a *section* of the sheaf $\mathfrak{F}$ over U.

Thus a section s maps x into the stem F_x; it has exactly one point in common with every stem over U. Two sections that coincide for a point $x \in X$ also coincide for some neighborhood of x.

Furthermore, since $s \cdot x$ belongs to F_x we can define addition and subtraction of two sections over U by $(s_1 \pm s_2)x = s_1 x \pm s_2 x \in F_x$. In this way we obtain the group of sections $\Gamma(U, \mathfrak{F})$.

The sheaves that we shall use will always have a base $X = Z$, where Z is the polycylinder introduced earlier. Furthermore they will always be analytic.

Take $X = Z$. A sheaf of ideals I_Z in the rings O_Z of germs of functions that are holomorphic at a point $z \in Z$ is called an *analytic sheaf* $\mathfrak{F}$ over Z. We take the usual topology on Z; on $F = \{I_Z\}$ we take the topology described above. Here the mapping $(f, \alpha) \to f\alpha$, where $f \in O_Z$ and $\alpha \in F_x$, is continuous. (In a more general situation this property is a requirement over a complex analytic variety.)

In particular, the sheaf O of germs of all holomorphic functions over Z is analytic. The sheaf $\mathfrak{F}$ mentioned above can be defined as a subsheaf of O. The section s of the analytic sheaf $\mathfrak{F}$ over the set U can be identified in a natural manner with the function $f \in S_U$ which is holomorphic over U.

Very important for what follows is the concept of a *coherent* analytic sheaf.

An analytic sheaf $\mathfrak{F}$ over Z is said to be *coherent* if for every point $z \in Z$ there exists a finite number of functions (sections of the sheaf O) that are holomorphic over some open neighborhood U of the point z and have the property that for an arbitrary point $z' \in Z$ the stem I_Z is contained in the ideal generated by this finite number of functions in the ring of germs $O_{Z'}$.

Now we can state the theorem of H. Cartan that we shall need. In what follows we shall base our study of a large class of tests and unbiased estimates on this theorem (Theorem 5 in [28]). We formulate it under more stringent restrictions than in its original presentation since this formulation will be sufficient for our purposes.

Theorem 1.3.3. *Let Z denote a polycylinder with open simply-connected* *bounded components and let $\mathfrak{F}$ denote a coherent analytic sheaf over Z. Let $u_1, u_2, \cdots, u_M$ denote finitely many sections of $\mathfrak{F}$ over Z such that for an arbitrary point $z \in Z$, the stem I_Z is generated by the germs of these functions in O_Z. Then the sections $u_1, \cdots, u_M$ in the ring O_Z of all functions that are holomorphic over Z generate an ideal containing the entire group of sections $\Gamma(Z, \mathfrak{F})$.*

From this theorem we derive a corollary that we shall have occasion to use later.

Corollary. *Let $f_1, \cdots, f_r$, where $r < s$, denote functions that are holomorphic on $\overline{Z}$. Suppose that the analytic set of points $(\theta_1, \cdots, \theta_s)$ defined by the conditions*

$$f_1 = 0, \ \ldots, \ f_r = 0; \ \operatorname{rank} \begin{Vmatrix} \frac{\partial f_1}{\partial \theta_1} & \cdots & \frac{\partial f_1}{\partial \theta_s} \\ \cdot & \cdot & \cdot \\ \frac{\partial f_r}{\partial \theta_1} & \cdots & \frac{\partial f_r}{\partial \theta_s} \end{Vmatrix} < r, \tag{1.3.1}$$

has no points inside Z. Then every function F that is holomorphic in Z and vanishes on the analytic set $f_1 = 0, \cdots, f_r = 0$ can be represented in the form

$$F = f_1 G_1 + \cdots + f_r G_r, \tag{1.3.2}$$

where the functions $G_1, \cdots, G_r$ are holomorphic in Z.

We note that this theorem deals essentially with the theory of ideals (in general, reducible) of analytic varieties. However, since the analytic set

$$f_1 = 0, \cdots, f_r = 0, \tag{1.3.3}$$

is not in general closed in Z, we cannot apply the corresponding theorems of Cartan in [28]. Consider a point $\theta \in Z$ and the germs of the functions $f_1, \cdots, f_r$ at it. If the functions $f_1, \cdots, f_r$ have no common zeros at the point θ, at least one of them, let us say f_1, has no zeros in some neighborhood U of the point z. Setting $G_1 = 1/f_1$, so that $1 = f_1 G_1$ in U, we see that the functions $f_1, \cdots, f_r$ generate at the point z the ring of all germs O_z, that is, the ideal $I_z = O_z$. Of course, the germ of the function F at z belongs to this ideal.

Suppose now that the functions $f_1, \cdots, f_r$ have a common zero at the point θ. Then by the hypotheses of the theorem we have at that point $\operatorname{rank} \|\partial f_i / \partial \theta_j\| = r$. Without loss of generality we may assume that the left corner minor of order r is nonzero. Then in a neighborhood of the point θ we may express $F = F(\theta_1, \cdots, \theta_s)$ in the form of a holomorphic function

$$F_1(f_1, f_2, \ldots, f_r, \theta_{r+1}, \ldots, \theta_s)$$
$$= F_1(0, \ldots, 0, \theta_{r+1}, \ldots, \theta_s) + f_1F_{11} + f_2F_{12} + \ldots + f_rF_{1r},$$

where the F_{1j} $(j = 1, \cdots, r)$ are holomorphic functions of $f_1, \cdots, f_r, \theta_{r+1}, \cdots, \theta_s$. If we set $f_1 = \cdots = f_r = 0$, we see that $F_1(0, \cdots, 0, \theta_{r+1}, \cdots, \theta_s)$ vanishes identically in that neighborhood; hence in that neighborhood we have $F = f_1G_1 + \cdots + f_rG_r$, where the $G_j = F_{1j}$ are holomorphic functions of $\theta_1, \cdots, \theta_s$.

Consider the sheaf $\mathfrak{F}$ of ideals generated on Z by the germs of the functions $f_1, \cdots, f_r$. Obviously it will be coherent. In accordance with Theorem 1.3.3, the functions $f_1, \cdots, f_r$, being an ideal in the ring of all functions that are holomorphic on Z, will generate all sections $\mathfrak{F}$ on Z, including the function F. From this we obtain (1.3.2), where all the G_j $(j = 1, \cdots, r)$ are holomorphic on Z $(G_j \in S_Z)$. In particular, if $f_1, \cdots, f_r$ have no common zeros in Z, then $1 = f_1G_1 + \cdots + f_rG_r$, where each $G_j \in S_Z$ $(j = 1, \cdots, r)$ (Cartan's example; see [28]).

We now present an important consequence of Theorems 1.3.1–1.3.3, discovered by Cartan in 1950 (see [27]).

Theorem 1.3.4. *Let A denote a compact polycylinder contained in Z and let I denote an ideal of functions that are holomorphic inside Z. Then I has a finite basis on A in the sense that there exists a set of functions $g_1, \cdots, g_s$ that are holomorphic in Z and an open neighborhood V such that $Z \supset V \supset A$ and for every function $f \in I$ there exists a representation of the form*

$$f|_V = h_1g_1 + h_2g_2 + \ldots + h_sg_s,$$

where $h_1, \cdots, h_s$ are functions that are holomorphic in V.[1)]

Proof. For every point $\theta \in A$, let us consider a polycylindrical neighborhood U_θ in which the ideal is finitely generated. Theorem 1.3.1 asserts that such a neighborhood exists. From the open covering of the compact set $\mathfrak{A}$ that we have obtained, let us choose a finite covering $\{u_{\theta_1}, \cdots, u_{\theta_l}\}$.

We may assume that $\bigcup_{i=1}^{l} u_{\theta_i} = Z'$ is a polycylinder (containing A). Consider the system of generating ideals $I_{\theta_i}\{U_1, \cdots, U_s\}$ and the ring of functions

1) The assertion that I has a finite basis on A can also be understood in the following sense. Let us choose elements of I as generators and construct from them an ideal $\widetilde{I}$ in the ring of functions that are holomorphic in Z. Then the ideal $\widetilde{I}$ is finitely generated. We note that if I has a finite basis on A in the sense indicated in the statement the theorem, it will also have a finite basis on A in the sense just explained.

that are holomorphic in Z'. Let $\widetilde{I}$ denote the ideal generated by I in that ring. From the coherence of the sheaf of ideals I_θ that are locally generated by the ideal $\widetilde{I}$ on Z and from Theorem 1.3.3, we may assert that the system $\{U_1, \cdots, U_s\}$ generates the entire group of sections. This means that for every function $t \in I$ there exists a set $h_1, \cdots, h_{l_s}$ of functions, holomorphic in Z', such that $f|_{Z'} = \Sigma h_i U_i|_{Z'}$. The theorem is proved.

CHAPTER II

SUFFICIENT STATISTICS AND EXPONENTIAL FAMILIES

§1. GENERAL INFORMATION ON SUFFICIENT STATISTICS

Consider a family of probability measures $\mathcal{P} = \{P_\theta, \theta \in \Omega\}$ defined on a single measurable space $(\mathfrak{X}, \mathfrak{A})$, the sample space. We shall call the index θ a *parameter* and the set Ω of indices θ the *parameter space*.

A statistic T is said to be *sufficient* for θ (or for the family $\mathcal{P}$) if for every measurable set $A \in \mathfrak{A}$ there exists a definition of conditional probability $P_\theta(A \mid t)$ that does not depend on the parameter θ.

In what follows we shall encounter numerous examples of sufficient statistics. At the moment, let us prove the following theorem (see [36]).

Theorem 2.1.1. *Suppose that $\mathfrak{X}$ is the Euclidean space E_n and let T denote a statistic that is sufficient for $\mathcal{P}$. Then there exists a definition of conditional probability $P_\theta(A \mid t)$ that does not depend on θ and that for every fixed t is a probability measure over $\mathfrak{A}$.*

This theorem resembles Theorem 0.3.1 in its content, and their proofs follow the same lines. For simplicity we shall prove it only for a one-dimensional space.

Let $r_1, r_2, r_3, \cdots$ denote the set of rational numbers arranged in some order and let $P_\theta(A \mid t)$ denote some value of the conditional probability of the set $(-\infty, x)$ for given t. For each value of $x = r_i$ (where $i = 1, 2, \cdots$), let us take a definition $P_\theta(r_i \mid t) = F(r_i \mid t)$ that does not depend on θ (as can be done, since the statistic T is sufficient). If $r_i < r_j$, then, by the properties of conditional probabilities (see for example [36]), we have $F(r_i \mid t) \le F(r_j \mid t)$ for all t except possibly for $t \in N_{ij}$, where N_{ij} is a set of measure zero with respect to the measure induced by the statistic. The function $F(x \mid t)$ is a nondecreasing function of x for rational values of x and for all $t \notin N' = \bigcup_{i,j} N_{ij}$

(which is also a set of measure 0). Furthermore, by the properties of conditional probabilities (*ibid.*), for $t \notin N''$, where N'' is a set of measure 0, we have

$$\lim_{n\to\infty} F\left(r_i - \frac{1}{n}\Big| t\right) = F(r_i | t) \qquad (i = 1, 2, \ldots);$$
$$\lim_{n\to\infty} F(n|t) = 1; \quad \lim_{n\to\infty} F(-n|t) = 0.$$

For $t \notin N' \cup N''$, the function $F(x|t)$ behaves like the distribution function $P(X < x | t)$, where X is a random variable. Now suppose that $F_1(x|t)$ is continuous from the left with respect to x and that it coincides with $F(x|t)$ at rational points. It is a distribution function and it defines a probability measure $P_1(A|t)$ that is independent of the parameter θ for $A \in \mathfrak{A}$. Let us show that $P_1(A|t)$ is the conditional probability of the set A for given t. Specifically, let us show that

$$P_1(A|t) = P(A|t) \tag{2.1.1}$$

for all $A \in \mathfrak{A}$ and for all t except for values of t in a set of corresponding measure 0. By what has been said above this is true for sets A that are intervals with rational or infinite end-points. Hence on the basis of elementary set-theoretic considerations it is true for the entire Borel σ-algebra $\mathfrak{A}$. On the other hand, on the set $N' \cup N''$ we can choose the measures $P(A|t)$ arbitrarily without violating the definition of conditional probabilities. This completes the proof of Theorem 2.1.1.

In what follows, we shall confine ourselves to admissable probability measures $\mathcal{P} = \{P_\theta, \theta \in \Omega\}$, that is, probability measures with probability density p_θ with respect to the same σ-finite measure μ on $(\mathcal{X}, \mathfrak{A})$. (Most often $\mathcal{X}$ will be a Euclidean space and μ will be a Lebesgue measure.)

We shall need the following theorem from measure theory.

Theorem 2.1.2. *The family $\mathcal{P}$ of probability measures is dominated by a σ-finite measure if and only if $\mathcal{P}$ has a countable equivalent family.*

By equivalent families of measures, we mean that vanishing of the measures of one family for any set $A \in \mathfrak{A}$ implies vanishing of the measures of the other family.

For the proof of the theorem see [36] or [18]. As a dominating measure for $\mathcal{P} = \{P_\theta, \theta \in \Omega\}$ we may take $\lambda = \sum_{i=1}^\infty c_i P_{\theta_i}$, where each c_i is positive, where

$\Sigma_{i=1}^{\infty} c_i = 1$, and where the P_{θ_i} constitute a suitably chosen countable family.

We shall now prove some extremely important factorization theorems that play a fundamental role in the theory of sufficient statistics.

Consider a family of probability measures $\mathcal{P} = \{P_\theta, \theta \in \Omega\}$ dominated by the measure $\lambda = \Sigma_{i=1}^{\infty} c_i P_{\theta_i}$, introduced above, which is equivalent to the family $\mathcal{P}$. Suppose that a statistic T maps a sample space $(\mathcal{X}, \mathfrak{A})$ onto another space $(\mathcal{T}, \mathfrak{B})$.

Theorem 2.1.3. *The statistic T is sufficient for $\mathcal{P}$ if and only if there exists a nonnegative $\mathfrak{B}$-measurable function $g(t, \theta)$ such that for all $\theta \in \Omega$,*

$$dP_\theta(x) = g(T(x), \theta)\, d\lambda(x). \tag{2.1.2}$$

Proof (cf. [36]). We first prove the necessity. Suppose that T is a sufficient statistic. Let A_0 denote the σ-subalgebra of $\mathfrak{A}$ induced by T. Then for all $\theta \in \Omega$, $A_0 \in \mathfrak{A}_0$, and $A \in \mathfrak{A}$ we have

$$\int_{A_0} P(A|T(x))\, dP_\theta(x) = P_\theta(A \cap \mathcal{A}_\theta). \tag{2.1.3}$$

Since $\lambda = \Sigma_{i=1}^{\infty} c_i P_{\theta_i}$, it follows from (2.1.3) that $\int_{A_0} P(A \mid T(x)) \times d\lambda(x) = \lambda(A \cap A_0)$, so that $P(A \mid T(x))$ is also a conditional probability for the measure λ.

Suppose that $g(T(x), \theta)$ is the Radon-Nikodym derivative $dP_\theta/d\lambda$ corresponding to the σ-algebra $\mathfrak{A}_0$ and the measure λ. To prove (2.1.2) it will be sufficient to show that it is the derivative $dP_\theta/d\lambda$ for the σ-algebra $\mathfrak{A}$ with the measure λ. Setting $A_0 = \mathcal{X}$ in equation (2.1.3), we find that

$$P_\theta(A) = \int_{\mathcal{X}} P(A|T(x))\, dP_\theta(x) = \int_{\mathcal{X}} E_\lambda[I_A(x)|T(x)]\, dP_\theta(x),$$

because, as we have just shown, $P(A \mid T(x))$ is also a conditional probability for the measure λ. Furthermore the function $E_\lambda[I_A(x) \mid T(x)]$ is $\mathfrak{A}_0$-measurable and $dP_\theta/d\lambda = g(T(x), \theta)$ for the σ-algebra $\mathfrak{A}_0$ and the measure λ. Consequently

$$\int_{\mathcal{X}} E_\lambda[I_A(x)|T(x)]\, dP_\theta(x)$$

$$= \int_{\mathcal{X}} E_\lambda[I_A(x)|T(x)]\, g(T(x), \theta)\, d\lambda(x) =$$

$$= \int_{\mathfrak{X}} E_\lambda [g(T(x), \theta) I_A(x) | T(x)] \, d\lambda(x)$$

$$= \int_{\mathfrak{X}} g(T(x), \theta) I_A(x) \, d\lambda(x) = \int_A g(T(x), \theta) \, d\lambda(x)$$

by virtue of the definition of conditional mathematical expectation. Thus $g(T(x), \theta)$ is the Radon-Nikodym density both for the σ-algebra $\mathfrak{A}$ and the measure λ, which completes the proof of the necessity of equation (2.1.2).

Let us prove its sufficiency. Suppose that equation (2.1.2) holds. We show that the conditional probability $P_\lambda(A \mid t)$ is a conditional probability for all measures $P \in \mathcal{P}$. For the σ-algebra $\mathfrak{A}$ we have $g(T(x), \theta) = dP_\theta(x)/d\lambda$. For given $A \in \mathfrak{A}$ and θ we define a measure ν on $\mathfrak{A}$ by $d\nu = I_A \, dP_\theta$. For the σ-algebra $\mathfrak{A}_0$ we have $d\nu(x)/dP_\theta(x) = E_\theta[I_A(x) \mid T(x)]$, so that $d\nu(x)/d\lambda(x) = P_\theta[A \mid T(x)] g(T(x), \theta)$ for the σ-algebra $\mathfrak{A}_0$. Furthermore, $d\nu(x)/d\lambda(x) = I_A(x) g(T(x), \theta)$ for the σ-algebra $\mathfrak{A}$, so that

$$\frac{d\nu(x)}{d\lambda(x)} = E_\lambda [I_A(x) g(T(x), \theta) | T(x)] = P_\lambda [A | T(x)] \, g(T(x), \theta)$$

for the σ-algebra $\mathfrak{A}_0$. Therefore

$$P_\lambda (A | T(x)) \, g(T(x), \theta) = P_\theta (A | T(x)) \, g(T(x), \theta)$$

for the σ-algebra $\mathfrak{A}_0$ and the measure λ and hence for the σ-algebra $\mathfrak{A}_0$ and the measure P_θ. Thus $P_\lambda(A \mid T(x))$ is one of the values of the conditional probability $P_\theta(A \mid T(x))$ and the statistic $T(x)$ is sufficient.

Theorem 2.1.3 has an important consequence.

Theorem 2.1.4. *Suppose that the probability measures* $P_\theta \in \mathcal{P}$ *have probability densities* $p_\theta = dP_\theta/d\mu$ *with respect to the* σ*-finite measure* μ. *Then the statistic* $T(x)$ *is sufficient for* $\mathcal{P}$ *if and only if there exists a nonnegative* $\mathfrak{B}$*-measurable function* $g(T(x), \theta)$ *and a nonnegative* $\mathfrak{A}$*-measurable function* h *defined on* $\mathfrak{X}$ *such that*

$$p_\theta(x) = g(T(x), \theta) \, h(x) \qquad (2.1.4)$$

for almost all x *corresponding to the* σ*-algebra* $\mathfrak{A}$ *and the measure* μ.

To prove this, consider the measure $\lambda = \sum_{i=1}^{\infty} c_i P_{\theta_i}$, which by Theorem 2.1.2 is equivalent to the family $\mathcal{P}$. If the statistic T is sufficient, then (2.1.4) follows from (2.1.2), where we set $h = d\lambda/d\mu$. Conversely, if (2.1.4) holds, then

$$d\lambda(x) = \sum_{i=1}^{\infty} c_i g(T(x), \theta_i) h(x) d\mu(x) = V(T(x)) h(x) d\mu(x),$$

where V is a measurable function. We define

$$g_1 = (T(x), \theta) = \frac{g(T(x), \theta)}{V(T(x))}, \quad \text{if} \quad V(T(x)) > 0,$$

and

$$g_1 = 0, \quad \text{if} \quad V(T(x)) = 0.$$

Then

$$dP_\theta(x) = g(T(x), \theta) h(x) d\mu(x) = g_1(T(x), \theta) d\lambda(x), \tag{2.1.5}$$

and thus $T(x)$ is a sufficient statistic.

At this point we make the obvious remark that if a mapping of a sufficient statistic T defined in a space $\mathfrak{T}$ onto another statistic T' defined in the same space is one-to-one and measurable in both directions, then such a mapping yields a sufficient statistic provided the function $g(T(x), \theta)$ in formula (2.1.4) is a measurable function of T'.

§2. EXAMPLES OF SUFFICIENT STATISTICS

We shall now examine a number of examples that illustrate the finding of sufficient statistics. Some of these examples will be needed later when we pose statistical problems with nuisance parameters; others are instructive as regards the general theory.

Example 1. In the Euclidean space E_n consider the family $\mathcal{P}$ of all densities $f(x_1) \cdots f(x_n)$ that are continuous with respect to Lebesgue measure and that correspond to a repeated sample $x_1, \cdots, x_n$, that is, to a vector with independent identically distributed components having the same probability density $f(x)$. (In what follows we shall use the term "repeated sample" without explanation.) If we arrange the elements of the sample $x_1, \cdots, x_n$ in order of size we obtain the variational series

$$x_1' \leqslant x_2' \leqslant \cdots \leqslant x_n'.$$

The mapping $T\colon (x_1, \cdots, x_n) \to (x_1', \cdots, x_n')$, which for brevity we shall identify with the expression $(x_1', \cdots, x_n')$, is a sufficient statistic. Indeed,

$$f(x_1)\cdot \ldots \cdot f(x_n) = f(x_1')\cdot \ldots \cdot f(x_n')$$

for the entire family $\mathcal{P}$. By Theorem 2.1.4, T is a sufficient statistic.

Example 2 (see Dynkin [19]). Suppose that $\mathcal{G}$ is a region in Euclidean space E_n. Let x denote an element of E_n and let S denote a set of values of the parameter θ. Suppose that for arbitrary $\theta \in S$ the probability density $p(x \mid \theta)$ is continuous and positive for $x \in \mathcal{G}$. We consider the mapping

$$x \to g(x, \theta) = \ln p(x \mid \theta) - \ln p(x \mid \theta_0),$$

which assigns to each $x \in \mathcal{G}$ the function $g(x, \theta)$, where θ ranges over S and θ_0 is any fixed value of the parameter θ. This mapping is a sufficient statistic, since

$$p(x \mid \theta) = \exp(g(x, \theta)) \cdot p(x \mid \theta_0),$$

and this meets the conditions of Theorem 2.1.4.

Example 3. In a Bernoulli scheme (of n repeated independent trials), let us treat the probability p of success in a single trial as the parameter p generating the family $\mathcal{P}$ of probability measures. The results of the sample can be coded by the row $x_1, \cdots, x_n$, where $x_i = 1$ if the ith trial succeeds and $x_i = 0$ if it fails. For the probability of the row $x_1, \cdots, x_n$ with given value of the parameter p, we have the probability density with respect to the "counting measure"

$$C_n^m p^m (1-p)^{n-m},$$

where m is the number of successes in n trials. The number m or the ratio m/n (the relative frequency) is a sufficient statistic for $\mathcal{P}$.

A more general example is the scheme of independent trials with k outcomes $E_1, \cdots, E_k$. The family of distributions $\mathcal{P}$ has $k-1$ parameters $p_1, \cdots, p_{k-1}$ (the probabilities of the outcomes $E_1, \cdots, E_{k-1}$). For n repeated trials, the absolute frequencies $m_1, \cdots, m_{k-1}$ of occurrence of the outcomes $E_1, \cdots, E_{k-1}$ or the relative frequencies $m_1/n, \cdots, m_{k-1}/n$ will be sufficient statistics.

Example 4. Suppose that independent trials with constant probability of success p are carried out until m successes occur. Let $Z+m$ denote the number of trials necessary for this result. Then the probability of the value $Z=z$ is

$$P(Z=z\,|\,p)=C^{z}_{m+z-1}p^{m}(1-p)^{z}.$$

If the results of $z+m$ trials are written as rows $(x_1, \cdots, x_{z+m})$ (for $z = 0, 1, 2, \cdots$) as described in Example 2, the number z of experiments such that m out of $z+m$ trials are successful is a sufficient statistic for the parameter p.

Example 5. For a repeated normal sample $x_1, \cdots, x_n \in N(a, 1)$, the probability density with respect to Lebesgue measure is

$$\frac{1}{(2\pi)^{n/2}}\exp\left(-\frac{1}{2}\sum_{i=1}^{n}(x_i-a)^2\right)$$
$$=\exp n\left(\bar{x}a-\frac{a^2}{2}\right)\frac{1}{(2\pi)^{n/2}}\exp\left(-\frac{1}{2}\sum x_i^2\right),$$

where as usual $\bar{x}$ denotes the quantity $\bar{x}=(1/n)\Sigma_{i=1}^{n}x_i$. From this we see that $\bar{x}$ is a sufficient statistic for the parameter a. For a sample from a more general family $N(a, \sigma^2)$, the probability density can be represented in the form

$$\frac{1}{(2\pi)^{n/2}\sigma^n}\exp\left[-\frac{1}{2\sigma^2}\sum_{i=1}^{n}(x_i-a)^2\right],$$

and $\bar{x}$ and $a_2=(1/n)\Sigma_{i=1}^{n}x_i^2$ or $\bar{x}$ and $s^2=(1/n)\Sigma_{i=1}^{n}(x_i-\bar{x})^2$, are sufficient statistics for the parameter $\theta=(a, \sigma^2)$.

Example 6. Let us consider a sequence of k independent repeated normal samples

$$(x_{i1}, \ldots, x_{in_i})\in N(a_i, \sigma_i^2).$$

For the parameter $\theta=(a_1, \sigma_1^2; \cdots; a_k, \sigma_k^2)$, the set $(\bar{x}_1, \cdots, \bar{x}_k;\ s_1^2, \cdots, s_k^2)$, where

$$\bar{x}_i=\frac{1}{n_i}\sum_{l=1}^{n_i}x_{il};\quad s_i^2=\frac{1}{n_i}\sum_{l=1}^{n_i}(x_{il}-\bar{x}_i)^2.$$

is a sufficient statistic. Here, the number of "scalar" sufficient statistics is equal to the number of scalar parameters. In particular, for $k=2$, corresponding to the two samples $x_1, \cdots, x_{n_1}\in N(a_1, \sigma_1^2)$; $y_1, \cdots, y_{n_2}\in N(a_2, \sigma_2^2)$

and the four parameters a_1, a_2, σ_1^2, and σ_2^2 are the four scalar statistics $\bar{x}, \bar{y}, s_1^2$, and s_2^2 (in the accepted notation).

Consider the probability density for a given pair of samples

$$\frac{1}{(2\pi)^{\frac{n_1+n_2}{2}}\sigma_1^{n_1}\sigma_2^{n_2}} \times \exp\left[-\left(\frac{1}{2\sigma_1^2}\sum_{i=1}^{n_1}(x_i-a_1)^2+\frac{1}{2\sigma_2^2}\sum_{i=1}^{n_2}(y_i-a_2)^2\right)\right]. \tag{2.2.1}$$

Let us set

$$a_1 = a_2 = a. \tag{2.2.2}$$

We thus have three independent scalar parameters a, σ_1^2 and σ_2^2 or the triple $\theta = (1/\sigma_1^2, a/\sigma_1^2, 1/\sigma_2^2)$, which is equivalent to them (in the sense explained). As before the sufficient statistics constitute a quadruple $(\bar{x}, \bar{y}, s_1^2, s_2^2)$. Here there are more scalar sufficient statistics than parameters. (This corresponds to the "incompleteness" of the system of sufficient statistics. We shall speak in greater detail of this later.)

If $\sigma_1^2 = \sigma_2^2 = \sigma^2$, then corresponding to the triple $\theta = (a_1, a_2, \sigma^2)$ is the triple of sufficient statistics $(\bar{x}, \bar{y}, \sum_{i=1}^{n_1} x_i^2 + \sum_{j=1}^{n_2} y_j^2)$.

Finally, if $a_1 = a_2 = a$ and $\sigma_1^2 = \sigma_2^2 = \sigma^2$, we have a homogeneous repeated sample of $n_1 + n_2$ normal variables from $N(a, \sigma^2)$. Of course,

$$\frac{1}{n_1+n_2}\left(\sum_{i=1}^{n_1} x_i + \sum_{j=1}^{n_2} y_j\right) \text{and} \frac{1}{n_1+n_2}\left(\sum_{i=1}^{n_1} x_i^2 + \sum_{j=1}^{n_2} y_j^2\right).$$

will be sufficient statistics.

Example 7. Consider a repeated sample $(x_1, y_1), \cdots, (x_n, y_n)$ of values of the two-dimensional normal vector (X, Y) with probability density

$$\frac{1}{2\pi\sigma_1\sigma_2\sqrt{1-\rho^2}} \exp - \frac{1}{2(1-\rho^2)}\left[\left(\frac{(X-a_1)}{\sigma_1^2} - 2\rho\frac{(X-a_1)(Y-a_2)}{\sigma_1\sigma_2} + \frac{(Y-a_2)^2}{\sigma_2^2}\right)\right].$$

The parameter $\theta = (a_1, a_2, \sigma_1^2, \sigma_2^2, \rho)$ includes the two mean values a_1 and a_2, two values σ_1^2 and σ_2^2 of the variances, and the coefficient of correlation ρ. As a sufficient statistic we may take the vector

$$\left(\bar{x},\ \bar{y},\ s_1^2,\ s_2^2,\ \frac{1}{n}\sum_{i=1}^{n}(x_i-\bar{x})(y_i-\bar{y})\right).$$

We can also replace the sample covariance $(1/n)\Sigma_{i=1}^n(x_i-\bar{x})(y_i-\bar{y})$ with the quantity $(1/n)\Sigma_{i=1}^n x_i y_i$ and we can replace the sample variances s_1^2 and s_2^2 with the quantities $(1/n)\Sigma_{i=1}^n x_i^2$ and $(1/n)\Sigma_{i=1}^n y_i^2$.

Example 8. Consider a repeated sample $x_1, \cdots, x_n$ from a distribution of the Pearson-III type with probability density

$$p(x)=\frac{\gamma^m}{\Gamma(m)}(x-a)^{m-1}\exp[-\gamma(x-a)] \quad \text{for } x\geqslant a,$$
$$p(x)=0 \ \text{ for } \ x<a;\ m>0,\ \gamma>0.$$

Here we have three scalar parameters: a, γ, and m. If we know the value of a, so that the family of distributions depends only on γ and m, then we have the sufficient statistics

$$\left(\sum_{i=1}^{n}(x_i-a),\ \prod_{i=1}^{m}(x_i-a)\right).$$

Now let us suppose that a is an unknown parameter. For the probability density of the sample we have the expression $p(x_1, \cdots, x_n) = p(x_1)\cdots p(x_n)$, or

$$p(x_1, \ldots, x_n)=\frac{\gamma^{mn}}{(\Gamma(m))^n}\prod_{i=1}^{n}(x_i-a)^{m-1}\exp\left[-\gamma\sum_{i=1}^{n}(x_i-a)\right],$$

if $\min(x_1, \cdots, x_n) \geq a$ and

$$p(x_1, \ldots, x_n)=0, \quad \text{if} \quad \min(x_1, \ldots, x_n)<a.$$

If $m \neq 1$, we have only the trivial sufficient statistics $T = (x_1', \cdots, x_n')$ (the variational series; cf. Example 1). However, if $m = 1$, so that we have the exponential distribution

$$p(x)=\gamma e^{-\gamma(x-a)} \quad (x\geqslant a),$$
$$p=0 \quad (x<a),$$

then the pair

$$\left(\sum_{i=1}^{n}x_i,\ \min_{i=1,\ldots,n}x_i\right)$$

obviously constitutes a sufficient statistic.

Example 9 (Kolmogorov [32]). Consider the family of uniform distributions where the carrier of the probability density is $[\theta - \frac{1}{2}, \theta + \frac{1}{2}]$. Here θ can be any real number. For a repeated sample $x_1, \cdots, x_n$ the probability density $p(x_1, \cdots, x_n)$ is equal to 1 if $\theta - \frac{1}{2} \leq x_i \leq \theta + \frac{1}{2}$ (for $i = 1, 2, \cdots, n$) but equal to 0 otherwise. Thus, $p(x_1, \cdots, x_n) = 1$ if and only if $\min_i (x_i + \frac{1}{2}) \geq \theta$ and $\max_i (x_i - \frac{1}{2}) \leq \theta$ and it is equal to 0 otherwise. Here, the parameter θ has two sufficient statistics: $\min_i x_i$ and $\max_i x_i$.

Example 10. In the preceding example suppose that the parameter θ cannot assume all real values but only the values $2m + \frac{1}{2}$, where m ranges over all integers (positive, negative, and zero). Suppose that we take a sample of two independent observations x_1 and x_2. Obviously each observation completely determines the parameter θ, so that x_1 and x_2 are each sufficient statistics.

§3. INFORMATIONAL PROPERTIES OF SUFFICIENT STATISTICS

The last, quite simple, example (cf. also the examples of this type in Basu's article [2]) is instructive for a discussion of information-theoretic properties of sufficient statistics. The relationships between statistics and information theory are well expounded in the famous book of Kullback [35]. Here we shall touch on only one point in that field.

Let $\mathcal{G}$ denote a region contained in Euclidean space E_n and let x denote an element of $\mathcal{G}$. Let θ denote a parameter assuming values in a region $\mathcal{H}$ contained in E_m. Let $p(x \mid \theta)$ denote the probability density with respect to Lebesgue measure. Suppose that this probability density is continuous and positive on $\mathcal{G} \times \mathcal{H}$. Let $E_{n_1} \subset E_n$, $n_1 < n$, denote a Euclidean space contained in E_n. Suppose that there exists a sufficient statistic $T \in E_{n_1}$ for the parameter θ. In the region $\mathcal{G}$ we set up a system of local coordinates (T, ξ), where ξ is an $(n - n_1)$-dimensional vector. Suppose that the Jacobian $\partial(T, \xi)/\partial(x)$ of the transformation exists, is continuous, and does not vanish for $x \in \mathcal{G}$.

Suppose that the parameter θ is also a random variable with distribution density $q(\theta) \neq 0$ in its domain of definition $\mathcal{H}$. Then we can consider the common distribution (x, θ) and we can define a common distribution density

$p(x, \theta)$. If the integral

$$I(x, \theta) = \int\int_{(\mathfrak{G}, \mathfrak{H})} p(x, \theta) \log_2 \frac{p(x, \theta)}{p(x)\, q(\theta)}\, dx\, d\theta, \tag{2.3.1}$$

where $p(x) = \int_{\mathfrak{H}} p(x \mid \theta)\, q(\theta)\, d\theta$ exists, it is called the *quantity of information in the sense of Shannon* contained in x with respect to θ or vice versa (see for example [35]). Here dx and $d\theta$ are the elements of volume. Let us shift to the coordinates T and ξ. We have

$$p(x \mid \theta) = p(T, \xi \mid \theta) \left| \frac{\partial(T, \xi)}{\partial(x)} \right|,$$

where $p(T, \xi \mid \theta)$ is the probability density for (T, ξ) corresponding to a given value of the parameter θ. From (2.3.1) we obtain

$$I(x, \theta) = \int\int_{(\mathfrak{G}, \mathfrak{H})} q(\theta)\, p(T, \xi \mid \theta) \left| \frac{\partial(T, \xi)}{\partial(x)} \right| \times \log_2 \frac{p(T, \xi \mid \theta)}{\int_{\mathcal{H}} p(T, \xi \mid \theta)\, q(\theta)\, d\theta} \left| \frac{\partial(x)}{\partial(T, \xi)} \right| dT\, d\xi. \tag{2.3.2}$$

If T is a sufficient statistic, then by Theorem 2.1.4,

$$p(T, \xi \mid \theta) = g(T, \theta)\, r(T, \xi) \neq 0, \tag{2.3.3}$$

where $r(T, \xi)$ is independent of θ. In this case (2.3.2) yields

$$I(x, \theta) = \int_{(T, \mathcal{H})} q(\theta)\, p_1(T \mid \theta) \log_2 \frac{g(T, \theta)}{\int_{\mathcal{H}} g(T, \theta)\, q(\theta)\, d\theta}\, dT\, d\theta, \tag{2.3.4}$$

where T is the range of the statistic T and $p_1(T \mid \theta)$ is the distribution density of the statistic T for a given value of θ. If we integrate both sides of equation (2.3.3) with respect to ξ, we obtain

$$p_1(T \mid \theta) = g(T, \theta)\, r_1(T), \tag{2.3.5}$$

where $r_1(T) = \int r(T, \xi)\, d\xi$, this integral being taken over all values of ξ.

If we multiply the numerator and denominator of the fraction in the argument of the logarithm in (2.3.4) by $r_1(T)$ and make the substitution (2.3.5), we obtain

$$I(x, \theta) = \int_{(T, \mathcal{H})} q(\theta) p_1(T \mid \theta) \log_2 \frac{p_1(T \mid \theta)}{p_1(T)} dT\, d\theta$$

$$= \int_{(T, \mathcal{H})} p_1(T, \theta) \log_2 \frac{p_1(T, \theta)}{p_1(T) q(\theta)} dT\, d\theta = I(T, \theta). \qquad (2.3.6)$$

Here $p_1(T, \theta)$ denotes the common density of T and θ, and $p_1(t) = \int_{\mathcal{H}} p_1(T, \theta) q(\theta)\, d\theta$ denotes the density of T. The expression $I(T, \theta)$ is the amount of information in the statistic T with respect to θ. Thus, under the conditions listed above regarding $p(x, \theta)$ and the statistic T, this statistic contains all the information concerning the parameter θ that is to be found in the observation of x. Conversely one can show that if

$$I(x, \theta) = I(T, \theta)$$

under the preceding assumption regarding $p(x \mid \theta)$ for a sufficiently broad class of *a priori* distributions of the parameter $q(\theta)$ and the statistic T, then the statistic T is sufficient. We note that in our derivation we could have used not the amount of information $I(x, \theta)$ but the functional

$$\sigma(x, \theta) = \int \int p(x, \theta) \varphi\left(\frac{p(x, \theta)}{p(x) q(\theta)}\right) dx\, d\theta, \qquad (2.3.7)$$

where ϕ is a rather arbitrary continuous function of a single variable. The conclusion regarding the sufficient statistics would remain valid. If a statistic U is stochastically independent of a sufficient statistic T, then under rather general assumptions the distribution U is independent of the parameter θ and contains no information about the parameter. However, this situation does not always obtain. Interesting "pathological" phenomena in this field were described by Basu [2].

Let $(\mathfrak{X}, \mathfrak{A})$ denote a measurable space and let $\mathcal{P} = \{P_\theta\}$ denote a family of measures defined on it. Then for every value $T = t$ and every $A \in \mathfrak{A}$ there exists a definition of conditional mathematical expectation $P(A \mid t)$ that does not depend on the parameter θ. For arbitrary $A \in \mathfrak{A}$ and $B \in \mathfrak{B}$ we have

$$P(A \cap T^{-1}(B)) = \int_B P(A \mid t)\, dP_\theta T^{-1}, \qquad (2.3.8)$$

where $P_\theta T^{-1}$ is the measure induced on $\mathfrak{B}$ by the statistic T.

Now let A denote a set belonging to $\mathfrak{A}$ whose probability is independent of the values of the statistic T for all values of the parameter θ. Thus

$$P_\theta(A \cap T^{-1}(B)) = P_\theta(A)\, P_\theta(T^{-1}(B)) \qquad (2.3.9)$$

for arbitrary $B \in \mathfrak{B}$. Since

$$P_\theta(A)\, P_\theta(T^{-1}(B)) = \int_B P_\theta(A)\, dP_\theta T^{-1}, \qquad (2.3.10)$$

when we compare this relation with (2.3.8), which is also valid for arbitrary $B \in \mathfrak{B}$, we conclude that for every value of the parameter θ,

$$P_\theta(A) = P(A \mid t) \qquad (2.3.11)$$

for almost all t with respect to the measure $P_\theta T^{-1}$.

Let θ_1 and θ_2 denote two values of the parameter θ. If there exists a set $B \in \mathfrak{B}$ such that $P_{\theta_1} T^{-1}(B) > 0$ and $P_{\theta_2} T^{-1}(B) > 0$, then we conclude from (2.3.11) that

$$P_{\theta_1}(A) = P_{\theta_2}(A).$$

If any two values of the parameter can be "coupled" in the above manner by means of the set $B \in \mathfrak{B}$, we see that the value of $P_\theta(A)$ is independent of θ and that the set A does not contain any information concerning the parameter θ. However, if the measures $P_{\theta_1} T^{-1}$ and $P_{\theta_2} T^{-1}$ do not "overlap", we cannot draw any such conclusion. Example 10 shows that in this case the set A, which is stochastically independent of the sufficient statistic, can even contain "complete information" on the parameter θ.

§4. SUFFICIENT STATISTICS FOR A REPEATED SAMPLE. EXPONENTIAL FAMILIES

Almost all the examples of sufficient statistics that we have given dealt with a repeated sample of random vectors in a Euclidean space. Such samples play an extremely important role in mathematical statistics. The simplest case is a sample of one-dimensional random variables $(x_1, \cdots, x_n)$. A number of works have been devoted to the study of sufficient statistics for this case. We mention Darmois [15], Koopman [34], Dynkin [19], Bahadur [1], and Brown [7]. Here we shall discuss certain theorems from [19] and [7].

Consider a region $\mathcal{G}$ contained in E_m and the family of distributions $\mathcal{P} = \{P_\theta\}$ with Borel σ-algebra in it. Consider the set of values of the parameter Ω as an abstract set.

If two functions $X_1(x)$ and $X_2(x)$ are defined in $\mathcal{G}$, we shall say that X_1 is *subordinate* to X_2 if for an arbitrary pair of values x' and x'' belonging to $\mathcal{G}$ the relation $X_2(x') = X_2(x'')$ implies the relation $X_1(x') = X_1(x'')$.

By Theorem 2.1.4, a statistic subordinate to a sufficient statistic is itself sufficient. If each of two statistics X_1 and X_2 is subordinate to the other statistic, we shall say that they are *equivalent*.

The statistic $\epsilon(x) = x$ (the identity mapping) is always sufficient. We shall call a statistic that is equivalent to $\epsilon(x)$ in a subregion $\tilde{\mathcal{G}} \subset \mathcal{G}$ a *trivial statistic*. If a statistic $X(x)$ is subordinate to every sufficient statistic, we shall call it a *necessary statistic*. Of course, it will be of interest to find statistics that are both necessary and sufficient.

Let us take a repeated sample of size n. Here we obtain points $(x_1, \cdots, x_n)$ in

$$E_{mn} = \underbrace{E_m \times \cdots \times E_m}_{n \text{ times}}.$$

In the region

$$\mathcal{G}^{(n)} = \underbrace{\mathcal{G} \times \cdots \times \mathcal{G}}_{n \text{ times}}$$

the distribution of the repeated sample P^n is induced. We shall call its statistics the *statistics of the repeated sample.*

Now suppose that for all values of $\theta \in \Omega$ the distributions $\mathcal{P}$ have a positive probability density $p(x \mid \theta)$ that is continuous with respect to x. In Example 2 of §2 we saw that the function $g(x, \theta) = \ln p(x \mid \theta) - \ln p(x \mid \theta_0)$, where θ_0 is any value of the parameter θ, is a sufficient statistic for $\mathcal{P}$. This statistic is also necessary; if $X(x)$ is a sufficient statistic, then by (2.1.4)

$$p(x \mid \theta) = g_1(X(x), \theta) h(x),$$

where $h(x)$ is independent of θ. Therefore $g(x \mid \theta) = \ln g_1(X(x), \theta) - \ln g_1(X(x), \theta_0)$ is, as a function of $\theta \in \Omega$, subordinate to $X(x)$.

If we replace $\mathcal{G}$ with $\mathcal{G}^n$ and $\mathcal{P}$ with $\mathcal{P}^n$, then we need to replace $g(x, \theta)$ with $g(x_1, \cdots, x_n, \theta) = g(x_1, \theta) + \cdots + g(x_n, \theta)$.

Now consider the family $\mathcal{P}$ of one-dimensional distributions defined on an interval $\Delta \subset E_1$. Let us suppose that for arbitrary $\theta \in \Omega$ there exists a continuous derivative $dp(x \mid \theta)/dx$. Furthermore let us suppose that $p(x \mid \theta) > c(\theta) > 0$ for $x \in \Delta$ and for every value of θ. Consider the minimum linear space (over the set of real numbers) $L(\mathcal{P}, \Delta)$ of functions defined on Δ that contains all constants and all functions $g(x, \theta)$ for all $\theta \in \Omega$. Denote the dimension of $L(\mathcal{P}, \Delta)$ by $r + 1$ (here r may be ∞).

Under the assumptions made above on $p(x \mid \theta)$ we have (see [19], [7]).

Theorem 2.4.1. *If the functions* $1, \phi_1(x), \cdots, \phi_r(x)$ *constitutes a basis in* $L(\mathcal{P}, \Delta)$, *then for arbitrary* $n \geq r$ *the system of functions*

$$\chi_i(x_1, \ldots, x_n) = \varphi_i(x_1) + \ldots + \varphi_i(x_n) \quad (i = 1, 2, \ldots, r) \tag{2.4.1}$$

is functionally independent and constitutes a necessary and sufficient statistic for a sample of size n.

We shall show that the statistic $(\chi_1, \cdots, \chi_r)$ is equivalent to the necessary and sufficient statistic $g(x_1, \cdots, x_n, \theta)$.

First let us show that the first statistic is subordinate to the second. Let $\phi(x)$ denote a function in $L(\mathcal{P}, \Delta)$ and suppose that $\chi(x_1, \cdots, x_n) = \Sigma_{j=1}^n \phi(x_j)$. We have $\phi(x) = c_0 + \Sigma_{q=1}^l c_q g(x, \theta_q)$, where the c_i are constants. Therefore $\chi(x_1, \cdots, x_n) = nc_0 + \Sigma_{q=1}^l c_q g(x_1, \cdots, x_q, \theta)$, so that χ is subordinate to the necessary statistic $g(x_1, \cdots, x_n, \theta)$ and hence is itself necessary. Furthermore, the converse subordination also holds: $g(x, \theta) = a_0(\theta) + \Sigma_{j=1}^n a_j(\theta)\phi_j(x)$, where the $a_j(\theta)$ are constants depending only on θ. Therefore $g(x_1, \cdots, x_n, \theta) = na_0(\theta) + \Sigma_{j=1}^r a_j(\theta)\chi_j(x_1, \cdots, x_n)$, so that $g(x_1, \cdots, x_n, \theta)$ is subordinate to $\{\chi_1, \cdots, \chi_r\}$. Thus this last statistic is a sufficient statistic equivalent to $g(x_1, \cdots, x_n, \theta)$.

The functions $1, \phi_1(x), \cdots, \phi_r(x)$ constitute a basis in $L(\mathcal{P}, \Delta)$ and consequently are linearly independent. Let us show now that, if s $(\leq r)$ arbitrarily chosen functions $1, \phi_1(x), \cdots, \phi_s(x)$ are linearly independent (for example, if they constitute a part of the basis), then for $n \geq s$ the system of functions

$$\chi_i(x_1, \ldots, x_n) = \sum_{j=1}^n \varphi_i(x_j)$$

(for $i = 1, \cdots, s$) are functionally dependent.

Each of the functions $\phi_i(x_j)$ is a linear combination of a constant and functions of the form

$$g(x_j, \theta_q) = \ln p(x_j|\theta_q) - \ln p(x_j|\theta_0).$$

By the hypothesis of the theorem this last function is continuously differentiable with respect to x_j in the interval Δ. The same applies to $\phi_i(x_j)$. If the functions $\chi_1, \cdots, \chi_s$ were functionally dependent in $\Delta^{(n)}$, we would have

$$\frac{D(\chi_1, \ldots, \chi_s)}{D(x_1, x_2, \ldots, x_s)} = \begin{vmatrix} \varphi_1'(x_1) & \varphi_1'(x_2) & \ldots & \varphi_1'(x_s) \\ \varphi_2'(x_1) & \varphi_2'(x_2) & \ldots & \varphi_2'(x_s) \\ \cdot & \cdot & \cdot & \cdot \\ \varphi_s'(x_1) & \varphi_s'(x_2) & \ldots & \varphi_s'(x_s) \end{vmatrix} \equiv 0. \qquad (2.4.2)$$

Let us show that this would imply linear dependence of the system $1, \phi_1, \cdots, \phi_s$. We prove this by induction. If $s = 1$, then $\phi_1'(x_1) = 0$, so that $\phi_1(x_1) = \text{const}$ and our assertion is trivial. Suppose now that it is true for $s - 1$. If $D(\chi_1, \cdots, \chi_{s-1})/D(x_1, \cdots, x_{s-1}) \equiv 0$, our assertion holds. Consequently we may assume that there exists a point $(x_1^{(0)}, \cdots, x_{s-1}^{(0)})$ at which this Jacobian is nonzero. Consider a point $(x_1^{(0)}, \cdots, x_{s-1}^{(0)}, x_s)$, where $x_s \in \Delta$ assumes arbitrary values. At such a point let us write the left-hand member of (2.4.2), expanding the determinant in terms of elements of the last column. We then obtain

$$\varphi(x_s) = A_1\varphi_1'(x_s) + \ldots + A_s\varphi_s'(x_s) \equiv 0. \qquad (2.4.3)$$

Here $A_1, \cdots, A_s$ are constants and $A_s \neq 0$. Thus we see that $\phi(x_s) = \text{const}$ in the interval Δ, so that we have linear dependence.

Thus the system of functions (2.4.1) is linearly independent and constitutes a necessary and sufficient statistic for $\mathcal{P}$ with respect to a sample of size n. This completes the proof of the theorem.

Theorem 2.4.2. *Under the conditions of Theorem 2.4.1, for arbitrary finite $n \leq r$, every sufficient statistic with respect to a sample of size n is trivial.*

Suppose that $n \leq r$. Then $L(\mathcal{P}, \Delta)$ contains n functions $\phi_1, \phi_2, \cdots, \phi_n$ such that the system $1, \phi_1, \phi_2, \cdots, \phi_n$ is linearly independent. It follows from the proof of Theorem 2.4.1 that the functions $\chi_1(x_1, \cdots, x_n), \cdots$

$\ldots, \chi_n(x_1, \ldots, x_n)$ of the type (2.4.1) corresponding to them are functionally independent. Then there exist $x_1^{(0)}, \ldots, x_n^{(0)} \in \Delta$ such that $D(\chi_1, \ldots, \chi_n)/D(x_1, \ldots, x_n) \neq 0$. In a neighborhood of the point $(x_1^{(0)}, \ldots, x_n^{(0)})$ we have

$$x_i = x_i(\chi_1, \ldots, \chi_n), \quad i = 1, 2, \ldots, n.$$

Thus by the definition given above the trivial sufficient statistic $(x_1, \ldots, x_n)$ is subordinate to $\chi_1, \ldots, \chi_n$ in some interval $\tilde{\Delta} \subset \Delta$ and hence is necessary.

Following Dynkin [19] the *rank* of a system of distributions $\mathcal{P}$ in a region $\mathcal{G}$ is defined as the largest number R such that, for every finite $n \leq R$, the family $\mathcal{P}$ has no nontrivial sufficient statistics with respect to a sample of size n in the region $\mathcal{G}$.

Suppose now that a system $\mathcal{P} = \{p(x \mid \theta)\}$ of distributions on an interval Δ is subject to the conditions described above: for every θ, there exists a continuous derivative $dp(x \mid \theta)/dx$ and we have the inequality $p(x \mid \theta) > c(\theta) > 0$ for $x \in \Delta$.

Theorem 2.4.3. *Suppose that the system of distributions $\mathcal{P}$ satisfies the conditions stated above and has finite rank R in the interval Δ. Then the density $p(x \mid \theta)$ can be represented in the form*

$$p(x \mid \theta) = \exp\left[\sum_{i=1}^{R} \varphi_i(x) c_i(\theta) + c_0(\theta) + \varphi_0(x)\right]. \tag{2.4.4}$$

Here $x \in \Delta$ and $\theta \in \Omega$; the functions $\phi_i(x)$ are continuously differentiable in Δ $(i = 1, \ldots, R)$, and the systems of functions $(1, \phi_1, \ldots, \phi_R)$ and $(1, c_1(\theta), \ldots, c_R(\theta))$ are linearly independent.

Families of the form (2.4.4) are called *exponential families*. Let us note, first, that the dimension $r + 1$ of the space $L(\mathcal{P}, \Delta)$ is equal to $R + 1$. To see this, note that this dimension is finite because otherwise, by Theorem 2.4.2, the sufficient statistics would be trivial for an arbitrary sample of size n (including for $n = R + 1$). This shows that $r + 1 \leq R + 1$. Furthermore, if $r + 1$ were less than $R + 1$, then, by taking $n = r + 1 \leq R$, we would obtain a sufficient statistic $(\chi_1, \ldots, \chi_r)$ of the form (2.4.1) that is nontrivial, whereas the size of the sample does not exceed R. Thus $r + 1 = R + 1$, so that $r = R$ and we take a continuously differentiable basis in Δ for the space $L(\mathcal{P}, \Delta)$ consisting of the $r + 1$ functions $1, \phi_1(x), \ldots, \phi_R(x)$. Then

$$g(x, \theta) = \sum_{i=1}^{R} c_i(\theta)\varphi_i(x) + c_0(\theta), \tag{2.4.5}$$

where the $c_i(\theta)$ are constants depending only on θ. Since $g(x, \theta) = \ln p(x \mid \theta) - \ln p(x \mid \theta_0)$, we obtain formula (2.4.4) with $\phi_0(x) = \ln p(x \mid \theta_0)$. Let us now show that the functions $1, c_1(\theta), \cdots, c_R(\theta)$ are linearly independent. Suppose that this is not the case. Then without loss of generality we may set $c_R(\theta) = b_0 + b_1 c_1(\theta) + \cdots + b_{R-1} c_{R-1}(\theta)$, where the b_i are constants. Substituting this expression into (2.4.5) we see that

$$g(x, \theta) = \sum_{i=1}^{R-1} c_i(\theta)\psi_i(x) + c_0(\theta) + b_0\varphi_R(x),$$

where $\psi_i(x) = \phi_i(x) + b_i\phi_R(x)$ (where $i = 1, 2, \cdots, R-1$). Therefore

$$g(x, \theta) = g(x, \theta) - g(x, \theta_0)$$
$$= \sum_{i=1}^{R-1} [c_i(\theta) - c_i(\theta_0)]\psi_i(x) + c_0(\theta) - c_0(\theta_0),$$

so that the dimension of $L(\mathcal{P}, \Delta)$ does not exceed R, which contradicts the hypothesis.

The following theorem is in a certain sense the converse of Theorem 2.4.3.

Theorem 2.4.4. *Suppose that*

$$p(x \mid \theta) = \exp\left(\sum_{i=1}^{R} c_i(\theta)\varphi_i(x) + c_0(\theta) + \varphi_0(x)\right), \tag{2.4.6}$$

where the $\phi_i(x)$ are continuously differentiable functions, $x \in \Delta$, $\theta \in \Omega$, and the systems $(1, \phi_1, \cdots, \phi_R)$ and $(1, c_1(\theta), \cdots, c_R(\theta))$ are linearly independent systems of functions. Then the rank of the system of distributions $\mathcal{P}$ is equal to R, and for $n \geq r$ the system of functions $\chi_i(x_1, \cdots, x_n)$ $(i = 1, \cdots, R)$ of the form (2.4.1) constitutes a necessary and sufficient statistic for a sample of size n.

We have

$$g(x, \theta) = \ln p(x \mid \theta) - \ln p(x \mid \theta_0)$$
$$= \sum_{i=1}^{R} (c_i(\theta) - c_i)\varphi_i(x) + c_0(\theta) - c_0; \quad c_i = c_i(\theta_0). \tag{2.4.7}$$

From this it is clear that the dimension of $L(\mathcal{P}, \Delta)$ does not exceed $R+1$, and consequently the rank of $\mathcal{P}$ does not exceed R. Furthermore, if the functions $1, c_1(\theta), \cdots, c_R(\theta)$ are linearly independent, so are the R functions $c_1(\theta) - c_1, \cdots, c_R(\theta) - c_R$. An elementary argument leads to the existence of R numbers $\theta_1, \cdots, \theta_R$ in Ω such that

$$\varphi_i(x) = b_{i0} + \sum_{j=1}^{R} b_{ij} g(x, \theta_j), \tag{2.4.8}$$

where the b_{ij} are constants. Thus the functions $1, \phi_1, \cdots, \phi_k$ belong to $L(\mathcal{P}, \Delta)$ and generate it. If they are linearly independent they constitute a basis in $L(\mathcal{P}, \Delta)$, and the present theorem follows from Theorems 2.4.1 and 2.4.2.

These theorems naturally lead us to consideration of exponential families of a general form, the study of one aspect of which constitutes the content of this book.

§5. EXPONENTIAL FAMILIES

The preceding section indicates the importance of the study of families of distributions of the form (2.4.4), i.e. exponential families. We shall need to generalize this concept, which was defined in §4 only for an interval Δ of the straight line E_1 and under special hypotheses. In making this generalization we shall follow Lehmann [36]. Let $\mathfrak{X}$ denote the Euclidean space E_k with Borel σ-algebra $\mathfrak{A}$. Suppose that a σ-finite measure μ is defined on the measurable space $(\mathfrak{X}, \mathfrak{A})$. Consider the family of densities $\{p_\theta(x)\}$ of the form

$$p_\theta(x) = C(\theta) \exp\left(\sum_{j=1}^{s} Q_j(\theta) T_j(x) \right) h(x), \tag{2.5.1}$$

where θ belongs to an abstract set Ω, where $T_j(x)$ $(j = 1, 2, \cdots, k)$ and $h(x)$ are measurable functions, and where the $Q_j(\theta)$ are arbitrary functions. Later we shall give representative examples.

If $x_1, \cdots, x_n$ is a repeated sample of a distribution of the type (2.5.1), then it follows immediately from the factorization theorem (2.1.4) that the statistics $\sum_{i=1}^n T_j(x_i)$ $(j = 1, 2, \cdots, s)$ constitute a sufficient statistic with respect to a sample of volume n.

We can simplify the writing of the probability density (2.5.1) if we assign

$h(x)$ to the measure $\mu(x)$ (that is, if we consider the measure $\nu(x)$ with $d\nu(x) = h(x)\,d\mu(x)$) and set $Q_j(\theta) = \theta_j$ $(j = 1, 2, \cdots, s)$. Here the θ_j have numerical values and the point $\theta = \theta_1, \cdots, \theta_s$ belongs to the Euclidean space E_s. We obtain the expression

$$p_\theta(x) = C(\theta)\exp\left[\sum_{j=1}^{s}\theta_j T_j(x)\right] \tag{2.5.2}$$

where $C(\theta)$ is a normalizing constant.

In this notation the parameters $(\theta_1, \cdots, \theta_s)$ in the space E_s may assume arbitrary sets of values related to one another (for example, they may be dependent on some base parameters that vary in E_r, where $r < s$).

However, we can also consider a space of parameters $\Omega \subset E_s$ that consists of those values of $\theta = (\theta_1, \cdots, \theta_s)$ for which

$$\underset{E_k}{\int \cdots \int} \exp\left(\sum_{j=1}^{s}\theta_j T_j(x)\right) d\nu(x) < \infty. \tag{2.5.3}$$

Here $\nu(x)$ must be a σ-finite measure on $\mathfrak{X} = E_k$ that is given in advance and that is independent of the parameter θ. The space Ω defined above is called the *natural parameter space*. Let us give some theorems of Lehmann [36] on exponential families under natural parametrization.

Theorem 2.5.1. *The natural parameter space Ω is a convex set.*

To prove this, we use Hölder's inequality (see for example [71], pp. 21–26), which may be written in the form

$$\int_{E_m} (a(x))^\alpha (b(x))^{1-\alpha}\, d\mu(x) \leqslant \left(\int_{E_m} a(x)\, d\mu(x)\right)^\alpha \left(\int_{E_m} b(x)\, d\mu(x)\right)^{1-\alpha}, \tag{2.5.4}$$

where $a(x)$ and $b(x)$ are measurable functions defined on E_m, μ_x is a σ-finite measure, $\alpha \in (0, 1)$, and all the integrals converge.

Let $\theta = (\theta_1, \cdots, \theta_s)$ and $\theta' = (\theta_1', \cdots, \theta_s')$ denote two points in the natural parameter space Ω and let α denote a number in the interval (0, 1). We need to show that the point $\alpha\theta + (1 - \alpha)\theta'$ belongs to Ω. In formula (2.5.4) we set

$\mu(x) = \nu(x)$, $a(x) = \exp(\sum_{j=1}^{s} \theta_j T_j(x))$, $b(x) = \exp(\sum_{j=1}^{s} \theta'_j T_j(x))$, and $m = n$. Then, on the basis of (2.5.4) and (2.5.3), the desired result follows for the values of the parameters θ and θ'.

In studying the family $\{p_\theta(x)\}$ in a natural parametrization, we need an extension into the complex domain.

Let $\xi = (\xi_1, \cdots, \xi_s)$ denote a point in the natural parameter space Ω, so that the integral in (2.5.3) converges when $\theta_j = \xi_j$ $(j = 1, 2, \cdots, s)$. Then this integral will obviously converge also for complex values $\theta_j = \xi_j + i\tau_j$, where the τ_j are arbitrary real numbers. Admittedly, the values of $p_\theta(x)$ given by formula (2.5.2) will not be positive and they will not in general even be real, so that we do not obtain extensions of the family of distributions. However such a procedure for extending the values of the parameters to the complex domain will be useful to us on several accounts in what follows. In many cases it allows us to apply the theory of functions of a complex variable.

Let $\phi(x)$ denote a bounded complex-valued measurable function defined on E_n. Consider the integral

$$\int_{E_n} \varphi(x) \exp\left[\sum_{j=1}^{s} \theta_j T_j(x)\right] d\nu(x). \tag{2.5.5}$$

Obviously this integral converges at every point $\xi = (\xi_1, \cdots, \xi_s)$ of the natural parameter space Ω and also at every point of the cylindrical subset of the complex space $E^{(2s)}$ consisting of points $\xi_j + i\tau_j$ (for $j = 1, 2, \cdots, s$), where the τ_j are arbitrary real numbers.

If the natural space Ω contains an interior point $(\xi_1^{(0)}, \cdots, \xi_s^{(0)})$, we may assert that the integral (2.5.5) is an analytic function of the parameters $\theta_1, \cdots$ $\cdots, \theta_s$ at that point. To see this, note that our integral can then be represented as the limit to which a sequence of finite sums converges uniformly with respect to the complex parameter $\theta = (\xi_1 + i\tau_1, \cdots, \xi_s + i\tau_s)$, where $(\xi_1, \cdots, \xi_s)$ lies in the neighborhood $(\xi_1^{(0)}, \cdots, \xi_s^{(0)})$. By Weierstrass' theorem on functions of several variables (see for example [5]) we may conclude that (2.5.5) is analytic at the point $(\xi_1^{(0)}, \cdots, \xi_s^{(0)})$ and at all complex points with the same abscissa.

The derivatives with respect to the θ_j can be calculated at these points by differentiating under the integral sign. This is true because the derivatives of such analytic functions can be expressed as integrals of the Cauchy type over small polycylinders around points θ of the type described, and by virtue of the absolute and uniform convergence with respect to θ of the integral (2.5.5) in a sufficiently small neighborhood of the given point we can transform the integral with respect to θ into an integral of the Cauchy type with respect to the measure $d\nu(x)$.

These considerations can be applied to the derivation of certain useful formulas (see [36]). From (2.5.2), we have, at interior points of Ω,

$$\int_{E_m} \exp\left(\sum_{j=1}^{s} \theta_j T_j(x)\right) d\nu(x) = (C(\theta))^{-1}. \qquad (2.5.6)$$

Differentiating under the integral sign with respect to θ_j and multiplying both sides of the equation by $C(\theta)$, we obtain

$$E(T_j(x)) = -\frac{\partial \ln C(\theta)}{\partial \theta_j}. \qquad (2.5.7)$$

Analogously,

$$E(T_i(x) T_j(x)) = -\frac{C''_{ij}}{C} - 2\frac{C'_i C'_j}{C^2},$$

where the subscripts denote the variables of differentiation. Since $\partial^2 \ln C/\partial\theta_i \partial\theta_j = C''_{ij}/C - C'_i C'_j/C^2$ it follows from (2.5.7) that

$$\begin{aligned} \operatorname{cov}(T_i(x), T_j(x)) &= E(T_i(x) T_j(x)) - E(T_i(x)) E(T_j(x)) \\ &= -\frac{\partial^2}{\partial\theta_i \partial\theta_j} \ln C(\theta). \end{aligned} \qquad (2.5.8)$$

Examples of exponential families can be obtained by repeated samples from the distributions exhibited in the different examples of §2 of the present chapter. A normal distribution is illustrated by Example 5 (for the one-dimensional case) and Example 7 (for the two-dimensional case). Example 3 illustrates a multinomial distribution. Normal distributions (of arbitrary dimension n) and binomial and multinomial distributions belong to this class. It will be instructive to consider Examples 8 and 9 of §2 (the Pearson-III distribution and uniform distribution with variable mean). If we consider the number a in Example 8 as given, then for a repeated sample we obtain an exponential family with parameters

$\theta_1 = \gamma$ and $\theta_2 = (m - 1)$ and with sufficient statistics

$$T_1 = \sum_{i=1}^{n} (x_i - a); \quad T_2 = \sum_{i=1}^{n} \ln (x_i - a).$$

On the other hand, if a is also a parameter, we do not obtain an exponential family. (In this and more general cases, it would be proper to speak of a generalization of exponential families, namely, "families with movable carriers", which are not exponential families.)

In Example 9, where the density is of a very simple form, assuming only the values 1 and 0, we also do not obtain an exponential family.

§6. SUFFICIENT STATISTICS AND UNBIASED ESTIMATES

Let $\mathcal{P} = \{P_\theta\}$, $\theta \in \Omega$, denote a family of distributions that is defined on a measurable space $(\mathfrak{X}, \mathfrak{A})$ and let $F(P_\theta) = f(\theta)$ denote a functional defined on $\mathcal{P}$.

A function $\phi(x)$ defined on $\mathfrak{X}$ is called an *unbiased estimate* for $f(\theta)$ if the mathematical expectation $E_\theta \phi(x) = \int_x \phi(x)\, dP_\theta(x)$ exists for all θ and if

$$E_\theta \varphi(x) = f(\theta), \qquad \theta \in \Omega. \tag{2.6.1}$$

In [32] Kolmogorov introduced upper estimates $\phi_+(x)$ and lower estimates $\phi_-(x)$ for $f(\theta)$, which are characterized by the properties

$$E_\theta \varphi_+(x) \geqslant f(\theta); \quad E_\theta \varphi_-(x) \leqslant f(\theta), \tag{2.6.2}$$

so that an estimate which is both upper and lower is unbiased, and conversely.

We shall study below the behavior of unbiased estimates in certain classes of exponential families. At the moment, however, what we need to do is establish connections between the theory of unbiased estimates and the theory of sufficient statistics.

Let $\phi(x)$ denote an unbiased estimate for $f(\theta)$ and let $\chi(x)$ denote a sufficient statistic for θ with an arbitrary set of values. Consider the conditional mathematical expectations $E_\theta[\phi(x) | \chi(x)]$. Since $\chi(x)$ is a sufficient statistic, this expression is independent of θ, so we may write $E_\theta[\phi(x)|\chi(x)] = E_\chi(\phi)$. The function $E_\chi(\phi)$ is a function only of $\chi(x)$ for given $\phi(x)$. Since $E(E_\chi(\phi)) = E_\phi = f(\theta)$, we see that the function $E_\chi(\phi)$ is an unbiased estimate of $f(\theta)$ (see Rao [61], Blackwell [4], and Kolmogorov [32]). We also have the following theorem ([16], [17], [15]).

Theorem 2.6.1. *If for some* $\theta \in \Omega$ *and an unbiased estimate* ϕ, *the variance* $D_\theta(\phi) = \int_x (\phi - f(\theta))^2 dP_\theta$, *exists, then* $D_\theta(E_\chi(\phi))$ *exists also and*

$$D_\theta(E_\chi(\varphi)) \leqslant D_\theta\varphi. \tag{2.6.3}$$

Proof. We have

$$\begin{aligned} D_\theta\varphi = E_\theta(\varphi - f(\theta))^2 &= E_\theta(\varphi - E_\chi(\varphi) + E_\chi(\varphi) - f(\theta))^2 \\ &= E_\theta(\varphi - E_\chi(\varphi))^2 + E(E_\chi(\varphi) - f(\theta))^2 \\ &\quad + 2E[(\varphi - E_\chi(\varphi))(E_\chi(\varphi) - f(\theta))]. \end{aligned}$$

Furthermore,

$$\begin{gathered} E[(\varphi - E_\chi(\varphi))E_\chi(\varphi)] = E\{E_\chi(\varphi - E_\chi(\varphi))E_\chi(\varphi)\} = 0; \\ E(\varphi - E_\chi(\varphi))f(\theta) = 0, \end{gathered}$$

so that

$$D_\theta\varphi = D_\theta(E_\chi(\varphi)) + E_\theta(\varphi - E_\chi(\varphi))^2,$$

which proves (2.6.3).

Theorem 2.6.1 can be generalized if we consider not the variance ϕ but "defects" of a more general form. Let us replace $D(\phi) = E(\phi - f(\theta))^2$ with the expression $Eg(\phi - f(\theta))$, where g is an arbitrary continuous convex (downward) function. We have (see Doob [18])

$$g(E(\varphi - f(\theta))\,|\,\chi) = g(E_\chi(\varphi) - f(\theta)) \leqslant E(g(\varphi - f(\theta))\,|\,\chi),$$

where we assume the existence of the corresponding integrals.

Again taking the mathematical expectations, we obtain

Theorem 2.6.2.

$$Eg(E_\chi(\varphi) - f(\theta)) \leqslant Eg(\varphi - f(\theta)). \tag{2.6.4}$$

This provides a generalization of Theorem 2.6.1.

Thus the operation of taking the conditional mathematical expectation of an unbiased statistic ϕ for a given value of a sufficient statistic leads to a new unbiased statistic the "convex defects" of which do not exceed those of the preceding one. In what follows, we shall frequently make use of this fact.

Now let us look at some examples of unbiased estimates.

Example 1. Let $x_1, \cdots, x_n$ denote a repeated sample from a distribution with parameters $Ex_i = a$ and $D(x_k) = \sigma^2$. Then the quantity $\bar{x} = (1/n)\sum_{i=1}^n x_i$ is an unbiased estimate of general mean a. An arbitrary linear form $\lambda_1 x_1 + \cdots$ $\cdots + \lambda_n x_n$, where $\sum_{i=1}^n \lambda_i = 1$, is such an estimate. The quantity

$$\bar{x}^2 - \frac{1}{n(n-1)} \sum_{i=1}^{n} (x_i - \bar{x})^2,$$

can serve as an unbiased estimate of the functional a^2, as can numerous other estimates.

Example 2. Suppose that the distribution is normal, so that $x_i \in N(a, \sigma_0^2)$, where σ_0 is known. Then we may consider the integral

$$\frac{1}{\sqrt{2\pi}\sigma_0} \int_A \exp\left[-\frac{(x-a)^2}{2\sigma_0^2}\right] dx = f_A(a)$$

over an arbitrary Borel set A on the real line. Suppose, for example, that x_i $(i = 1, \cdots, n)$ denotes the dimensions of an arbitrary manufactured object and that A constitutes a zone lying outside the tolerated range $[a_0 - 3\sigma_0, a_0 + 3\sigma_0]$. Then $f_A(a)$ denotes the portion of rejects for a general mean a. An unbiased estimate of $f_A(a)$ is the statistic (see [32])

$$\varphi_A(\bar{x}) = \frac{1}{\sqrt{2\pi}\sigma_0} \int_A \exp\left[-\frac{(x-\bar{x})^2}{2\sigma_0^2}\right] dx.$$

Analogous estimates are made in [32] when a and σ are unknown.

Example 3. Let $x_1, \cdots, x_n \in N(a, \sigma^2)$ denote a repeated sample in a normal set with unknown variance σ^2. Let us suppose that $n \geq 2$. We know that under these conditions $\bar{x}$ and s^2, where $s^2 = (1/n)\sum_{i=1}^n (x_i - \bar{x})^2$, are stochastically independent and $\bar{x} \in N(a, \sigma^2/n)$; $ns^2/\sigma^2 = \chi^2_{n-1}$, where χ^2_{n-1} is a χ^2-distribution with $(n-1)$ degrees of freedom. (See Cramér [33]. See Chapter IV regarding the independence of the statistics in connection with the theory of Neyman structures.) If we set $ns^2/\sigma^2 = Z_n$ we have, for arbitrary $\rho > 0$,

$$EZ_n^\rho = \frac{1}{2^{\frac{n-1}{2}}\Gamma\left(\frac{n-1}{2}\right)} \int_0^\infty x^{\frac{n}{2}-\frac{3}{2}+\rho} e^{-\frac{x}{2}} dx = 2^\rho \frac{\Gamma\left(\frac{n-1}{2}+\rho\right)}{\Gamma\left(\frac{n-1}{2}\right)}.$$

Thus for arbitrary positive ρ the statistic

$$\left(\frac{n}{2}\right)^{\rho} \frac{\Gamma\left(\frac{n-1}{2}\right)}{\Gamma\left(\frac{n-1}{2}+\rho\right)} s^{2\rho}$$

is an unbiased estimate for the parameter $\sigma^{2\rho}$.

Example 4 ([30] and [72]). Let $(x_1, y_1), \cdots, (x_n, y_n)$ denote a two-dimensional normal sample (see Example 7 of §2) with coefficient of correlation ρ and let

$$r = \frac{\sum_{i=1}^{n} (x_i - \bar{x})(y_i - \bar{y})}{\left(\sum_{i=1}^{n} (x_i - \bar{x})^2 \sum_{i=1}^{n} (y_i - \bar{y})^2\right)^{1/2}}$$

denote the sample coefficient of correlation. Then

$$E \arcsin r = \arcsin \rho, \tag{2.6.5}$$

so that (unexpectedly) $\arcsin \rho$ has the unbiased estimate $\arcsin r$.

Example 5. Let $x_1, \cdots, x_n$ denote a repeated sample from the family of all distributions on E_1 with continuous density $f(x)$. Let $T(x_1, \cdots, x_n)$ denote a continuous bounded statistic. This statistic is an unbiased estimate of the functional $T_f = E_f T(x_1, \cdots, x_n)$ for given f. Since it is bounded, it has a variance $D_f(T) = E(T - T_f)^2$. A sufficient statistic of this family is the variational series $x'_1 \leq \cdots \leq x'_n$. For a fixed sufficient statistic, the conditional distribution of the sample consists of $n!$ points constituting $n!$ permutations of the variational series. Thus for a given value of a sufficient statistic the conditional mathematical expectation of T is equal to

$$\frac{1}{n!} \sum_{G} GT(x_1, \ldots, x_n) = \widetilde{T},$$

where G is the set of all $n!$ permutations of $x_1, \cdots, x_n$ and the notation $GT(x_1, \cdots, x_n)$ denotes an arbitrary member of G. By Theorem 2.6.1 the variance of the symmetrized statistic $\widetilde{T}$ does not exceed the variance of T.

CHAPTER III

NUISANCE PARAMETERS. TESTS WITH INVARIANT POWER FUNCTIONS

§ 1. NUISANCE PARAMETERS

Suppose that on a measurable space $(\mathfrak{X}, \mathfrak{A})$ there is defined a family of measures $\{P_\theta\}$, for $\theta \in \Omega$, that are dominated by a σ-finite measure μ with respect to which there is a probability density $p_\theta(x)$. We shall assume that the values of the parameter $\theta = (\theta_1, \cdots, \theta_s)$ lie in a Borel set Σ_1 (usually a parallelepiped) contained in the Euclidean space E_s. Furthermore, we shall assume that for every value of x the density $p_\theta(x) = p(x; \theta_1, \cdots, \theta_s)$ is a sufficiently smooth function of the parameters $\theta_1, \cdots, \theta_s$. The degree of smoothness required will vary from problem to problem and will be specified in the individual cases.

Instead of the given parameters $\theta_1, \cdots, \theta_s$, we may introduce other parameters $\gamma_1(\theta_1, \cdots, \theta_s), \cdots, \gamma_s(\theta_1, \cdots, \theta_s)$. For suitable choice of these functions $\gamma_1, \cdots, \gamma_s$, the property of smoothness of $p_\theta(x)$ for given x is retained for the new parameters $\gamma_1, \cdots, \gamma_s$. Let us now suppose that we subject a hypothesis H_0 regarding the parameters $\theta_1, \cdots, \theta_q$ (where $q < s$) to statistical verification. The hypothesis may be of the form $(\theta_1, \cdots, \theta_q) \in \omega$, where ω is a subset of the space of the parameters $(\theta_1, \cdots, \theta_q)$. We make no hypothesis of any sort regarding the remaining parameters $\theta_{q+1}, \cdots, \theta_s$. However, since these parameters appear in the expression for the probability density $p(x; \theta_1, \cdots, \theta_s)$, they may play a part in all calculations of the distributions of the statistics that we may make in verifying the hypothesis H_0; for this reason we have to keep them in mind. Following H. Hotelling, we call such parameters $\theta_{q+1}, \cdots, \theta_s$ *nuisance parameters* of the given problem.

Frequently, the hypothesis H_0 deals with the behavior of certain functions $\gamma_1(\theta_1, \cdots, \theta_s), \cdots, \gamma_q(\theta_1, \cdots, \theta_s)$ $(q < s)$ of the given parameters and is of the form $(\gamma_1, \cdots, \gamma_q) \in \omega_q$, where ω_q is a subset of E_q. Let us assume that the functions $\gamma_1(\theta_1, \cdots, \theta_s), \cdots, \gamma_q(\theta_1, \cdots, \theta_s)$ are functionally independent and sufficiently smooth and that there exist $s - q$ additional functions $\gamma_{q+1}(\theta_1, \cdots, \theta_s), \cdots$ $\cdots, \gamma_s(\theta_1, \cdots, \theta_s)$, such that the parameters $\gamma_1(\theta_1, \cdots, \theta_s), \cdots, \gamma_s(\theta_1, \cdots, \theta_s)$ constitute a new system of parameters with the same properties as the original system. Then it is natural to consider $\gamma_{q+1}(\theta_1, \cdots, \theta_s), \cdots, \gamma_s(\theta_1, \cdots, \theta_s)$ as the nuisance parameters of the hypothesis H_0.

Let us look at some examples of the verification of a hypothesis with nuisance parameters.

Example 1. Verification of a linear hypothesis in the one-dimensional case (see for example [36]). Let $x_1, \cdots, x_n$ denote independent normal variables with means $\xi_1, \cdots, \xi_n$ and common variance σ^2. We know that the vector of the means $(\xi_1, \cdots, \xi_n)$ lies in the t-dimensional linear subspace $\Pi_\Omega \subset E_n$. Suppose that we wish to verify the hypothesis H_0 that this vector lies in the subspace Π_ω of dimension $t - r < t$.

Here the coordinates in t-dimensional space Π_Ω and the standard deviation σ serve as the parameters $(\theta_1, \cdots, \theta_s)$, where $s = t + 1$. Let us assume that there are r linear relationships among the first t parameters, so that there will be $r + 1$ nuisance parameters in the problem.

Suppose, in particular, that we have a linear regression scheme: $\xi_j = \alpha + \beta t_j$ (for $j = 1, 2, \cdots, n$), where the t_j are known. We wish to verify the hypothesis H_0: $\beta = 0$. Here the space Π_Ω consists of vectors of the form $\alpha(1, \cdots, 1) + \beta(t_1, \cdots, t_n)$, so that we have the three parameters α, β, and σ for $t = 2$. With the hypothesis H_0: $\beta = 0$, we have two nuisamce parameters α and σ.

The hypothesis H'_0 that $\alpha = \alpha_0$ and $\beta = \beta_0$, which fixes the value of the regression parameters, reduces to a linear hypothesis in the sense of the above definition if we take the new variables $y_j = x_j - \alpha_0 - \beta_0 t_j$ (for $j = 1, 2, \cdots, n$). For the new variables, we have the hypothesis H'_0: $\alpha = 0$, $\beta = 0$ and a single nuisance parameter σ.

Example 2. An important special case of a linear hypothesis in the somewhat more general sense described at the end of the preceding example is Student's problem: For a repeated normal sample $x_1, \cdots, x_n \in N(a, \sigma^2)$, verify the hypothesis H_0: $a = a_0$ regarding a given value of the general mean a. The nuisance parameter is σ.

Example 3. For the same sample, verify the hypothesis H_0: $a/\sigma = \gamma_0$. Making a suitable change of parameters $\gamma_0 = \gamma_0(a, \sigma) = a/\sigma$ and $\gamma_1 = \gamma_1(a, \sigma)$, we can treat γ_1 as a nuisance parameter. In particular, we may set $\gamma_1 = a$ or $\gamma_1 = \sigma$.

Example 4 (the Behrens–Fisher problem). Let $x_1, \cdots, x_{n_1}$ and $y_1, \cdots, y_{n_1}$ denote two repeated samples taken respectively from normalized sets $N(a_1, \sigma_1^2)$ and $N(a_2, \sigma_2^2)$. Verify the hypothesis H_0: $a_1 - a_2 = 0$. We may treat a_1, σ_1, and σ_2 as nuisance parameters. Basically, this problem is a special case of a generalized linear hypothesis when not only the mean values but also the variances are assumed variable. The present problem is one of the best-known problems with nuisances parameters and it will be treated in detail at the end of the book.

Example 5. Under the condition of Example 4, verify the hypothesis H_0: $\gamma_0 = \sigma_1^2/\sigma_2^2 = \gamma_1$. Here the nuisance parameters are, for example, a_1, a_2, and σ_2.

§2. TESTS WITH INVARIANT POWER FUNCTIONS

Let us again look at the measurable space $(\mathfrak{X}, \mathfrak{A})$ and the family of measures with probability density $p(x; \theta_1, \cdots, \theta_s)$ introduced in the preceding section defined on it. Let us consider a null statistical hypothesis H_0 of the form $\theta_1 = \theta_1^{(0)}, \cdots, \theta_q = \theta_q^{(0)}$, where $q < s$. As has already been noted, the hypothesis H'_0 that there are q relationships of the form $\gamma_1(\theta_1, \cdots, \theta_s) = 0, \cdots, \gamma_q(\theta_1, \cdots, \theta_s) = 0$ among the parameters when certain analytic properties are imposed on these relations can be reduced to such a form. An alternative to H_0 is the hypothesis that the parameters fall in the complement (with respect to Ω) of the set defined by the hypothesis H_0. We denote the hypothesis alternative to H_0 by H_1.

Simple hypothesis H'_0 that the parameters $\theta_1, \cdots, \theta_s$ assume given values satisfying the relationships indicated will be called "special cases of the hypothesis H_0."

The hypothesis H_0 is a complicated statistical hypothesis since it does not fix the values of the nuisance parameters $\theta_{q+1}, \cdots, \theta_s$ and hence does not give the probability density $p(x; \theta_1, \cdots, \theta_s)$.

Let us assume that verification of the hypothesis H_0 is done with the aid of a *critical function* $\Phi(x)$, which we shall also call a *test.* If $\Phi(x)$ assumes only the values 0 and 1, we shall say that the test is *unrandomized* and we shall call the zone of values of x such that $\Phi(x) = 1$ the *critical zone* Z, so that $\Phi(x)$ is the characteristic function of the critical zone. If $x \in Z$, then H_0 is rejected and we take H_1. If $\Phi(x)$ assumes other values in the interval $[0, 1]$ we say that the test is *randomized*. If an observation yields the value x, we need to have a "play-off" with two outcomes: H_0 with probability $1 - \Phi(x)$ and H_1 with probability $\Phi(x)$, and we should accept or reject H_0 in accordance with these outcomes.

We know that, even in the simplest case of testing a simple hypothesis H_0 against a simple alternative H_1, to obtain the most powerful test we need to introduce randomization (see for example [36]).

Let us denote by $E_\theta\Phi(x)$ the mathematical expectation of $\Phi(x)$ for given $\theta = (\theta_1, \cdots, \theta_s)$. Those values $(\theta_1, \cdots, \theta_s)$ for which the hypothesis H_0: $\theta_1 = \theta_1^{(0)}, \cdots, \theta_q = \theta_q^{(0)}$ is satisfied correspond to the value of the *level* of the test $\Phi(x)$:

$$E_\theta\Phi(x) = \alpha(\theta_1^{(0)}, \ldots, \theta_q^{(0)}, \theta_{q+1}, \ldots, \theta_s) \tag{3.2.1}$$

and they characterize the probability of rejecting H_0 when it is valid. If θ assumes values corresponding to H_1, then

$$E_\theta\Phi(x) = \beta(\theta_1, \ldots, \theta_s) \tag{3.2.2}$$

gives the power of the test Φ. We shall refer to both the values of (3.2.1) and the values of (3.2.2) as the values of the *power function* $\phi(\theta_1, \cdots, \theta_s)$ of the test $\Phi(x)$, keeping in mind their different statistical connotations depending on whether they correspond to the values of the parameter H_0 or the parameter H_1. Thus we have

$$E_\theta\Phi(x) = \varphi(\theta_1, \ldots, \theta_s); \quad (\theta_1, \ldots, \theta_s) \in \Omega. \tag{3.2.3}$$

To a significant degree, the power function ϕ characterizes the properties of a test although there are other characteristics that are important for many measurable spaces, for example those spaces encountered in sequential analysis.

To calculate the power function $\phi(\theta_1, \cdots, \theta_s)$ we need to know the values of the parameters. If we wish to determine the behavior of a test for simple particular cases H_0: $(\theta_1^{(0)}, \cdots, \theta_q^{(0)}, \theta'_{q+1}, \cdots, \theta'_s)$ and the alternative H_1: $(\theta_1^{(1)}, \cdots, \theta_q^{(1)}, \theta''_{q+1}, \cdots, \theta''_s)$, then we need to know not only the values of the "basic" parameters $\theta_1, \cdots, \theta_q$ but also the values of the nuisance parameters $\theta_{q+1}, \cdots, \theta_s$. The question naturally arises: is it possible to construct tests for which the power function $\phi(\theta_1, \cdots, \theta_s)$ is independent of the nuisance parameters $\theta_{q+1}, \cdots, \theta_s$ and is determined only by the values of the "basic" parameters $\theta_1, \cdots, \theta_q$ with which the hypothesis H_0 and H_1 deal? We shall call tests of such a kind *tests with invariant power function* (with respect to the nuisance parameters).

There exist trivial examples of such tests: The test $\Phi \equiv \alpha$, where α is a constant belonging to the interval (0, 1), is obviously a test with invariant power function. However, one can easily see that this test is completely useless for verifying the hypothesis H_0.

In general, if a test Φ with invariant power function $\phi(\theta_1, \cdots, \theta_s)$ that is independent not only of the nuisance parameters $\theta_{q+1}, \cdots, \theta_s$ but also of the basic parameters $\theta_1, \cdots, \theta_q$, that test will be useless for verifying H_0. In what follows we shall be concerned with tests $\Phi(x)$ with invariant power function that are not useless. Such tests might be of particular interest if, for a sufficiently broad class of problems, we could obtain sufficiently broad classes of such tests and examine their properties, seeking tests that are optimum in some respect or other. Their advantage over other tests would consist in the fact that these other (noninvariant) tests do not in general admit calculation of the power function $\phi(\theta_1, \cdots, \theta_s)$ for particular cases of H_0 and H_1 and hence their qualities are to a considerable degree unclear.

At the present time however, we have extremely few results in this direction, and the results that we do have deal primarily with particular cases of testing hypothesis (of the type of Examples 1–5 in § 1). Since these results are interesting however, we shall present them in the following section.

§ 3. SOME RESULTS DEALING WITH TESTS WITH INVARIANT POWER FUNCTIONS

The question of unrandomized tests with invariant power function for Student's problem (Example 2 of § 1) was first studied by Dantzig [14] in 1940. Dantzig showed that all such tests are useless. Here we shall look at a somewhat more general statement of this problem.

Let $x_1, \cdots, x_n \in N(a, \sigma^2)$ denote a repeated normal sample. Sufficient statistics for the parameters a and σ^2 are $\bar{x}$ and $s^2 = (1/n) \sum_{i=1}^n (x_i - \bar{x})^2$. Let $\gamma(a, \sigma)$ denote a piecewise continuous function of the parameters a and σ. Suppose that we are verifying a hypothesis H_0: $\gamma(a, \sigma) = \gamma_0$ (where γ_0 is one of the possible values of γ). To verify it we take a randomized test $\Phi(x)$. Consider the conditional mathematical expectation

$$E(\Phi(x) \mid \bar{x}, s^2) = \Phi_1(\bar{x}, s^2). \tag{3.3.1}$$

Since $\bar{x}$ and s^2 are sufficient statistics, the expression $\Phi_1(\bar{x}, s^2)$ will not depend on the parameters and will also be a test, because with a suitable definition of conditional mathematical expectation we have $0 \leq \Phi_1 \leq 1$. Furthermore, Theorems

2.6.1 and 2.6.2 enable us to assume that the properties of the new test $\Phi_1(\overline{x}, s^2)$, which depends only on sufficient statistics, will be no worse in several respects than the properties of the original test. In all cases, for arbitrary values of a and σ,

$$E_{a,\sigma}\Phi(x) = E_{a,\sigma}\Phi_1(\overline{x}, s^2), \tag{3.3.2}$$

that is, the power function of the test $\Phi(x)$ coincides with the power function of the test $\Phi_1(\overline{x}, s^2)$, so that investigation of the tests with invariant power function can be conducted only in a region of tests that are dependent only on the sufficient statistics.

In what follows we shall frequently use an operation of the type (3.3.1), which we shall call the operation of *projection* onto the σ-algebra of sufficient statistics. For now, let us consider the question of verifying the hypothesis H_0: $\gamma(a, \sigma) = \gamma_0$ by using tests with invariant power function. It turns out that the existence of such tests that are not useless depends in a very real way on the form of the function $\gamma(a, \sigma)$. For example, let us take $\gamma(a, \sigma) = \sigma$, so that the hypothesis H_0 is of the form $\sigma = \sigma_0$. The unrandomized test Φ with critical zone $s^2 > C$ (for any $C > 0$) will obviously possess a power function that is independent of a but dependent on σ, so that it has an invariant power function but still is not useless. A similar situation is true of a test with critical zone $f(x_i - x_j) > C$, where f is an arbitrary continuous function of the differences in observations. Its power function is independent of a but (in general) dependent on σ. We can express this situation very concisely by saying that the hypothesis H_0: $\sigma = \sigma_0$ admits an invariant verification. However, as we shall see later, the hypothesis H_0: $a = a_0$ does not have an invariant verification. We can prove this and a more general theorem.

Theorem 3.3.1. *The hypothesis* H_0:

$$\frac{a}{\sigma^\rho} = \gamma_0$$

does not admit an invariant verification for $\rho < 1$.

To prove this theorem we need only consider randomized and unrandomized tests that depend only on sufficient statistics. By virtue of (3.3.1) and (3.3.2), the remaining tests reduce to these. Let us set $\overline{x} = X$ and $s^2 = V$. As we know, $\overline{x}$ and s^2 are stochastically independent (for phenomena of this sort see Chapter IV).

Furthermore (see for example [33]) the statistic X has probability density

$$p_1(x) = \left(\frac{n}{2\pi}\right)^{1/2} \frac{1}{\sigma} \exp\left[-\frac{n}{2\sigma^2}(x-a)^2\right]. \tag{3.3.3}$$

The statistic V has probability density

$$p_2(v) = \left(\frac{n}{2}\right)^{\frac{n-1}{2}} \frac{1}{\Gamma\left(\frac{n-1}{2}\right)} \sigma^{-n+1} v^{\frac{n}{2}-\frac{3}{2}} \exp\left(-\frac{nv}{2\sigma^2}\right). \tag{3.3.4}$$

Thus the common probability density is

$$p(x, v) = C_n \sigma^{-n} v^{\frac{n-3}{2}} \exp\left(-\frac{n}{2\sigma^2}(v + (x-a)^2)\right)$$
$$= C_n \sigma^{-n} \exp\left(-\frac{na^2}{2\sigma^2}\right) v^{\frac{n-3}{2}} \exp\left[-\frac{n}{2\sigma^2}(v + x^2) + \frac{na}{\sigma^2} x\right],$$

где $C_n = \dfrac{n^{n/2}}{\sqrt{2\pi}\, 2^{\frac{n-1}{2}} \Gamma\left(\frac{n-1}{2}\right)}$. (3.3.5)

Let us also set $n/2\sigma^2 = \lambda$ and $na/\sigma^2 = \mu$ (this corresponds to a natural parametrization of our exponential family). From (3.3.5) we obtain

$$p(x, v) = C_n' \lambda^{\frac{n}{2}} \exp\left(-\frac{\mu^2}{4\lambda}\right) v^{\frac{n-3}{2}} \exp[-\lambda(v + x^2) + \mu x], \tag{3.3.6}$$

where $C'_n > 0$ is a new constant.

Let $\Phi_1(X, V)$ denote a test for the hypothesis H_0: $a/\sigma^\rho = \gamma_0$ that depends only on the sufficient statistics X and V. In accordance with (3.3.6). we have an expression for its power function:

$$\varphi(\lambda, \mu) = E_{\lambda,\mu}\Phi_1(X, V)$$
$$= C_n' \lambda^{\frac{n}{2}} \exp\left(-\frac{\mu^2}{4\lambda}\right) \int_{-\infty}^{\infty} dx \int_0^{\infty} dv \Phi_1(x, v) v^{\frac{n-3}{2}}$$
$$\times \exp[-\lambda(v + x^2) + \mu x]. \tag{3.3.7}$$

Thus

$$\int_{-\infty}^{\infty} dx \int_0^{\infty} dv \Phi_1(x, v) v^{\frac{n-3}{2}} \exp[-\lambda(v + x^2) + \mu x]$$
$$= C_n'' \lambda^{-\frac{n}{2}} \exp\left(\frac{\mu^2}{4\lambda}\right) \varphi(\lambda, \mu). \tag{3.3.8}$$

Since $\Phi_1(x, v)$ is a Lebesgue-measurable function such that $0 \le \Phi_1(x, v) \le 1$, we see from (3.3.7) that $\phi(\lambda, \mu)$ is an analytic function of λ and μ for $\lambda > 0$, $-\infty < \mu < \infty$.

The hypothesis H_0:

$$\frac{a}{\sigma^\rho} = \gamma_0$$

can be represented in terms of the parameters λ and μ as follows:

$$\frac{\mu}{\lambda^{1-\rho/2}} = 2\left(\frac{n}{2}\right)^{\rho/2} \gamma_0 = \gamma_1. \tag{3.3.9}$$

Thus, it is a question of the existence of a test $\Phi_1(X, V)$ such that in formula (3.3.8),

$$\varphi(\lambda, \mu) = \psi\left(\frac{\mu}{\lambda^{1-\rho/2}}\right) \tag{3.3.10}$$

for $\lambda > 0$ and $-\infty < \mu < \infty$. Here the function $\psi = \psi(t)$ must not be a constant if the test is to be of any use. Because of the analyticity of $\phi(\lambda, \mu)$ in the region defined above, the function $\psi(t)$ is differentiable for $t > 0$. Let us make use of this fact. If $\psi(t)$ is not a constant, there exists a point $t = t_0$, where $\psi'(t_0) \neq 0$. In equation (3.3.8) we set $\phi(\lambda, \mu) = \psi(\mu/\lambda^{1-\rho/2})$ and differentiate both sides with respect to μ. Since $\lambda > 0$, we can differentiate under the integral signs on the left-hand side of the equation. We obtain

$$\int_{-\infty}^{\infty} dx \int_0^{\infty} dv \Phi_1(x, v) v^{\frac{n-3}{2}} x \exp[-\lambda(v + x^2) + \mu x]$$
$$= C_n'' \lambda^{-n/2} \exp\left(\frac{\mu^2}{4\lambda}\right)\left(\frac{\mu}{2\lambda}\psi\left(\frac{\mu}{\lambda^{1-\rho/2}}\right) + \psi'\left(\frac{\mu}{\lambda^{1-\rho/2}}\right)\frac{1}{\lambda^{1-\rho/2}}\right). \tag{3.3.11}$$

Now, let us set $\mu = t_0\lambda^{1-\rho/2}$ and consider the behavior of the two sides of equation (3.3.11) for small positive values of λ. Let us find an upper bound for the left-hand side of that equation by replacing the factor x with $|x|$ and the function $\Phi_1(x, v)$ with its maximum value, namely 1. For the left-hand side of equation (3.3.11) we obtain the upper bound

$$2\int_0^{\infty} v^{\frac{n-3}{2}} \exp(-\lambda v)\, dv \int_0^{\infty} x \exp(-\lambda x^2 + |\mu| x)\, dx$$
$$= 2\Gamma\left(\frac{n-1}{2}\right)\lambda^{-\frac{n-1}{2}} \int_0^{\infty} x \exp(-\lambda x^2 + |\mu| x)\, dx. \tag{3.3.12}$$

To find a bound for the remaining integral, we remember that $\mu = t_0\lambda^{1-\rho/2}$, where $\rho < 1$. Let us break this integral into two integrals: $\int_0^{\lambda_1}(\ )\,dx + \int_{\lambda_1}^{\infty}(\ )\,dx$, where $\lambda_1 = (\ln 1/\lambda)/\sqrt{\lambda}$. For the values of μ indicated, we see that

$$\int_{\lambda_1}^{\infty} (\quad)\, dx = O\left(\exp - \frac{1}{2}\left(\ln \frac{1}{\lambda}\right)^2\right).$$

Thus a trivial bound for $\int_0^{\lambda_1}(\ )\, dx$ is $\lambda_1^2 = O((\ln(1/\lambda)^2)/\lambda)$. By substituting this value into (3.3.12) we obtain a bound for (3.3.12) as $\lambda \downarrow 0$:

$$O\left(\lambda^{-\frac{n}{2}-\frac{1}{2}}\left(\ln \frac{1}{\lambda}\right)^2\right). \tag{3.3.13}$$

Here the right-hand side of equation (3.3.11) is of the form

$$C_n'' \lambda^{-n/2} \exp\left(\frac{t_0^2}{4}\lambda^{1-\rho}\right)\left(\frac{t_0}{2}\lambda^{-\rho/2}\cdot\psi(t_0) + \psi'(t_0)\,\lambda^{\frac{\rho}{2}-1}\right).$$

Since $\rho < 1$, we have $\rho/2 - 1 < -1/2$. Since $\psi'(t_0) \neq 0$, we obtain a contradiction by letting λ approach 0 from above; thus the proof of Theorem 3.3.1 is complete.

In particular, we see that the hypothesis H_0: $a = a_0$ does not admit an invariant verification.

We note now that for $\rho = 1$ and invariant verification of the hypothesis H_0: $\gamma = a/\sigma = \gamma_0$ is possible. To carry out this verification we may take, for example, the statistic $U = X/\sqrt{X_1^2 + \cdots + X_n^2}$, where

$$X = \bar{x}\sqrt{n+1},\ X_j = \left(1 - \frac{1}{n+1}\right)^{-1/2}(x_j - \bar{x})\ (j = 1, 2, \ldots, n).$$

The distribution of this statistic depends only on a/σ, so that the unrandomized test with critical zone $|U| \geq C$ will have an invariant power function.

A result analogous to the last one can be formulated for the general case of a linear hypothesis (Example 1 of § 1). A very simple case of a more general hypothesis admitting inequality of the variances of the elements of the sample is the Behrens–Fisher problem (Example 4 of § 1). In the notation of that example, suppose that we are verifying the hypothesis H_0: $a_1 - a_2 = 0$. We shall take a_1, σ_1, and σ_2 as the nuisance parameters.

Theorem 3.3.2. *The hypothesis of equality of means in the Behrens–Fisher problem does not admit an invariant verification.*

Proof. Corresponding to the four parameters a_1, a_2, σ_1, and σ_2 in the problem are the four sufficient statistics $X_1 = \bar{x}$, $X_2 = \bar{y}$, $V_1 = s_1^2$, and $V_2 = s_2^2$. Remembering the laws of distribution of the form (3.3.3) and (3.3.4) for two samples of sizes n_1 and n_2, we obtain an expression for the common probability density

$$p(x_1, x_2, v_1, v_2,) = C_n\sigma_1^{-n_1}\sigma_2^{-n_2}\exp\left(-\frac{n_1a_1^2}{2\sigma_1^2} - \frac{n_2a_2^2}{2\sigma_2^2}\right)$$

$$\times v_1^{\frac{n_1-3}{2}} v_2^{\frac{n_2-3}{2}} \exp\left[-\frac{n_1}{2\sigma_1^2}(v_1 + x_1^2) - \frac{n_2}{2\sigma_2^2}(v_2 + x_2^2)\right.$$

$$\left. + \frac{n_1a_1}{\sigma_1^2}x_1 + \frac{na_2}{\sigma_2^2}x_2\right]; \quad C_n > 0. \tag{3.3.14}$$

Just as in the preceding derivation, we introduce the natural parameters

$$\frac{n_1}{2\sigma_1^2} = \lambda_1; \quad \frac{n_2}{2\sigma_2^2} = \lambda_2; \quad \frac{n_1a_1}{\sigma_1^2} = \mu_1; \quad \frac{n_2a_2}{\sigma_2^2} = \mu_2.$$

From these equations we get $2a_1 = \mu_1/\lambda_1$ and $2a_2 = \mu_2/\lambda_2$, so that the null hypothesis takes the form H_0: $\mu_1/\lambda_1 - \mu_2/\lambda_2 = 0$. The more general hypothesis H_0: $a_1 - a_2 = \delta$, where δ is given, reduces to this one when we substitute, for example, $x_i + \delta/2$ for x_i (where $i = 1, 2, \cdots, n_1$) and $y_j - \delta/2$ for y_j (where $j = 1, 2, \cdots, n_2$). Thus, as in the preceding derivation, the question of an invariant verification of H_0 reduces to the question of existence of the equation

$$\int_{-\infty}^{\infty} dx_1 \int_{-\infty}^{\infty} dx_2 \int_0^{\infty} dv_1 \int_0^{\infty} dv_2 v_1^{\frac{n_1-3}{2}} v_2^{\frac{n_2-3}{2}}$$

$$\times \exp[-\lambda_1(v_1 + x_1^2) - \lambda_2(v_2 + x_2^2) + \mu_1x_1 + \mu_2x_2]$$

$$= C_n'\lambda_1^{-\frac{n_1}{2}}\lambda_2^{-\frac{n_2}{2}}\exp\left(\frac{\mu_1^2}{4\lambda_1} + \frac{\mu_2^2}{4\lambda_2}\right)\psi\left(\frac{\mu_1}{\lambda_1} - \frac{\mu_2}{\lambda_2}\right). \tag{3.3.15}$$

Here $C_n' > 0$ and ψ is a power function not depending on the nuisance parameters. The left-hand member of the equation is an analytic function of λ_1, λ_2, μ_1, and μ_2 for positive λ_1 and λ_2 and arbitrary μ_1 and μ_2. Therefore $\psi = \psi(t)$ is a twice differentiable function. Let us show that $\psi''(t) \equiv 0$. Let us suppose, to the contrary, that there exists a point t_0 such that $\psi''(t_0) \neq 0$. We multiply both sides of equation (3.3.15) by the quantity $\exp[-(\mu_1^2/4\lambda_1 + \mu_2^2/4\lambda_2)]$ and apply the differential operator $\partial^2/\partial\mu_1\partial\mu_2$ to both sides of the resulting equation. Then we set $\mu_1/\lambda_1 = 2t_0$ and $\mu_2/\lambda_2 = t_0$. On the right-hand side of the equation obtained by multiplying (3.3.15) by the exponential expression indicated above, we obtain the expression

$$-C_n'\lambda_1^{-\frac{n_1}{2}}\lambda_2^{-\frac{n_2}{2}}\psi''(t_0)\cdot\frac{1}{\lambda_1\lambda_2}. \tag{3.3.16}$$

On the left side we obtain an expression that by virtue of (3.3.13) takes the form

$$O\left(\lambda_1^{-\frac{n}{2}}\lambda_2^{-\frac{n}{2}}\right) \tag{3.3.17}$$

as $\lambda_1 \downarrow 0$ and $\lambda_2 \downarrow 0$. If $\psi''(t_0) \neq 0$, we obtain a contradiction between (3.3.16) and (3.3.17). Thus $\psi''(t) \equiv 0$ and $\psi(t) = \alpha + \beta t$, where α and β are constants. Furthermore, the argument $t = \mu_1/\lambda_1 - \mu_2/\lambda_2$ can assume arbitrary values and $\psi(t)$, as power function, satisfies the condition $0 \leq \psi(t) \leq 1$. Therefore $\beta = 0$ and $\psi(t) \equiv \alpha$, so that the test is useless. This completes the proof of the theorem.

However, the methods that we have expounded do not enable us to answer the general question of hypothesis of the form H_0: $\gamma(a, \sigma) = \gamma_0$ that admit an invariant verification in a problem of the type of Student's problem, to say nothing of more general problems. To solve such problems, we need finer analytical methods.

§4. STEIN'S TEST

The above derivations had to do with samples of constant size, normal distributions defined on a Euclidean space. In a well-known article, Stein [67] showed in 1945, in particular, that the situation as regards Student's problem (Example 2 of §1) changes if, instead of the tests based on samples of constant size, we use two-sample tests with samples of random size, i.e. if we take the point of view of sequential analysis. It then turns out that the hypothesis H_0: $a = a_0$ in Student's problem will admit an invariant verification. The present section is devoted to an exposition of this result of Stein.

Suppose that independent random variables $x_1, x_2, \cdots$ are distributed according to a normal law $N(a, \sigma^2)$. We wish to verify the hypothesis H_0: $a = a_0$ with the aid of a test whose power function depends only on $a - a_0$ and is independent of σ^2. Suppose that we have a sample of size n_0: $x_1, \cdots, x_{n_0}$. Suppose that

$$s^2 = \frac{1}{n_0 - 1}\left\{\sum_{i=1}^{n_0} x_i^2 - \frac{1}{n_0}\left(\sum_{i=1}^{n_0} x_i\right)^2\right\} \tag{3.4.1}$$

is an unbiased estimate of the variance. Furthermore, let us define the integer n by

$$n = \max\left\{\left[\frac{s^2}{z}\right] + 1,\ n_0 + 1\right\}, \tag{3.4.2}$$

where $z > 0$ is a prenamed constant and $[q]$ denotes the greatest integer not exceeding q. If we know the number s^2, we can determine the real numbers b_i

$(i = 1, \cdots, n)$ such that

$$\sum_{i=1}^{n} b_i = 1; \quad b_1 = b_2 = \ldots = b_{n_0}; \quad s^2 \sum_{i=1}^{n} b_i^2 = z. \tag{3.4.3}$$

Such a choice is possible since, when $\Sigma_{i=1}^{n} b_i = 1$ and $b_1 = b_2 = \cdots = b_{n_0}$, the minimum $\min \Sigma_{i=1}^{n} b_i^2$ is attained when $b_i = 1/n$ $(i = 1, 2, \cdots, n)$ and its value is $1/n$. By virtue of (3.4.2), we have $n \geq s^2/z$ and $s^2/n \leq z$. Now we define a random variable t' by

$$t' = \frac{\sum_{i=1}^{n} b_i x_i - a_0}{\sqrt{z}} = \frac{\sum_{i=1}^{n} b_i (x_i - a)}{\sqrt{z}} + \frac{a - a_0}{\sqrt{z}} = u + \frac{a - a_0}{\sqrt{z}}, \tag{3.4.4}$$

where

$$u = \frac{\sum_{i=1}^{n} b_i (x_i - a)}{\sqrt{z}}. \tag{3.4.5}$$

Let us show that u has a Student's distribution $t_{n_0 - 1}$ with $n_0 - 1$ degrees of freedom (see for example [33]). Specifically, by (3.4.3),

$$u = \frac{\sum_{i=1}^{n} b_i (x_i - a)}{s\sqrt{\sum_{i=1}^{n} b_i^2}} = \frac{b_1 \sum_{i=1}^{n_0} (x_i - a) + \sum_{j=n_0+1}^{n} b_j (x_j - a)}{s\sqrt{\sum_{i=1}^{n} b_i^2}}.$$

We note that the expression $b_1 \Sigma_{i=1}^{n_0} (x_i - a)$ is stochastically independent of s by the independence of $\bar{x}$ and s^2 in a normal sample. The expression $\Sigma_{j=n_0+1}^{n} b_j (x_j - a)$ is stochastically independent of s and of $b_1 \Sigma_{i=1}^{n_0} (x_i - a)$ since it contains only the observations of x_j with $j > n_0$, whereas the preceding expressions contain only the observations of x_i with $i \leq n$. Thus for given s and given constants b_i, the numerator of the fraction (3.4.5) is distributed independently of s according to a normal law $N(0, \sigma^2 \Sigma_{i=1}^{n} b_i^2)$. Furthermore, the quantity $(n_0 - 1)s^2/\sigma^2$ has the distribution $\chi^2_{n_0 - 1}$, so that s is distributed like $\sigma\sqrt{\chi^2_{n_0-1}/(n_0 - 1)}$. Thus the fraction u is distributed like $X/\sqrt{\chi^2_{n_0-1}/(n-1)}$ where $X \in N(0, 1)$ is independent of $\chi^2_{n_0-1}$. Therefore $u = t_{n_0-1}$ is distributed according to Student's law with $n_0 - 1$ degrees of freedom. The statistic u can be used to construct an unrandomized test of the hypothesis H_0: $a = a_0$ with power function independent of σ. Let α denote the desired level (the possibility of falling into the critical zone for H_0). Let $\xi_{\alpha/2}$ denote the abscissa at which

$$P\{t_{n_0-1} > \xi_{\alpha/2}\} = \frac{\alpha}{2}. \tag{3.4.6}$$

We determine the critical zone to be the zone

$$\left| \frac{\sum_{i=1}^{n} b_i x_i - a_0}{\sqrt{z}} \right| > \xi_{\alpha/2}. \qquad (3.4.7)$$

Then we obtain a test for H_0 with power function

$$P\left\{ \left| t_{n_0-1} + \frac{a - a_0}{\sqrt{z}} \right| > \xi_{\alpha/2} \right\}. \qquad (3.4.8)$$

Here z is a constant chosen in advance, and therefore the power function $\phi(a)$ depends only on a and is independent of σ.

We note also that a Student's distribution (for t_{n_0-1}) is even and unimodal and that it has a single vertex at the coordinate origin. For given a_0 and $\xi_{\alpha/2}$ the probability (3.4.8) attains a minimum at $a = a_0$ and its value at a_0 is equal to the level α. For $a \neq a_0$ it is higher than that level, so that the probability of rejecting H_0 when it is not valid is greater than when it is valid, i.e. this test is unbiased.

To test H_0: $a = a_0$ against a one-sided alternative $a > a_0$, we may follow a similar procedure. The critical zone of the level α can be defined as

$$\frac{\sum_{i=1}^{n} b_i x_i - a_0}{\sqrt{z}} > \xi_\alpha \qquad (3.4.9)$$

with power function

$$\varphi(a) = P\left\{ t_{n_0-1} > \xi_\alpha + \frac{a_0 - a}{\sqrt{z}} \right\}.$$

Of course one can also construct other critical zones with invariant power function by means of the statistic u.

Let us also stop to look at the distribution of the random variable n, i.e. the number of observations necessary for obtaining a solution. We have

$$n = \max\left\{ \left[\frac{s^2}{z}\right] + 1,\ n_0 + 1 \right\}.$$

Thus

$$\begin{aligned} P\{n = n_0 + 1\} &= P\left\{ \frac{s^2}{z} \leqslant n_0 \right\} \\ &= P\left\{ \frac{(n_0 - 1)s^2}{\sigma^2} < \frac{n_0(n_0 - 1)z}{\sigma^2} \right\} = P\{\chi^2_{n_0-1} < y\}, \end{aligned} \qquad (3.4.10)$$

where $y = n_0(n_0 - 1)z/\sigma^2$. For $\nu > n_0 + 1$ we have

$$P\{n=\nu\}=P\left\{\nu<\frac{s^2}{z}+1\leqslant\nu+1\right\}$$
$$=P\left\{\frac{(\nu-1)(n_0-1)z}{\sigma^2}<\chi^2_{n_0-1}<\frac{\nu(n_0-1)z}{\sigma^2}\right\}, \qquad (3.4.11)$$

and the necessary values of the probabilities $P\{n=\nu\}$ can be found from a table of the χ^2 distribution. From this result, we can find the mathematical expectation $E(n)$ of the necessary number of observations. We note that these last numbers depend on the nuisance parameter σ. It should be noted that our wish to make the power function strictly invariant with respect to the nuisance parameter can lead to a certain loss of information. Therefore it is expedient to use certain variations of this test that ensure weak dependence of the power function on the nuisance parameter but that raise the lower limit of the power (cf. Stein [67]). We shall not pursue these considerations since the primary aim of the present monograph is to give as complete a description as possible of certain forms of invariant tests, so as to make it possible to form an opinion of their qualities in comparison with other tests and on the loss of information in comparison with those situations in which the values of the nuisance parameters are known. For further information on tests of the Stein type as applied to various hypotheses associated with a normal law see the article by Chatterjee [78].

CHAPTER IV

SIMILAR TESTS AND STATISTICS

§1. SIMILARITY OF TESTS AND OF STATISTICS

In the preceding chapter we considered several unbiased tests. The principle of unbiasedness is often put forward as a desirable property of tests.

Suppose that we have a measurable space $(\mathfrak{X}, \mathfrak{A})$ with family of probabilistic measures $\{P_\theta\}$ (for $\theta \in \Omega$) that are dominated by a σ-finite measure μ and two (in general, complex) hypotheses H_0: $\theta \in \omega \subset \Omega$ and H_1: $\theta \in \Omega \setminus \omega$. Suppose that we are verifying the hypothesis H_0 with the aid of a test Φ (in general randomized) with power function $\phi_\Phi(\theta) = E_\theta(\Phi(x))$. The principle of unbiasedness of a test Φ consists in the requirement that the probability of rejecting the hypothesis H_0 when it is not valid be less than when it is valid. Thus we must have

$$\left.\begin{array}{ll} \varphi_\Phi(\theta) \leqslant \alpha & \text{for } \theta \in \omega, \\ \varphi_\Phi(\theta) \geqslant \alpha & \text{for } \theta \in \Omega \setminus \omega. \end{array}\right\} \tag{4.1.1}$$

A test Φ of such a form is said to be *unbiased*. Of course, this property in itself does not guarantee high quality of the test. For example the trivial test $\Phi \equiv \alpha$ is obviously unbiased. However, it is a rather natural requirement.

Following Sverdrup [68], we note that if the parameter space Ω is equipped with a topology according to which $\phi_\Phi(\theta)$ is a continuous function of θ and the common boundary $\partial\omega$ of the regions ω and $\Omega \setminus \omega$ is nonempty, it follows from (4.1.1) that $\phi_\Phi(\theta) = \alpha$ for $\theta \in \partial\omega$. Thus $\phi_\Phi(\theta)$ has a constant value α for $\theta \in \partial\omega$. If we compare this behavior of $\phi_\Phi(\theta)$ with the behavior of the trivial test $\Phi_1 \equiv 1$ for which $\phi_{\Phi_1}(\theta) = E_\theta\Phi_1 = 1$ for arbitrary $\theta \in \Omega$, we see that the critical function ϕ is similar in a certain sense to the entire space $\mathfrak{X}$ (the critical zone of the trivial test Φ_1). Therefore, the test Φ is said to be *similar* on $\partial\omega$.

These considerations indicate the procedure for studying tests $\Phi(x)$ for which

$$E_\theta \phi(x) = \alpha \tag{4.1.2}$$

for all values of θ in a set $\Omega_0 \subset \Omega$. Such tests are said to be similar with respect to the family of distributions $\{P_\theta\}$ (for $\theta \in \Omega_0$) or, more briefly, with respect to the parameter set Ω_0. The concept of similarity was first introduced by Neyman [56]. Let us pause to look at the properties of such tests for the scheme of families of distributions with nuisance parameters (cf. §1 of the preceding chapter). Suppose that we are testing the null hypothesis H_0: $\gamma_1(\theta_1, \cdots, \theta_s) = 0, \cdots, \cdots, \gamma_q(\theta_1, \cdots, \theta_s) = 0$, where $q < s$ and $\gamma_1, \cdots, \gamma_q$ are sufficiently smooth functions, for a family of densities $\{p(x; \theta_1, \cdots, \theta_s)\}$. Suppose that the s smooth functions $\gamma_1(\theta_1, \cdots, \theta_s), \cdots, \gamma_q(\theta_1, \cdots, \theta_s), \gamma_{q+1}(\theta_1, \cdots, \theta_s), \cdots, \gamma_s(\theta_1, \cdots, \theta_s)$ constitute a new parametrization and that $\gamma_{q+1}, \cdots, \gamma_s$ are the nuisance parameters. If we were to succeed in constructing a test Φ with invariant power function for the trial H_0, its power function $\phi_\Phi(\gamma_1, \cdots, \gamma_q)$ would depend only on the parameters $\gamma_1, \cdots, \gamma_q$ and would not contain nuisance parameters. However, we know from the preceding chapter that, in the simplest classical cases of testing a linear hypothesis and its generalizations, such tests do not exist for repeated samples of constant size. Therefore it is natural to weaken the requirement of invariance of the power function with respect to the nuisance parameters. We can require that the power function ϕ_Φ be independent of the nuisance parameters at least when the null hypothesis H_0: $\gamma_1 = 0, \cdots, \gamma_q = 0$ is realized. Then the values of ϕ_Φ will constitute a level α (the probability of rejecting H_0 when it is valid), and we arrive at the similarity condition

$$\phi_\Phi(\theta) = E_\theta \Phi(x) = \alpha, \tag{4.1.3}$$

provided that

$$H_0: \gamma_1(\theta_1, \ldots, \theta_s) = 0, \ldots, \gamma_q(\theta_1, \ldots, \theta_s) = 0; \; q < s. \tag{4.1.4}$$

The points $(\theta_1, \cdots, \theta_s)$ of the set (4.1.4) are limiting values of points that violate these relations. Therefore if $\phi_\Phi(\theta)$ is a continuous function of θ, then, in accordance with what was said above, the similarity condition (4.1.3) of a test with hypothesis H_0 follows from the requirement that the test be unbiased.

The general definition of similarity of a test for $\theta \in \Omega_0 \subset \Omega$ can be regarded as a special case of unbiasedness of statistics (cf. Chapter II, §6). Specifically, in the present case, we are considering tests, i.e. statistics $\Phi(x)$ satisfying the inequalities $0 \le \Phi(x) \le 1$ for $\theta \in \Omega_0 \subset \Omega$. Here the functional $f(\theta)$ (cf. (2.6.1)) is equal to the constant α. From an analytical point of view, the condition $0 \le \Phi(x) \le 1$ is equivalent to consideration of bounded statistics $\xi(x)$: if $|\xi(x)| < K$, where K is a constant, then the linear transformation $\xi(x) \to 1/2 + \xi(x)/2k$ makes the statistic $\xi(x)$ a test. In what follows, some of the results that we obtain for similar tests will be carried over to a certain degree to unbiased estimates.

If we require, in addition to the conditions for similarity of a test (4.1.3) and (4.1.4), that the test Φ with critical zone Z be unrandomized, then conditions (4.1.3) and (4.1.4) take the form

$$P_\theta(Z) = \alpha \quad \text{for} \quad H_0\colon \gamma_1 = 0,\ \gamma_2 = 0,\ \ldots,\ \gamma_q = 0; \qquad q < s. \quad (4.1.5)$$

Thus the conditional probability of falling in the ciritical zone Z under H_0 must be a constant and equal to the level α for each simple particular case H_0'. Study of similar unrandomized tests is quite difficult. The zone Z of a test of the form indicated is called the *similarity zone*. In the general case, in analogy with definition (4.1.2) we shall call the scalar statistic $T(x)$ *similar* if, for arbitrary $\theta \in \Omega_0 \subset \Omega$ and an arbitrary value of ξ, the probability $P_\theta(T(x) < \xi) = F_T(\xi)$ is independent of θ. An analogous definition can be made for the most general case of a statistic T. Suppose that the statistic T maps measurable spaces $(\mathfrak{X}, \mathfrak{A})$ into measurable spaces $(\mathfrak{T}, \mathfrak{B})$ and that a family of measures $\{P_\theta\}$ (for $\theta \in \Omega$) is defined on $\mathfrak{X}$. The statistic T is similar for $\theta \in \Omega_0 \subset \Omega$ if $P_\theta(T^{-1}B)$ is independent of θ for arbitrary $\theta \in \Omega_0$ and $B \in \mathfrak{B}$.

If $T(x)$ is a scalar similar statistic, then zones of the form $Z_{\xi_1\xi_2}\colon \xi_1 \le T(x) < \xi_2$ are similar for arbitrary ξ_1 and ξ_2 (with $\xi_1 < \xi_2$) and they can be used to construct unrandomized similar tests. As an approximation to (scalar) similar statistics one can consider statistics $T(x)$ with mean value $E_\theta T(x)$ and variance $D_\theta(T(x))$ that are independent of θ for $\theta \in \Omega_0 \subset \Omega$. However, their study involves great difficulties.

Together with similar zones, we sometimes consider *bisimilar zones*. These were also introduced by Neyman [56]. Suppose that k measurable real functions $f_1(x), \cdots, f_k(x)$ are defined on the space $(\mathfrak{X}, \mathfrak{A})$ with the system of measures $\{P_\theta\}$ (for $\theta \in \Omega$). Suppose also that

$$\int_{\mathfrak{X}} f_i(x)\, dP_\theta(x) = 0 \quad \text{for} \quad \theta \in \Omega_0 \subset \Omega, \qquad i = 1, \ldots, k.$$

Then the similar zone $Z \subset \mathfrak{X}$ is said to be *bisimilar* if it satisfies the additional conditions

$$\int_Z f_i(x)\, dP_\theta(x) = 0 \quad (i = 1, 2, \ldots, k). \qquad (4.1.6)$$

Bisimilar zones are useful for the construction of similar tests possessing additional desirable properties. It is natural to consider the bisimilar statistics $T(x)$ (including randomized tests $\Phi(x)$) defined by the equations

$$E_\theta T(x)=\alpha;\quad E_\theta f_i(x)T(x)=0;\quad i=1,\ 2,\ \ldots,\ k;\quad \theta\in\Omega_0\subset\Omega, \tag{4.1.7}$$

where the $f_i(x)$ $(i = 1, 2, \cdots, k)$ are the functions referred to above.

§ 2. NEYMAN STRUCTURES. LEHMANN'S AND SCHEFFÉ'S THEOREMS

Let $\Phi(x)$ denote a similar test of level α for $\theta \in \Omega_0 \subset \Omega$,

$$E_\theta\Phi(x)=\alpha;\quad \theta\in\Omega_0\subset\Omega. \tag{4.2.1}$$

Suppose that a family of probabilistic measures $\{P_\theta\}$, $\theta \in \Omega$, has a sufficient statistic T with respect to $\theta \in \Omega_0$. Then for an arbitrary value $T = t$ the conditional distribution $\Phi(x)$ is, for given t, independent of the parameter $\theta \in \Omega_0$. Consider the "projection" of the test $\Phi(x)$ onto the sufficient statistic T (cf. Chapter III, §3)

$$E_\theta(\Phi(x)\,|\,t)=\Phi_1(t). \tag{4.2.2}$$

The statistic $\Phi_1(t)$ is independent of the parameter $\theta \in \Omega_0$. If it is constant and equal to the level α for almost all values of t (with respect to all measures induced by T for $\theta \in \Omega_0$), a condition which we shall write in the form

$$E_\theta(\Phi(x)|t) = \Phi_1(t) = \alpha \quad \text{(for almost all } t), \tag{4.2.3}$$

then

$$E_\theta\Phi(x)=E\Phi_1(t)=\alpha,\quad \theta\in\Omega_0, \tag{4.2.4}$$

i.e. the test Φ is similar. A test Φ satisfying condition (4.2.3) is called a *Neyman structure* for the statistic T (cf. Neyman [58]). An analogous concept can be defined for the statistic $\xi(x)$.

An easily-grasped interpretation of Neyman structures is as follows. If T is a sufficient statistic and $\xi(x)$ is another statistic, then the conditional distribution of $\xi(x)$ for a given value $T = t$ is independent of the parameter $\theta \in \Omega_0$. If $\mathfrak{X}$ and the values of the statistic T are contained in Euclidean spaces, then under definite (and not excessively cumbrous) analytical conditions we may speak of the conditional distribution of the statistic $\xi(x)$ on the surface $T = t$ (see for example Cramér [33]). This distribution is independent of θ. Under the same conditions on the surface $T = t$, we can "cut out" a zone Z_t such that the conditional probability $P_\theta(\xi(x) \in Z_t | t)$ is equal to $\bar{\alpha}$, that is, equal to the level for $\theta \in \Omega_0$. If we succeed in "gluing" these zones Z_t "continuously" for all values of t into a single zone Z, this zone will obviously be similar to the zone for the level α. Of course, rigorous treatment of such considerations is rather laborious but it can be done without difficulty for many important classical tests.

Lehmann and Scheffé [37] have characterized an important class of families of measures $\mathcal{P}$ for which all similar tests are Neyman structures. This characterization involves the concept of completeness of a family of distributions. A family of probabilistic measures $\{P_\theta\}$, $\theta \in \Omega$, is said to be *complete* if an arbitrary measurable function $f(x)$ such that

$$E_\theta f(x) = 0, \quad \theta \in \Omega, \tag{4.2.5}$$

is equal to 0 almost everywhere. A family $\{P_\theta\}$ is said to be *boundedly complete* if an arbitrary bounded measurable function $f(x)$ satisfying the condition (4.2.5) is equal to 0 almost everywhere. We have (see Lehmann and Scheffé [37] and Lehmann [36])

Theorem 4.2.1. *Let $\{P_\theta\}$, $\theta \in \Omega$, denote a family of distributions and let T denote a sufficient statistic for $\theta \in \Omega_0 \subset \Omega$. For all similar tests $\Phi(x)$ for the hypothesis H_0: $\theta \in \Omega_0$ to be Neyman structures with respect to T, it is necessary and sufficient that the family of measures induced by T be boundedly complete.*

Proof of sufficiency. Let $\{P_\theta^T\}$ denote the boundedly complete family of distributions induced by the statistic T. Consider a similar test $\Phi(x)$ of level α. We have $E_\theta(\Phi(x) - \alpha) = 0$ for $\theta \in \Omega_0$. Consider the "projection" $E(\Phi(x|t)) = \Phi_1(t)$. Then $E(\Phi_1(t) - \alpha) = 0$ for all measures P_θ^T with $\theta \in \Omega_0$. Since the family is boundedly complete, $\Phi_1(t) = \alpha$ with probability 1, and hence the test $\Phi(x)$ is a Neyman structure with respect to T.

Proof of necessity. Let us suppose that the family $\{P_\theta^T\}$, $\theta \in \Omega_0$, is not boundedly complete. Then there exists a function f such that $f(T) \neq 0$ with positive probability, $|f(T)| \leq M < \infty$ (where M is some constant), and $Ef(T) = 0$ for all P_θ^T with $\theta \in \Omega_0$. Suppose that $c = (1/m) \min(\alpha, 1 - \alpha)$ and $\Phi(t) = \alpha + cf(t)$. Then $0 \leq \Phi(t) \leq 1$ and $E_\theta \Phi(t) = \alpha$ for $\theta \in \Omega_0$, so that $\Phi(t)$ is a similar test that is not a Neyman structure. This completes the proof of the theorem.

Exponential families (see Chapter II, §5) constitute an important special case of complete families. Consider the exponential family (2.5.2)

$$dP_\theta(x) = C(\theta)\exp\left[\sum_{j=1}^{s}\theta_j T_j(x)\right] d\mu(x), \tag{4.2.6}$$

where $\mu(x)$ is the σ-finite measure with respect to which the probability densities in (2.5.2) are taken. In this connection we have (see Lehmann [36])

Theorem 4.2.2. *Suppose that a parameter set Ω_0 contains an s-dimensional parallelepiped. Then the family of distributions induced by the sufficient statistic $T = (T_1(x), \cdots, T_s(x))$ for $\theta \in \Omega_0$ is complete.*

Proof. By means of a translation of the statistic T by a constant (vector) amount we may assume that the s-dimensional parallelepiped contained in Ω is of the form

$$I: -a \leqslant \theta_i \leqslant a, \quad a > 0, \quad j = 1, 2, \ldots, s.$$

Furthermore, suppose that $f(T)$ is a measurable function such that $E_\theta f(T) = 0$ for all $\theta \in \Omega_0$. Let $\nu(T)$ denote the measure induced by $\mu(x)$ on the space $\mathcal{T}$ of values of T. We have

$$\int_{\mathcal{T}} \exp(\theta_1 T_1 + \ldots + \theta_s T_s) f(T)\, d\nu(T) = 0 \tag{4.2.7}$$

for $(\theta_1, \cdots, \theta_s) \in I$. The integral (4.2.7) is assumed to converge absolutely. It is a multiple Laplace-Stieltjes integral. It can be extended to the complex "polystrip" Π: $\theta_j = \xi_j + i\tau_j$, where $-a \leq \xi_j \leq a$, and $-\infty < \tau_j < \infty$ for $j = 1, 2, \cdots \cdots, s$, in which it is an analytic function. By virtue of (4.2.7) it vanishes on the entire polystrip Π and hence (see Theorem 1.2.1) $f(T) = 0$ almost everywhere (with respect to the measure ν). The theorem is proved.

We make a slight addition to this theorem that will be useful in the theory of unbiased statistics for exponential families.

Suppose that an unbiased statistic $\chi(T)$ in such a family depends only on sufficient statistics and that it estimates the function $\rho(\theta_1, \cdots, \theta_s)$. Then it is the only unbiased estimate depending solely on sufficient statistics. To see this, suppose that $\chi_1(T)$ is another unbiased estimate for $\rho(\theta_1, \cdots, \theta_s)$. Then $E_\theta(\chi(T) - \chi_1(T)) = 0$ for $\theta \in I$, so that by Theorem 4.2.2 we have $\chi(T) = \chi_1(T)$ with probability 1.

Theorems 2.6.1 and 2.6.2 enable us to amplify this remark somewhat. Let $\xi(x)$ denote an arbitrary unbiased estimate of the function $\rho(\theta_1, \cdots, \theta_s)$ with variance $D(\xi(k))$ or, in general, the loss

$$Eg(\xi(x) - \rho(\theta)), \text{where } \rho(\theta) = \rho(\theta_1, \ldots, \theta_s),$$

and let g denote a given convex (downward) function. The conditional mathematical expectation $E(\xi(x)|T) = \chi(T)$ is also an unbiased estimate of $\rho(\theta)$. Also, it depends only on the sufficient statistic and from what was said above it is the unique unbiased estimate of this form. Thus the "projections" of arbitrary unbiased estimates onto sufficient statistics coincide. In accordance with Theorem 2.6.2, for an arbitrary unbiased estimate $\xi(x)$,

$$Eg(\xi(x)-\rho(\theta)) \geqslant Eg(\chi(T)-\rho(\theta)), \tag{4.2.8}$$

so that the statistic $\chi(T)$ yields the minimum of the loss.

Whereas the various forms of exponential families provide interesting examples of completeness, consideration of examples of incomplete families of a distribution is also instructive. Let us look again at the Behrens-Fisher problem (Chapter III, §3). In accordance with formula (3.3.14) the corresponding distribution has four natural parameters (denoted in the text following formula (3.3.14) by λ_1, λ_2, μ_1, and μ_2). If no relationships are imposed on these parameters, the set of their values constitutes a four-dimensional parallelepiped and the family defined by (3.3.14) is complete. However, in the Behrens-Fisher problem we further impose the condition $a_1 = a_2$, that is, $\lambda_1\mu_2 - \lambda_2\mu_1 = 0$, so that we do not have a parallelepiped of the type indicated. In formula (3.3.14) let us set $a_1 = a_2 = a$. Let $\psi(\bar{x} - \bar{y}) = \psi(X_1 - X_2)$ denote an arbitrary continuous odd function of $X_1 - X_2$ for which there exists a mathematical expectation when the probability density is that of (3.3.14). The quantity $X_1 - X_2$ is a normal variable with zero mean, so that the mathematical expectation of the odd function $\psi(X_1 - X_2)$ vanishes. This proves the incompleteness of the family (3.3.14) for $a_1 = a_2$. The majority of the questions considered below deal with the behavior of exponential families in which the conditions imposed on the natural parameters produce incompleteness. The description of similar tests (especially unrandomized tests) and of unbiased estimates under such conditions is an interesting problem of analytical statistics.

§3. SOME METHODS OF CONSTRUCTING SIMILAR ZONES

Let us consider the usual situation in which we have a measurable space $(\mathfrak{X}, \mathfrak{A})$ equipped with a family of measures $\{P_\theta\}$, $\theta \in \Omega$. Let Ω_0 denote a subset of Ω. (This corresponds to the null hypothesis H_0: $\theta \in \Omega_0$.) The zone A, which is similar with respect to Ω_0, has a constant probability $P_\theta(A) = \alpha$ for $\theta \in \Omega_0$. Let $\Phi(x)$ denote the characteristic function of the set A (that is, $\Phi(x) = 1$ for $x \in A$ and $\Phi(x) = 0$ for $x \notin A$). We then have an unrandomized test for the hypothesis H_0.

Now let $\Phi_1(x)$ denote a randomized test for H_0. We thus have the similarity condition $E_\theta \Phi_1(x) = \alpha$ for $\theta \in \Omega_0$. If the similar test $\Phi_1(x)$ were unrandomized, not only the mean value $E_\theta \Phi_1(x)$ but also the distribution of $\Phi_1(x)$ would depend on the parameter $\theta \in \Omega_0$. This is the basic difference between unrandomized and randomized tests. However, this difference is maintained only for the basic sample space $\mathfrak{X}$ and it can be removed in a certain sense by a suitable broadening of that space.

Let $\mathfrak{U}$ denote the interval $[0, 1]$ with Borel σ-algebra and uniform distribution. Consider the Cartesian product $\mathfrak{X} \times \mathfrak{U}$ with the corresponding product of the σ-algebras and the product measure. Let $\Phi_1 = \Phi_1(x)$ denote a randomized test on the space $(\mathfrak{X}, \mathfrak{A})$. Consider the statistic (Φ_1, U), where U is independent of $x \in \mathfrak{X}$ and is uniformly distributed on $\mathfrak{U}$. With the aid of this statistic let us define a new test Φ^* on $(\mathfrak{X}, \mathfrak{U})$ by

$$\Phi^* = \begin{cases} 1 & \text{for} \quad U - \Phi_1(x) < 0, \\ 0 & \text{for} \quad U - \Phi_1(x) \geqslant 0. \end{cases} \tag{4.3.1}$$

We see that the test Φ^* is unrandomized. Furthermore, for $\theta \in \Omega_0$,

$$E_\theta \Phi^* = E_\theta (E(\Phi^* | \Phi_1(x))) = E_\theta \Phi_1(x) = \alpha, \tag{4.3.2}$$

so that the test Φ^* is similar to the level α for $\theta \in \Omega_0$. Of course, construction of Φ^* is equivalent to the process of randomization as a supplementary observation over the statistic $\mathfrak{U}$.

Sometimes the procedure described can be applied without broadening the basic sample space $\mathfrak{X}$.

Suppose that $\mathfrak{X}$ is n-dimensional Euclidean space E_n and that we have the sufficient statistic $(T_1, \cdots, T_k)$ with range in the Euclidean space E_k $(k < m)$ for $\theta \in \Omega_0$. Let $(T_1, \cdots, T_k;\ \xi_1, \cdots, \xi_{m-k})$ denote a system of local coordinates in $\mathfrak{X}$. Here the σ-algebras are, as usual, assumed to be Borel sets, and we may speak of the conditional distribution $P_\theta\{\xi_1, \cdots, \xi_{m-k} | T_1, T_2, \cdots, T_s)$ for $\theta \in \Omega$ (see §3 of the Introduction).

Now suppose that we have constructed a randomized test $\Phi_1(T_1, \cdots, T_k)$ for testing a hypothesis H_0: $\theta \in \Omega_0$ that depends only on sufficient statistics. Let us take an arbitrary measurable scalar function $V(T_1, \cdots, T_k;\ \xi_1, \cdots, \xi_{m-k})$ and construct the conditional distribution

$$P(V < y | T_1, \ldots, T_k) = F(y;\ T_1, \ldots, T_k) \tag{4.3.3}$$

for an arbitrary real value of y. For $\theta \in \Omega_0$ this distribution does not depend on θ. Let us suppose that for almost all values of $(T_1, \cdots, T_k)$, the function $F(y; T_1, \cdots, T_k)$ is a strictly monotonic function of y. Then, as is well known (see [33]), the transformation $U = F(V(T_1, \cdots, T_k); (\xi_1, \cdots, \xi_{m-1}))$ yields a new function the conditional distribution of which is uniform on the interval $[0, 1]$ for almost all values of $T_1, \cdots, T_k$. We can now define an unrandomized test Φ^* by formula (4.3.1), where we replace $\Phi_1(x)$ with $\Phi_1(T_1, \cdots, T_k)$. This test Φ^* depends only on $x \in \mathfrak{X}$. Furthermore, for arbitrary $\theta \in \Omega$,

$$E_\theta \Phi^* = E_\theta (E\Phi^* | \Phi_1(T_1, \ldots, T_k)) = E_\theta \Phi_1(T_1, \ldots, T_k), \tag{4.3.4}$$

so that the power functions of the tests Φ^* and Φ_1 coincide. In particular, if the test $\Phi_1(T_1, \cdots, T_k)$ is similar for $\theta \in \Omega_0$, the same will be true of the test Φ^*. Its critical zone will be a similar zone.

When we have constructed a randomized similar test Φ that depends only on sufficient statistics, we can, generally speaking, use it to construct (choosing different functions V) various unrandomized tests of the same power and the corresponding similar zones. In Chapter V we shall be able, in the case of certain broad subclasses of exponential families, to give a rather complete description of randomized similar tests that depend only on sufficient statistics. Our discussion shows that these tests correspond, in general, to unrandomized tests with the same power function that depend on the sample point $x \in \mathfrak{X}$ and not on sufficient statistics. Construction of nontrivial unrandomized similar tests that depend only on sufficient statistics is not always possible and is a different problem. We shall speak further about this and shall give some important special cases (particularly as regards the Behrens-Fisher problem in Chapters VIII and IX).

Neyman structures (see §2) are a classical means for constructing similar zones (see Neyman [58]). Suppose that we have the situation just described: $\mathfrak{X} = E_m$ and $T_1, \cdots, T_k$ are sufficient statistics for $\theta \in \Omega_0 \subset \Omega \subset E_n$, where $n < m$. Suppose that the family of probability densities $\{P_\theta\}$, $\theta \in \Omega_0$, and the sufficient statistic $T = (T_1, \cdots, T_k)$ satisfy the condition formulated in §3 of Chapter II regarding the possibility of "coupling" any two values of the parameter $\theta \in \Omega_0$. Then we will not have the pathological situations described in that section. If $\chi(x)$ is a statistic independent of the sufficient statistic T, their distribution will not depend on the parameter θ if $\theta \in \Omega_0$. Consequently $\chi(x)$ is a sufficient statistic for $\theta \in \Omega_0$. From a geometrical point of view, construction of the similar statistic $\chi(x)$ that assumes only the values 0 and 1 (the characteristic function of the similar zone) amounts to delineating zones of given probability α on the level surfaces of the sufficient statistic $T = (T_1, \cdots, T_k)$ and "gluing" them as described in §2. The similar zones constructed as Neyman structures are quite different from those constructed above in accordance with formulas (4.3.1)–(4.3.3): the characteristic functions of the latter are conditional distributions depending on sufficient statistics.

We note that direct application of Neymann structures to the construction of similar tests of the hypothesis H_0: $\theta \in \Omega_0$ obviously leads to useless tests if the sufficient statistic $T = (T_1, \cdots, T_s)$ that is applied is sufficient for $\theta \in \Omega$ but not for $\theta \in \Omega_0$. Here replacement of the basic sample space $\mathfrak{X}$ with the space of values of a statistic Y constructed in such a way that the new statistics that are sufficient for $\theta \in \Omega_0$ but not for $\eta \in \Omega$ in the new sample space may be of help. Such constructions will be made in the following chapters.

Let us now look at other ways of constructing similar zones. It might be noted that nontrivial similar zones do not exist for all families. In fact, nontrivial randomized similar tests may fail to exist. Consider for example the family of probability densities (with respect to Lebesgue measure) of the form

$$p_\theta(x) = c_1(\theta)(1 + \cos 2\pi\theta x) \text{ and } p_\theta(x) = c_2(\theta)(1 + \sin 2\pi\theta x),$$

where $x \in [0, 1]$ and the $c_i(\theta)$ $(i = 1, 2)$ are normalizing constants. Let Φ denote the entire real line $-\infty < \theta < \infty$ and let Ω_0 consist of all integers $\theta = 0, \pm 1, \pm 2, \cdots$.

If $\phi(x)$ is a similar randomized test for the hypothesis H_0: $\theta \in \Omega_0$ of level α, then

$$\int_0^1 (\Phi(x) - \alpha)\cos 2\pi n x \, dx = 0; \quad \int_0^1 (\Phi(x) - \bar{\alpha})\sin 2\pi n x \, dx = 0$$

where $n = 0, 1, 2, \cdots$. (Note that $\int_0^1 (\Phi(x) - \bar{\alpha})\, dx = 0$ also by virtue of the similarity condition for $\theta = 0$.) Therefore $\Phi(\alpha) \equiv \alpha$ with probability 1, so that the test is trivial.

An even simpler example is provided by the family of distributions that is defined on $\mathfrak{X} = [0, 1]$ and that has probability density

$$p_\theta(x) = (\theta + 1)x^\theta, \quad \theta \in \Omega_0 = \{0, 1, 2, \ldots\}.$$

If $\Phi(x)$ is a similar test of the level α it follows from the equation

$$\int_0^1 (\Phi(x) - \alpha)\, x^\theta\, dx = 0,$$

$\theta \in \Omega_0$, that $\Phi(x)$ is equal to α almost everywhere.

Analogous examples are easily obtained from Theorem 4.2.2 for exponential families. Thus we see that construction of similar zones is possible only for special families of distributions. Here we shall exhibit some of these families.

Finite families of distributions (families with a finite set of parameters Ω_0) were investigated in this connection by Ljapunov [54] in 1940 and later by Neyman [57]. Ljapunov's article contains a number of important results regarding completely additive vector-valued set functions. We shall need only a few relatively weak results that can be proven more easily than can Ljapunov's results.

Theorem 4.3.1.[1)] *Suppose that n continuous probability measures* $\mu_1, \cdots$ $\cdots, \mu_n$ *are defined on a set* $\mathfrak{X}$ *with* σ*-algebra* Σ *of subsets. Then for arbitrary* λ *in* $[0, 1]$ *there exists a subspace* $A_\lambda \in \Sigma$ *such that* $\mu_1(A_\lambda) = \mu_2(A_\lambda) = \cdots$ $\cdots = \mu_n(A_\lambda) = \lambda$.

Proof. Without loss of generality we may assume that the measures are absolutely continuous with respect to one of them since we can define a new measure in terms of the n measures by $\mu_{n+1} = (\mu_1 + \cdots + \mu_n)/n$ and require that $\mu_{n+1}(A_\lambda) = \lambda$, which changes nothing. All the other measures are absolutely continuous with respect to this measure.

Furthermore, it will be sufficient to prove the theorem for $\lambda = ½$, since repeated application of the theorem enables us to construct sets A_λ, where λ is an arbitrary dyadic fraction (i.e. a fraction of the form $a/2^n$ with a and n integers), by carrying out the constructiions in the measurable space $\mathfrak{X} - A_\lambda$ with $\mu_i(\mathfrak{X} - A_\lambda) = 1 - \lambda$. Then for arbitrary λ we may consider a sequence of these dyadic fractions that approaches λ.

Suppose that the theorem is true for all $n \leq n_0$ and suppose that $\mu_1, \cdots, \mu_{n_0+1}$ satisfy the conditions of the theorem. From what was said above, to every λ in $[0, 1]$ is assigned a set $A_\lambda \in \Sigma$ such that $\mu_k(A_\lambda) = \lambda$ for $k = 1, \cdots, n_0$. Also, $A_{\lambda_1} \subset A_{\lambda_2}$ for $\lambda_1 < \lambda_2$.

Consider the Borel σ-algebra $\mathfrak{A}$ constructed on the sets A_λ. Here

---

1) Proof of this theorem, which is a consequence of Ljapunov's results, was communicated to the author by V. N. Sudakov.

$\mu_1(B) = \mu_2(B) = \cdots = \mu_{n_0}(B)$ for arbitrary $B \in \mathfrak{U}$. The measures $\mu_1(B)$ and $\mu_{n_0+1}(B)$ satisfy the conditions of the theorem. There are two of them, so that it remains for us to prove the theorem for $n = 2$.

For every $\lambda \in [0, 1]$, let us construct a set A_λ such that $\mu_1(A_\lambda) = \lambda$ (which is possible by virtue of the continuity of the measure μ_1). Let $\mathfrak{U}$ denote the Borel σ-algebra on the sets A_λ. Let $\{A_\lambda\}$ denote the family, constructed for the measure μ_1, of subsets of Σ with the properties described above. We may interpret λ as a point on a circle of unit circumference. To every Borel subset of the circle, in particular to every semicircle, there corresponds a subset in $\mathfrak{U}$. Let us now define a continuous function on the circle as follows: to every λ we assign the value μ_λ, the measure of the set corresponding to the semicircle beginning at λ. Obviously, if this function is not a constant it assumes values both greater and less than ½ and hence it must assume the value ½ somewhere on the semicircle. This completes the proof of Theorem 4.3.1.

Theorem 4.3.1, which, as stated above, is a particular case of Ljapunov's results [54], can be applied to the construction of certain classes of families with nontrivial similar zones.

Let $\mathcal{P}_\theta$, $\theta \in \Omega_0$, denote a family of measures defined on $(\mathfrak{X}, \mathfrak{A})$ that is dominated by a σ-finite measure μ. Then, in accordance with Theorem 2.1.4, for a sufficient statistic $T(x)$ we have the decomposition (2.1.4)

$$p_\theta(x) = g(T(x), \theta)\, h(x),$$

where g is measurable in the space of values of $T(x)$ and where $h(x)$ is a nonnegative $\mathfrak{A}$-measurable function. Conversely, (2.1.4) is a sufficient condition for sufficiency of $T(x)$.

In the articles [40] and [41] by Linnik, which are generalized in the article by Kagan and Linnik [23], there is introduced a class of distributions generalizing (2.1.4) for nontrivial $T(x)$. If $\mathfrak{X}$ is the Euclidean space E_m and if $T(x) \in E_k$ $(k < m)$ is a nontrivial sufficient statistic, then under rather general conditions it is possible to construct similar zones in the form of Neyman structures, as described at the beginning of this section. Let us now consider the family of probability densities of the form

$$p_\theta(\alpha) = R_1(T(x), \theta)\, r_1(x) + \ldots + R_q(T(x), \theta)\, r_q(x). \qquad (4.3.5)$$

Here $x \in \mathfrak{X} = E_m$ and $T(x) \in E_k$ $(k < m)$ is an arbitrary statistic. The density is with respect to a σ-finite dominating measure μ, and $\theta \in \Omega_0$ is an abstract parameter. Correspondingly, for given θ, the functions $R_i(T(x), \theta)$ $(i = 1, 2, \cdots, m)$

and the functions $r_i(x)$ are assumed measurable. We see that the family (4.3.5) generalizes the family (2.1.4) for nontrivial $T(x) \in E_k$ and $\mathfrak{X} = E_m$. Let us now show that under rather general assumptions this family has similar zones (obviously nontrivial) for an arbitrary level $\alpha \in (0, 1)$. First of all, let us put the probability densities (4.3.5) in the form of finite sums of sign-constant terms. We set

$$R_i^+(T, \theta) = \max\{R_i(T, \theta), 0\}; \quad r_i^+(x) = \max\{r_i(x), 0\},$$
$$R_i^-(T, \theta) = \min\{R_i(T, \theta), 0\}; \quad r_i^-(x) = \min\{r_i(x), 0\}.$$

Then we have

$$R_i(T, \theta) = R_i^+(T, \theta) + R_i^-(T, \theta),$$
$$r_i(x) = r_i^+(x) + r_i^-(x).$$

Substituting these expressions into (4.3.5), denoting by $\widetilde{R}_i(T, \theta)$ either $R_i^+(T, \theta)$ or $R_i^-(T, \theta)$ and by $\widetilde{r}_i(x)$ either $r_i^+(x)$ or $r_i^-(x)$, we obtain from (4.3.5)

$$p_\theta(x) = \widetilde{R}_1(T, \theta)\widetilde{r}_i(x) + \ldots + \widetilde{R}_n(T, \theta)\widetilde{r}_n(x), \tag{4.3.6}$$

where $\widetilde{R}_i(T, \theta)$ and $\widetilde{r}_i(x)$ have corresponding measurability and $q \leq n \leq 4q$. Now let us impose the condition

$$\pi_{i, \theta}(x) = \frac{1}{C_i(\theta)} |\widetilde{R}_i(T, \theta)\widetilde{r}_i(x)|, \quad i = 1, 2, \ldots, n \tag{4.3.7}$$

for all $\theta \in \Omega_0$ and $i = 1, 2, \cdots, n$. We may assume without loss of generality that $C_i(\theta) > 0$ for $\theta > \Omega_0$. We define

$$C_i(\theta) = \int_{\mathfrak{X}} |\widetilde{R}_i(T(x), \theta)\widetilde{r}_i(x)| d\mu(x) < \infty. \tag{4.3.8}$$

Then we set

$$p_\theta(x) = \sum_{i=1}^{n} \varepsilon_i C_i(\theta)\pi_{i, \theta}(x). \tag{4.3.9}$$

Here the $\pi_{i,\theta}(x)$ are normalized probability densities with respect to the measure μ, the quantity $\epsilon_i = \pm 1$, and

$$\sum_{i=1}^{n} \varepsilon_i C_i(\theta) = 1, \quad \theta \in \Omega_0. \tag{4.3.10}$$

Also, we denote by $\Pi_{i,\theta}$ the probability measures corresponding to the densities $\pi_{i,\theta}(x)$. By virtue of our assumptions (that the space $\mathfrak{X}$ and the range of $T(x)$ are Euclidean spaces and that we are using Borel σ-algebras), we obtain for every value of $T(x)$ the conditional distributions of the probabilities $\Pi_{i,\theta}(A|T)$ (cf. Theorem 2.1.1). Since T is a sufficient statistic we may assume for $\theta \in \Omega_0$ that

$$\Pi_{i, \theta}(A|T) = \Pi_{i, \theta_0}(A|T), \quad i = 1, 2, \ldots, n, \tag{4.3.11}$$

for any fixed value $\theta_0 \in \Omega_0$. In accordance with Theorem 4.3.1, for given $\alpha \in (0,1)$ and a given value of T there exists a set $A_T \in \mathfrak{A}$ such that

$$\Pi_{1,\theta_0}(A_T|T) = \Pi_{2,\theta_0}(A_T|T) = \ldots = \Pi_{n,\theta_0}(A_T|T) = \alpha,$$

so that, by virtue of (4.3.11),

$$\Pi_{1,\theta}(A_T|T) = \Pi_{2,\theta}(A_T|T) = \ldots = \Pi_{n,\theta}(A_T|T) = \alpha \tag{4.3.12}$$

for all $\theta \in \Omega_0$.

Suppose that for almost all T the level surfaces of the statistic T in the Euclidean space E_k are piecewise smooth and admit introduction of local coordinates (T, ξ) in the space $\mathfrak{X} = E_k$. Then (see Cramér [33]) for almost all T it is possible to assign to the sets $A_T \in \mathfrak{A}$ the sets C_T on the surfaces with given values of T (the traces of the A_T) such that

$$\Pi_{i,\theta}(A_T|T) = \Pi_{1,\theta}(C_T|T), \quad \theta \in \Omega_0, \quad i = 1, 2, \ldots, n.$$

Let us suppose that "gluing" with respect to the values of T is possible, i.e. that the set $A = \bigcup C_T$ is measurable. Then, from (4.3.11),

$$\Pi_{i\theta}(A) = \alpha, \quad \theta \in \Omega_0, \quad i = 1, 2, \ldots, n.$$

Thus $\int_A \pi_{i,\theta}(x)\, d\mu(x) = \alpha$ for the same values of θ and i. Let us integrate both sides of equation (4.3.9) over the set A keeping equation (4.3.10) and the equation just derived in mind. We obtain

$$\int_A p_\theta(x)\, d\mu(x) = \alpha, \quad \theta \in \Omega_\theta. \tag{4.3.13}$$

We have obtained a similar zone of level α. The construction composed generalizes Neyman structures. In the article [23] this construction is expounded in a more general form: the sample space and the range of the statistic are not assumed to be Euclidean spaces.

Let us give some examples of families of the form (4.3.5).

Example 1. Suppose that we have a probability density with respect to a dominating σ-finite measure μ on a space $(\mathfrak{X}, \mathfrak{A})$ of the form

$$p_\theta(x) = \exp\left\{\sum_{i=1}^{N} \varphi_i(x)\psi_i(\theta)\right\} \sum_{i=1}^{M} q_i(x) s_i(\theta) \tag{4.3.14}$$

with corresponding conditions of measurability. Consider a repeated sample of size $n > N$ for the sample space $\mathfrak{X}$. The corresponding probability density with respect to the product-measure for $\mu(x)$ is of the form

$$p_\theta(x_1, \ldots, x_n) = \exp\left\{\sum_{i=1}^{N} T_i(x)\psi_i(\theta)\right\}$$
$$\times \sum_{i_1, i_2, \ldots, i_n=1}^{M} q_{i_1}(x)\ldots q_{i_n}(x)s_{i_1}(\theta)\ldots s_{i_n}(\theta) \tag{4.3.15}$$

where $T_i(x) = \Sigma_{j=1}^n \phi_i(x_j)$. If the second sum in the product (4.3.14) degenerated to unity, then (4.3.15) would yield an exponential family with nontrivial sufficient statistics. The family (4.3.15) is of the form (4.3.5) and, in accordance with what was said above, it admits nontrivial similar zones. It can be regarded as a "degenerate exponential family".

Example 2 (cf. Example 8, Chapter II, §2). Consider a distribution of the Pearson-III type:

$$p(x;\ a, \gamma) = \frac{\gamma^m}{\Gamma(m)}(x-a)^{m-1}\exp\{-\gamma(x-a)\} \quad (x \geqslant a);$$

where $p(x;\ a, \gamma) = 0$ for $x < a$. Here $m > 0$ and $\gamma > 0$. Let m denote a positive integer ascertained from observation. Suppose that we wish to verify the hypothesis H_0: $\gamma = \gamma_0$ with the aid of a test that is similar with respect to the values of a from a repeated sample of size n.

We know (see Chapter II, §2) that for $m > 1$ there will be only trivial sufficient statistics and we cannot construct Neyman structures. However we can use the device described above. The probability density of the repeated sample $(x_1, \cdots, x_n)$ is equal to

$$p(x_1, \ldots, x_n;\ a, \gamma)$$
$$= \frac{\gamma^{mn}}{(\Gamma(m))^n}\prod_{i=1}^{n}(x_i-a)^{m-1}\exp\left\{-\gamma\sum_{i=1}^{n}(x_i-a)\right\}$$

for $\min_i x_i \geq a$. On the other hand,

$$p(x_1, \ldots, x_n;\ a, \gamma) = 0$$

for $\min_i x_i < a$.

Since $m-1$ is a nonnegative integer, we obtain representations of the form (4.3.5), where $T(x) = (\Sigma_{i=1}^n x_i, \min_i x_i)$, for $p(x_1, \cdots, x_n;\ a, \gamma)$, and the remaining functions are easily found. Consequently, for $\gamma = \gamma_0$ and an arbitrary level $\alpha \in (0, 1)$, we can construct zones that are similar with respect to a.

§4. APPROXIMATELY SIMILAR ZONES

We have seen that in general there are no similar zones or even randomized similar tests for infinite families of distributions. However, as Besicovitch discovered in 1961 (see [3]), there are approximately similar zones under extremely general conditions. Here we shall expound a portion of his results. To make the exposition simpler we have strengthened somewhat the hypotheses in his theorems.

Consider the family of distributions $\{P_\theta\}$, $\theta \in \Omega_0$, on a measurable space $(\mathfrak{X}, \mathfrak{A})$ admitting a scalar statistic $T(x)$ that has a probability density $\rho_\theta(x)$ with respect to Lebesgue measure for all values of θ. Also let us assume that $\rho_\theta(x)$ is an absolutely continuous function whose derivative exists everywhere and has an absolutely bounded majorant $M(x)$:

$$|\rho_\theta'(x)| \leqslant M(x), \quad M(x) \leqslant M_0. \tag{4.4.1}$$

Under these conditions we have

Theorem 4.4.1. *For* $\alpha \in (0, 1)$ *and* $\epsilon > 0$ *there exists a set* $A_\epsilon \in \mathfrak{A}$ *such that*

$$|P_\theta(A_\varepsilon) - \alpha| < \varepsilon \quad \textit{for} \quad \theta \in \Omega_0. \tag{4.4.2}$$

To prove this it will be sufficient to construct a Lebesgue-measurable set E_ϵ on the real axis such that

$$\left| \int_{E_\varepsilon} \rho_\theta(x)\,dx - \alpha \right| \leqslant \varepsilon \quad \text{for} \quad \theta \in \Omega_0. \tag{4.4.3}$$

Then we can set $A_\epsilon = T^{-1}(E_\epsilon)$.

To construct E_ϵ we take an interval $I_N = [-N, N]$ of the real line such that

$$\int_{I_N} \rho_\theta(x)\,dx \geqslant 1 - \frac{\varepsilon}{4}. \tag{4.4.4}$$

Then we partition I_N by means of the points $\cdots, x_{-1}, x_0, x_1, x_2, \cdots$ into subintervals $I_{n-1,1} = [x_{n-1}, x_n]$ such that

$$|x_n - x_{n-1}| \leqslant \eta, \tag{4.4.5}$$

where η is a small number to be chosen later.

Now, in each interval $I_{n-1,1} = [x_{n-1}, x_n]$, let us consider the subinterval $I_{n-1,\alpha} = [x_{n-1}, (1-\alpha)x_{n-1} + \alpha x_n]$. We now have, for $x \in I_{n-1,1}$,

$$\rho_0(x) = \rho_\theta(x_{n-1}) + (x - x_{n-1})\gamma(x, n),$$

where

$$|\gamma(x,\ n)| \leqslant \sup_{x \in I_{n-1,1}} M(x) = M_{n-1} \leqslant H_0.$$

Thus

$$\int_{x_{n-1}}^{x} \rho_\theta(x)\,dx = (x - x_{n-1})\,\rho_\theta(x_{n-1}) + \frac{(x - x_{n-1})^2}{2}\,\gamma_1(x,\ n), \tag{4.4.6}$$

where

$$|\gamma_1(x,\ n)| \le M_{n-1}. \tag{4.4.6a}$$

Therefore

$$\int_{I_{n-1,\alpha}} \rho_\theta(x)\,dx = \alpha(x_n - x_{n-1})\,\rho_\theta(x_{n-1}) + \frac{\alpha^2 (x_n - x_{n-1})^2}{2}\,\gamma_1(x,\ n), \tag{4.4.7}$$

$$\int_{I_{n-1,1}} \rho_\theta(x)\,dx = (x_n - x_{n-1})\,\rho_\theta(x_{n-1}) + \frac{(x_n - x_{n-1})^2}{2}\,\gamma_1(x,\ n). \tag{4.4.8}$$

Furthermore, by inequalities (4.4.5) and (4.4.6a) we have

$$\sum_{(n)} \frac{(x_n - x_{n-1})^2}{2}\,|\gamma_1(x,\ n)| \leqslant \eta \sum_{(n)} M_{n-1}(x_n - x_{n-1}) \leqslant 2NH_0\eta. \tag{4.4.9}$$

Now, by choosing η sufficiently small, we can get $2NH_0\eta \le \epsilon/4$. Then from (4.4.8) we have

$$\int_{-N}^{N} \rho_\theta(x)\,dx = \sum_{(n)} (x_n - x_{n-1})\,\rho_\theta(x_{n-1}) + \gamma\frac{\varepsilon}{4},$$

where $|\gamma| \le 1$. Thus

$$\sum_{(n)} (x_n - x_{n-1})\,\rho_\theta(x_{n-1}) = \int_{-N}^{N} \rho_\theta(x)\,dx + \gamma\frac{\varepsilon}{4}. \tag{4.4.10}$$

Let us set $E_\epsilon = \bigcup_n I_{n-1,\alpha}$. From (4.4.7) we obtain

$$\int_{E_\varepsilon} \rho_\theta(x)\,dx = \alpha\sum_{(n)} (x_n - x_{n-1})\,\rho_\theta(x_{n-1}) + \gamma_1\frac{\varepsilon}{4}, \tag{4.4.11}$$

where $|\gamma_1| \le 1$. From (4.4.10) we obtain

$$\int_{E_\varepsilon} \rho_\theta(x)\,dx = \alpha\int_{-N}^{N} \rho_\theta(x)\,dx + \frac{\gamma\alpha\varepsilon}{4} + \gamma_1\frac{\varepsilon}{4}$$

$$= \alpha\int_{-N}^{N} \rho_\theta(x)\,dx + \gamma_2\frac{\varepsilon}{2}, \quad |\gamma_2| \leqslant 1.$$

Then, keeping (4.4.4) in mind, we obtain (4.4.3).

In [3] this theorem is proven under the less restrictive hypothesis that a majorant $M(x)$ exists and is bounded at all points except points of a closed set of measure 0. The proof under this hypothesis becomes somewhat more complicated.

Consider a family of probability densities

$$\{p_\theta(x)\}; \quad p_\theta(x) = (\theta+1)\,x^\theta; \quad \theta = 0, 1, 2, \ldots; \quad x \in [0, 1]$$

(see §3) for which there are not even any randomized similar tests. By Theorem 4.4.1, there are still approximate similar zones A_ϵ for this family such that

$$\int_{A_\varepsilon} p_\theta(x)\,dx = \alpha + \gamma\varepsilon, \quad |\gamma| \leqslant 1.$$

The approximately similar zones constructed in accordance with Theorem 4.4.1 have characteristic structure of a "laminated threshold". If we decrease the number ϵ characterizing the approximation, the layers become ever finer and the number of them increases in a bounded region of space. From a statistical standpoint the use of approximately similar critical zones of this kind is inexpedient. Of course, in evaluating the expediency of such a method we must begin with a definite criterion, for example the behavior of the power function. However, we can indicate beforehand why it is not in general statistically expedient to use approximately similar zones constructed as indicated in the proof of Theorem 4.4.1.

For an observation $X \in \mathfrak{X}$ to belong to an approximately similar zone A_ϵ it is sufficient that the probability density $p_\theta(x)$ of a statistic $\Phi(x)$ be smooth and that it have a bounded derivative. If the space Ω of the parameters $(\theta_1, \cdots, \theta_s) \in E_s$ is a compact set, the case in which $|p'_\theta(x)|$ is bounded for $\theta \in \Omega_0 \subset \Omega$ but not bounded for $\theta \in \Omega \backslash \Omega_0$ is likely to be a pathological rather than a natural situation. The natural cases are those in which this quantity, if bounded for $\theta \in \Omega_0$, is also bounded for $\theta \in \Omega$. Then the approximately similar zones A_ϵ constructed for testing the hypothesis H_0: $\theta \in \Omega_0$ will be approximately similar with the same level for all $\theta \in \Omega$ and a test based on them is from a practical point of view useless.

§5. INDEPENDENT STATISTICS

For the construction of many tests, in particular in problems dealing with a normal law, it is helpful to use independent statistics. Thus we have several times used the independence of the statistics $\bar{x}$ and s^2 in a repeated normal sample. In many cases such properties of independence are connected with the properties of completeness of exponential families and with the Lehmann-Scheffé theorems. These relationships were studied in the articles by Hogg and Craig [73] and Feĭgel'son [74]. We shall use the notation of Theorem 4.2.1 and shall assume the completeness conditions mentioned in that theorem.

Let $\chi(x)$ denote a similar statistic for $\theta \in \Omega_0$. Consider zones of the form $\{\chi(x) < \xi\}$. They will be similar for arbitrary ξ. Suppose that $\Phi_\xi(x)$ is a test for the hypothesis H_0: $\theta \in \Omega_0$ with critical zone $\{\chi(x) < \xi\}$ of level α. By Theorem 4.2.1 the test $\Phi_\xi(x)$ is a Neyman structure. Thus for almost all values of $T = t$ we have $E(\phi_\xi(x)|t) = \alpha$ when $\theta \in \Omega_0$. This means that

$$P\{\chi(x) < \xi \mid t\} = P(\chi(x) < \xi)$$

for almost all values of t.

Obviously, similar reasoning can be applied to vector-valued similar statistics.

Thus the statistic $\xi(x)$ does not depend on the sufficient statistic $T(x)$. This circumstance enables us to prove the independence of rather complicated statistics immediately, without calculations that are often extremely laborious.

Let us look at some appropriate examples.

Example 1. In a repeated normal sample $x_1, \cdots, x_n \in N(a, \sigma_0^2)$, where σ_0 is known, a sufficient statistic for a is $\bar{x}$ (see Example 5, §2, Chapter II). For $a \in (-\infty, \infty)$, the family is exponential and complete (in accordance with Theorem 4.2.2). Let $g(x_i - x_j)$ denote a Lebesgue-measurable vector-valued function depending only on the observed differences $x_i - x_j$. Obviously this function is a complete statistic with respect to a. On the basis of what we showed above, $g(x_i - x_j)$ is stochastically independent of $\bar{x}$. In particular, s^2 is independent of $\bar{x}$.

Example 2. In the same normal sample $x_1, \cdots, x_n \in N(a, \sigma^2)$, let us suppose that the two parameters are independent. The statistics $\bar{x}$ and s^2 are sufficient statistics for (a, σ). Let $g(x_i - x_j)$ denote a Lebesgue-measurable vector-valued function of the observed differences. Suppose also that $g(x_i - x_j)$ is a homogeneous function of zero dimension with respect to its arguments, so that

$$g\left(\frac{x_i - x_j}{\sigma}\right) \equiv g(x_i - x_j) \quad \text{for arbitrary } \sigma > 0.$$

Then obviously $g(x_i - x_j)$ is similar with respect to (a, σ). Since the exponential family in question is complete, the statistic $g(x_i - x_j)$ is independent of the pair $(\bar{x}, s^2)$. In particular, statistics of the form

$$\frac{m_\nu}{m_2^{\nu/2}} = \sum_{i=1}^{n} \left(\frac{x_i - \bar{x}}{s} \right)^\nu,$$

where the notation is the convential one and $\nu \geq 2$ is an integer, are independent of the pair $(\bar{x}, s^2)$. (This property was derived in Cramér's book [33] by means of a rather complicated calculation.)

Example 3. Let $x_1, \cdots, x_n$ denote a repeated sample from a distribution of the Pearson-III type (see Example 8, §2, Chapter II) where $a = 0$. Thus

$$p(x;\ a,\ \gamma) = p(x;\ 0,\ \gamma) = \frac{\gamma^m}{\Gamma(m)} x^{m-1} \exp(-\gamma x) \quad (x \geqslant 0),$$

$$p(x;\ a,\ \gamma) = 0 \quad (x < 0).$$

Here the parameter $\gamma \in (0, \infty)$. A sufficient statistic is $\bar{x}$. Let $g(x_1, \cdots, x_n)$ denote a measurable vector-valued function of zero dimension with respect to its arguments. This function is independent of $\bar{x}$. In particular, the random vector

$$\left(\frac{x_1}{x_1 + \ldots + x_n}, \frac{x_2}{x_1 + \ldots + x_n}, \ldots, \frac{x_n}{x_1 + \ldots + x_n} \right)$$

is independent of $x_1 + \cdots + x_n$.

Example 4. Consider a repeated sample from a 2-dimensional normal set with independent components, i.e. with probability density (in the usual notation)

$$p_\theta(x,\ y) = \frac{1}{2\pi\sigma_1\sigma_2} \exp - \frac{1}{2} \left[\frac{(x - a_1)^2}{\sigma_1^2} + \frac{(y - a_2)^2}{\sigma_2^2} \right],$$

$$\theta = (a_1,\ a_2,\ \sigma_1,\ \sigma_2).$$

Here $a_i \in (-\infty, \infty)$ and $\sigma_i \in (0, \infty)$ $(i = 1, 2)$, and we have a complete exponential family.

Now consider a measurable vector-valued function $g(x_1, \cdots, x_n;\ y_1, \cdots, y_n)$ that is invariant under the transformation $x_i \to \alpha x_i + \beta$, $y_i \to \gamma y_i + \delta$, where α, β, γ, and δ are constants with α and γ nonzero. This transformation makes $p_\theta(x, y)$ a probability density, where $a_1 = a_2 = 0$ and $\sigma_1 = \sigma_2 = 1$. In view of this distribution, the function $g(x_1, \cdots, x_n;\ y_1, \cdots, y_n)$ is independent of the parameters a_1, a_2, σ_1, and σ_2, and hence is a similar statistic that is independent of the sufficient statistics of the family:

$$\bar{x},\ \bar{y},\ s_1^2 = \frac{1}{n} \sum_{i=1}^{n} (x_i - \bar{x})^2;\ s_2^2 = \frac{1}{n} \sum_{i=1}^{n} (y_i - \bar{y})^2.$$

In particular, the sample coefficient of correlation

$$r = \frac{\frac{1}{n}\sum_{i=1}^{n}(x_i - \bar{x})(y_i - \bar{y})}{s_1 s_2}$$

is a function of the type indicated. Thus the sample coefficient of correlation r is independent of the quadruple of sufficient statistics $(\bar{x}, \bar{y}, s_1^2, s_2^2)$. This fact will be useful to us in Chapter IX. For other examples see [74].

The properties of independence of statistics can in turn be used to characterize the distributions and to construct certain nonparametric tests. Numerous works, for example [28], [38] and [53], are devoted to problems of characterization in this sense.

§6. APPLICATIONS OF A THEOREM OF H. CARTAN TO THE STUDY OF FAMILIES OF STATISTICS

An important corollary of the theorems of H. Cartan given in Chapter I is Theorem 1.3.2, which was proved there under the assumption that, in the ring O_Z of complex functions $f(z_1, \cdots, z_k)$ that are holomorphic on a compact simply-connected polycylinder, all the ideals are finitely generated on that polycylinder.

Following the article [48] we shall apply this result to the study of families of independent identically distributed statistics. Here we shall see the utility of considering ideals of functions for studying the behavior of a statistic. In Chapter V we shall use ideals of functions to construct (in a certain sense) complete families of similar tests. Let us begin by considering families of identically distributed statistics, keeping in mind the fact that the occurrence of identical distribution of statistics is closely related to their independence (see [38]). Specifically, if (X, Y) is a two-dimensional random vector, then a necessary and sufficient condition for their components X and Y to be independent is that

$$E \exp i(t_1 X + t_2 Y) = E \exp(itX) E \exp(itY) \tag{4.6.1}$$

for arbitrary real t_1 and t_2. Let (X', Y') denote a random vector with components that are independent and distributed like X and Y respectively. We denote this fact by writing

$$X \cong X', \quad Y \cong Y'.$$

Then for arbitrary real t_1 and t_2 equation (4.6.1) can be written in the form

$$E \exp i(t_1 X + t_2 Y) = E \exp i(t_1 X' + t_2 Y').$$

i.e.

$$t_1X+t_2Y\cong t_1X'+t_2Y'$$

for arbitrary real t_1 and t_2. Thus the independence of X and Y is equivalent to identical distribution of t_1X+t_2Y and $t_1X'+t_2Y'$ for arbitrary real values of t_1 and t_2.

Let $(\mathfrak{X}, \mathfrak{A})$ denote a measurable space equipped with a probability measure μ. Let $Q(X; a_1, \cdots, a_s)$ denote a family of statistics that are dependent on the real parameters $a_1, \cdots, a_s$ and that vary in some closed bounded parallelepiped Ω. To shorten our notation we set

$$(a_1, \ldots, a_s)=a, \quad Q(X; a_1, \ldots, a_s)=Q(X, a).$$

If two statistics $Q(X, a)$ and $Q(X, a')$ of the family are identically distributed, i.e. if $Q(X, a)\cong Q(X, a')$, then existence of the moments $E(Q(X, a))^r$ and $E(Q(X, a'))^r$, $r=0, 1, 2, \cdots$, would mean that the moments are equal:

$$E(Q(X, a))^r=E(Q(X, a'))^r, \quad r=0, 1, 2, \ldots, \tag{4.6.2}$$

Conversely, if the problem of moments were determinate in the present case, then a countable set of equations (4.6.2) would imply identical distribution of the statistics $Q(X, a)$ and $Q(X, a')$. The conditions under which the problem of moments is determinate are well-known:

$$\sum_{r=1}^{\infty} \alpha_{2r}^{-\frac{1}{2r}}=\infty,$$

where α_{2r} is the $2r$th moment (see [80]).

Although these conditions are only sufficient, they are almost necessary. In this connection, these conditions are not satisfied even with a normal distribution of polynomial families of statistics (see [39]). Therefore we shall consider not the moments (4.6.2) but their generalizations which we shall call the Φ-moments. Specifically, suppose that we have a family of continuous functions $\{\Phi_r(U)\}$ $(r=1, 2, \cdots)$ of a real argument U. Suppose that the family $\{\Phi_r(U)\}$ has property (D):

(D) *Determinacy of the analogue of the moment problem*: The mathematical expectations (Φ-moments)

$$E\Phi_r(Q(X, a)), \quad a\in\Omega, \tag{4.6.3}$$

exist, and their value determines the distribution $Q(X, a)$ up to μ-measure zero.

Furthermore, let us suppose that, relative to the family $\{\Phi_r(U)\}$, the statistics $Q(X, a)$ possess property (Γ):

(Γ) *Holomorphic connectedness*. Let us define

$$E(\Phi_r(Q(X, a))) = \varphi_r(a_1, \ldots, a_s). \tag{4.6.4}$$

Then the region of the s complex variables $a_1, \cdots, a_s$ contains a simply-connected compact polycylinder $\Omega_1 \supset \Omega$ on which all the functions (4.6.4), $r = 1, 2, \cdots$, exist and are homomorphic. (This applies to boundary as well as interior points.)

By assumption, giving (4.6.3) determines the distribution $Q(X, a)$. Therefore if

$$E\Phi_r(Q(X, a)) = E\Phi_r(Q(X, a')), \quad r = 1, 2, \ldots, \tag{4.6.5}$$

for two values of the parameter $a, a' \in \Omega$, then $Q(X, a)$ and $Q(X, a')$ are identically distributed statistics. The converse is obvious. If $Q(X, a) \cong Q(X, a')$, then (4.6.5) holds for $r = 1, 2, \cdots$.

It turns out that condition (Γ) leads to the fact that the property of identical distribution of any two statistics of the family follows from a finite number of relations of the form (4.6.5).

Theorem 4.6.1. *For given* $(\mathfrak{X}, \mathfrak{A}, \mu)$, *and given families* $\{Q(X, a)\}$ *and* $\{\Phi_r\}$, *if the condition* (Γ) *of holomorphic connectedness is satisfied, then there exists a finite constant* R_0 *such that the relations*

$$E\Phi_r(Q(X, a)) = E\Phi_r(Q(X, a')), \quad a' \in \Omega, \tag{4.6.6}$$

where $r \le R_0$, *imply that*

$$Q(X, a) \cong Q(X, a'). \tag{4.6.7}$$

To prove this we note that (4.6.7) follows from a countable set of relations of the form (4.6.5). These may be put in the form

$$\varphi_r(a_1, \ldots, a_s) = \varphi_r(a'_1, \ldots, a'_s). \tag{4.6.8}$$

Consider the polycylinder $\Omega_1 \times \Omega_1$ of points of the form (a, a'). Let us construct an ideal J generated by the differences

$$\varphi_r(a_1, \ldots, a_s) - \varphi_r(a'_1, \ldots, a'_s) \quad (r = 1, 2, \ldots).$$

On the basis of the corollary of Cartan's theorem mentioned above, the ideal J must have a finite basis and must be generated by the differences

$$\varphi_r(a_1, \ldots, a_s) - \varphi_r(a'_1, \ldots, a'_s)$$

for $r \leq R_0$. Thus, if equations (4.6.8) are satisfied for $r \leq R_0$, where $a = (a_1, \cdots, a_s)$ and $a' = (a'_1, \cdots, a'_s)$ are two points belonging to Ω, then they are also satisfied for all $r > R_0$, so that equations (4.6.5) are satisfied for all r and hence (4.6.7) holds.

An analogous assertion can be made for conditions of independence of two statistics of a given family. Consider the same measurable space $(\mathcal{X}, \mathcal{A})$ with measure μ and two families of statistics $\{Q(X, a)\}$, for $a \in \Omega$, and $\{T(X, a')\}$, for $a' \in \Omega'$. Furthermore, let $\{\Phi_r(U)\}$ and $\{\Psi_t(U)\}$ $(r, t = 1, 2, \cdots)$ denote two families of continuous functions such that the functions

$$E[\Phi_r(Q(X, a))\Psi_t(T(X, a'))] = \varphi_{r,t}(a, a'), \; r, t = 1, 2, \ldots, \qquad (4.6.9)$$

and the functions

$$E\Phi_r(Q(X, a)) = \chi_r(a), \quad E\Psi_t(T(X, a')) = \tau_t(a'), \qquad (4.6.10)$$
$$r, t = 1, 2, \ldots$$

are both defined in the polycylinder $\Omega \times \Omega'$ of values of (a, a').

Furthermore let us suppose that a countable set of equations of the form

$$\varphi_{r,t}(a, a') = \chi_r(a)\tau_t(a'), \quad r, t = 1, 2, \ldots, \qquad (4.6.11)$$

for points $a \in \Omega$ and $a' \in \Omega'$ imply independence of the statistics $Q(X, a)$ and $T(X, a')$. In analogy with condition (Γ) above, let us formulate a new condition (Γ') of "holomorphic connectedness" for this situation:

(Γ'). There exist simply-connected compact polycylinders $\Omega_1 \supset \Omega$ and $\Omega'_1 \supset \Omega'$ such that the functions $\chi_r(a)$, $r = 1, 2, \cdots$, are holomorphic for $a \in \Omega$, the functions $\tau_t(a')$, $t = 1, 2, \cdots$, are holomorphic for $a' \in \Omega'_1$, and the functions $\phi_{r,t}(a, a')$ $(r, t = 1, 2, \cdots)$ are holomorphic for $(a, a') \in \Omega_1 \times \Omega'_1$. (Again, this applies to boundary as well as interior points.)

Under these conditions we have

Theorem 4.6.2. *For given families* $\{Q(X, a)\}$, $\{T(X, a')\}$, $\{\Phi_r\}$, *and* $\{\Psi_t\}$, *there exists a constant* R_1 *such that finitely many uncorrelatedness relations*

$$E[\Phi_r(Q(X, a))\Psi_t(T(X, a'))]$$
$$= E\Phi_r(Q(X, a))E\Psi_t(Q(X, a')) \qquad (4.6.12)$$

for $r \leq R_1$ *and* $t \leq R_1$ *and for arbitrary given values of* a *and* a', *imply independence of the statistics* $Q(X, a)$ *and* $T(X, a')$.

The proof is analogous to the preceding proof. On the polycylinder $\Omega_1 \times \Omega'_1$ we consider the ideal I generated by the functions

$$\varphi_{r,t}(a, a') - \chi_r(a)\tau_t(a'). \qquad (4.6.13)$$

By virtue of the compactness and simple connectedness of the polycylinder $\Omega_1 \times \Omega_1'$, the ideal I has a finite basis, by the above-mentioned corollary of Cartan's theorem, so that it is generated by functions of the form (4.6.13) for $r \leq R_1$ and $t \leq R_1$. If equations (4.6.12) are satisfied for $r \leq R_1$ and $t \leq R_1$ and for any particular values $a \in \Omega$ and $a' \in \Omega'$, then, by virtue of what has been said, they are satisfied for all values of r and t. Hence the statistics $Q(X, a)$ and $T(X, a')$ are then independent for given values of a and a'.

Thus we see that independence of two statistics in the families of statistics described above is equivalent to the existence of a finite number of uncorrelatedness relations of the form (4.6.12).

Let us give some examples of the application of Theorems 4.6.1 and 4.6.2.

Example 1. Let $\{Q(X, a)\}$ denote a family of positive polynomials, for example the squares of other polynomials. Suppose that $a \in \Pi$, where Π is a compact parallelepiped in E_s. Let us take for the family $\{\Phi_r(U)\}$ the set of functions

$$\Phi_r(U) = U^{r-1} \exp(-G(U)), \quad r = 1, 2, \ldots, \tag{4.6.14}$$

where $G(U)$ is a nonnegative polynomial of degree ≥ 2 for $U > 0$. An arbitrary compact polycylinder Z contains Π, and the family (4.6.14) gives a basis for using Theorem 4.6.1. Identical distribution of the statistics $Q(X, a)$ and $Q(X, a')$ for $a \in \Pi$ and $a' \in \Pi$ is equivalent to coincidence of their Φ-moments (of which there are $r \leq R_0$) constructed with the aid of (4.6.14).

Example 2. To apply Theorem 4.6.2 we may take two families of nonnegative polynomials $\{Q(X, a)\}$ and $\{T(X, a')\}$ for $a \in \Pi$ and $a' \in \Pi$ and, in addition to the family (4.6.14), an analogous family of functions with the same or a different polynomial $G(U)$.

For the families $\{Q(X, a)\}$ and $\{T(X, a')\}$ we may take the set of rational functions and many other sets of functions.

CHAPTER V

COTEST IDEALS FOR EXPONENTIAL FAMILIES

§1. SIMILAR TESTS AND COTEST IDEALS

In the present chapter we shall examine the construction of classes, complete in a certain sense, of similar tests for exponential families under certain restrictions. Such a construction is equivalent to finding a basis of an ideal. In this section we consider the question from a general point of view, without involving exponential families.

Suppose that we have a measurable space $(\mathfrak{X}, \mathfrak{A})$. Let $\{P_\theta\}$, $\theta \in \Omega$, denote a family of measures defined on it that are dominated by a σ-finite measure μ. Suppose that we are testing the null hypothesis H_0: $\theta \in \Omega_0 \subset \Omega$. To simplify the statement of the problem, let us consider a Bayes situation as an alternative: a probability measure $B(\theta)$ is defined on the set $\Omega \setminus \Omega_0$; if we know $B(\theta)$ then the alternative becomes the simple hypothesis. To convert the null hypothesis H_0 into a simple hypothesis let us consider randomized tests that are similar with respect to Ω_0 for testing it. Such a situation may seem somewhat artificial, but it is possibly the least complicated one for an initial description of similar tests and the choosing of tests that are optimum in some sense or other.

Let $\Phi = \Phi(x)$ denote a randomized similar test of level $\alpha \in (0, 1)$ for testing the null hypothesis H_0: $\theta \in \Omega_0$, so that

$$E(\Phi \mid \theta) = \alpha \text{ for } \theta \in \Omega_0. \tag{5.1.1}$$

Under such conditions we shall refer to an arbitrary statistic of the form $\psi = A(\Phi - \alpha)$, where A is a constant, as a *cotest*. Obviously

$$E(\psi \mid \theta) = 0 \text{ for } \theta \in \Omega_0. \tag{5.1.2}$$

For every cotest $\psi = \psi(x)$ there exists a constant C such that

$$-\alpha \leq C\psi(x) \leq 1 - \alpha \tag{5.1.3}$$

for all x. The number α is called the *level* of the cotest. A *precotest* of a level α is defined to be any statistic $\xi(x)$ for which $E(\xi(x)|\theta)$ whenever $\theta \in \Omega$ and

$$E(\xi(x)|\theta) = 0 \text{ for } \theta \in \Omega_0. \tag{5.1.4}$$

To describe similar tests of level α we need only know how to describe the precotests. If we have a description of the precotests, for example a parametric description with arbitrary functions belonging to some class serving as parameters, we need only choose from them the precotests such that $-\alpha \leq \xi(x) \leq 1-\alpha$ for all x. Similar tests are of the form

$$\Phi(x) = \xi(x) + \alpha.$$

Let us now look at the less general scheme introduced in §1 of Chapter II. The parameter $\theta = (\theta_1, \cdots, \theta_s)$ lies in some compact Borel subset Ω (usually a parallelepiped) of the Euclidean space E_s. The probability density $p(x; \theta_1, \cdots, \theta_s)$ with respect to the dominating measure μ is assumed to be a continuous function of the parameter $\theta \in \Omega$ for almost all values of x.

Consider the precotests belonging to the class $\mathfrak{K}$ of complex-valued statistics $f(x)$ satisfying the following condition: the space $\overline{E}_{2s}$ of complex s-tuples $\{\theta_1, \cdots, \theta_s\}$ contains a simply-connected compact polycylinder $Z \supset \Omega$ such that the mathematical expectations

$$E(f(x)|\theta) = \phi(\theta) \tag{5.1.5}$$

exist and are holomorphic on Z. The "images" $\phi(\theta)$ of the statistics $f(x)$ constitute a ring $\overline{K}$ over the field of complex numbers.

Of course the class $\mathfrak{K}$ of statistics of this sort may be trivial and reduce to the statistic $f(x) \equiv 0$. Even if it is not trivial it may prove too small to contain the cotests that are optimum in some sense or other. However, for the case of exponential families the class $\mathfrak{K}$ is sufficiently broad. Furthermore, finding the preimages $f(x)$ from their images $\phi(\theta)$ is comparatively simple for these families.

If $f(x) \in \mathfrak{K}$ is a precotest, the image $\phi(\theta)$ vanishes for $\theta \in \Omega_0$. The converse assertion is also obvious. The set of functions $\phi(\theta) \in \overline{K}$ that vanish for $\theta \in \Omega_0$ constitutes an ideal of the ring $\overline{K}$. We shall call this ideal the *cotest ideal* of the ring $\overline{K}$ and denote it by I_{H_0}. (It would be more accurate to call it the "precotest ideal" but such a term would be too cumbersome.) Thus description of the cotests in the class $\mathfrak{K}$ reduces to description of the ideals of functions in the ring $\overline{K}$ that are generated by functions that vanish on $\Omega_0 \subset Z$ (see Chapter I). It should be noted, however, that the theory of ideals in rings of holomorphic functions was developed (by Cartan and others in the publications cited in §3 of Chapter I) primarily for rings of all functions that are holomorphic on a compact polycylinder and not on their subrings. However, even the information that we already have is quite useful for investigating similar tests in certain families of distributions.

Under the conditions of a given Bayes measure $B(\theta)$ on simple alternatives to H_0, it is natural to treat as the optimum similar test out of all similar tests that one Φ that maximizes the integral $\int_{(\Omega\setminus\Omega_0)} E(\Phi\mid\theta)\,dB(\theta)$. However, if such a test exists it is usually hard to find. Therefore it may prove expedient to find an ϵ-optimum similar test Φ_ϵ satisfying the condition

$$W(\Phi_\varepsilon\mid B)=\int\limits_{(\Omega\setminus\Omega_0)} E(\Phi_\varepsilon\mid\theta)\,dB(\theta)\geqslant\sup_{\Phi}\int\limits_{(\Omega\setminus\Omega_0)} E(\Phi\mid\theta)\,dB(\theta)-\varepsilon, \qquad (5.1.6)$$

where the supremum is over all similar tests Φ of given level α and where ϵ is a positive number. Instead of an ϵ-optimum test, we need only find an ϵ-optimum cotest ψ_ϵ satisfying condition (5.1.6), with Φ_ϵ replaced by ψ_ϵ, and the condition

$$-\alpha\le\psi_\epsilon(T)\le 1-\alpha. \qquad (5.1.7)$$

For families (see above) with densities of the form $\{p(x;\theta_1,\cdots,\theta_s)\}$ which for all x are smooth functions of the parameter $\theta=(\theta_1,\cdots,\theta_s)$, we may consider optimization of the test from the standpoint of the local properties of its power function (see [36]). For example, if a hypothesis H_0 is of the form H_0: $\gamma(\theta_1,\cdots,\theta_s)=\gamma_0$, where γ is a smooth function, and if $\gamma_1=\gamma(\theta_1,\cdots,\theta_s),\cdots$ $\cdots,\gamma_s=\gamma_s(\theta_1,\cdots,\theta_s)$ is a local coordinate system, we may examine the behavior of the power function

$$E(\Phi\mid\theta)=E(\Phi\mid\gamma,\gamma_2,\cdots,\gamma_s). \qquad (5.1.8)$$

For $\gamma=\gamma_0$ and arbitrary admissible values of $\gamma_2,\cdots,\gamma_s$, the function (5.1.8) assumes a constant value α. If this function is twice continuously differentiable with respect to γ for arbitrary $\gamma_2,\gamma_3,\cdots,\gamma_s$, we may require that "the tests be unbiased in a neighborhood of H_0"; i.e. we may require that $(\partial/\partial\gamma)E(\Phi\mid\gamma,\gamma_2,\cdots,\gamma_s)=0$ for $\gamma=\gamma_0$ and arbitrary values of $\gamma_2,\cdots,\gamma_s$. It is desirable to have the value of the second derivative $(\partial^2/\partial\gamma^2)E(\Phi\mid\gamma,\gamma_2,\cdots,\gamma_s)$ as great as possible at $\gamma=\gamma_0$ for all values of $\gamma_2,\cdots,\gamma_s$. Of course this cannot in general be done simultaneously for all points $\gamma_2,\cdots,\gamma_s$.

If the family of distributions $\{P_\theta\}$ that we are studying has nontrivial sufficient statistics, then by what was said in §3 of Chapter III it is natural for us to confine ourselves to similar tests that depend only on sufficient statistics. Characteristics of the type (5.1.6) will be no worse for these.

§2. STATEMENT OF THE PROBLEM FOR INCOMPLETE EXPONENTIAL FAMILIES

Let us again look at exponential families of distributions. To make the exposition as simple as possible we assume that the sample space is the Euclidean space E_n and that the dominating measure is Lebesgue measure.

As we know (see Chapter II), a compatible distribution of sufficient statistics $(T_1, \cdots, T_s)$ in the space E_s also constitutes an exponential family with probability density which it will be convenient for us to write in the form

$$p_\theta(T_1, \ldots, T_s) = C(\theta) \exp[-(\theta_1 T_1 + \ldots + \theta_s T_s)]\, h(T_1, \ldots, T_s). \tag{5.2.1}$$

Let us assume that the set of points of discontinuity of $h(T_1, \cdots, T_s)$ has Lebesgue measure 0.

Many problems with tests of the "even type" involve only sufficient statistics $T_1, \cdots, T_s$ that retain sign. For example, such a problem is the problem of constructing similar tests for the Behrens-Fisher problem that depend only on the statistics $(\bar{x} - \bar{y})^2$, s_1^2, and s_2^2, where the notation is that of Chapter III, §3. We shall look at this problem again in Chapter VIII, §2. Cases of the type

$$T_i \geq c_i \quad \text{or} \quad T_i \leq c_i \quad (i = 1, 2, \cdots, s), \tag{5.2.2}$$

where the c_i are constants, reduce to this case if we replace the T_i with sufficient statistics $T_i - c_i$ $(i = 1, 2, \cdots, s)$. We shall call exponential families with sufficient statistics of this kind *one-sided exponential families*. Without loss of generality we can write (making a transformation of parameters if necessary) the probability densities for the sufficient statistics of a one-sided family in the form

$$p_\theta(T_1, \ldots, T_s) = C(\theta) \exp[-(\theta_1 T_1 + \ldots + \theta_s T_s)]\, h(T_1, \ldots, T_s), \tag{5.2.3}$$

where, in the case of a one-sided family, we assume $T_i \geq 0$ for $i = 1, 2, \cdots, s$. In a more general situation we assume that $h(T_1, \cdots, T_s)$ behaves like the function $m(T_1, \cdots, T_s)$ in §1 of Chapter I, specifically that $h(T_1, \cdots, T_s)$ vanishes for $T_j < 0$, where $j = 1, 2, \cdots, s_1$ and $s_1 \leq s$. The integral (5.2.3) converges absolutely in the Cartesian product $\mathcal{P}$ of the s_1 half-planes R_j: $\operatorname{Re} \theta_j > 0$

($j = 1, 2, \cdots, s_1$) and the $s - s_1$ strips $S_j: 0 < \operatorname{Re}\theta_j \le A_j$, where the A_j are finite numbers. (By a linear transformation of the parameters and the variables we can reduce arbitrary strips to this type.) The region $\mathcal{P}$ described above is the appropriate region of variation for natural complex parameters. In posing problems of testing statistical hypotheses we shall assume that the parameters θ_j ($j = 1, 2, \cdots, s$) are real. As a rule, the region Ω of variation of these parameters will be a closed bounded parallelepiped in the Euclidean space E_s. Let us assume that the null hypothesis H_0 in this region is of the form

$$\Pi_1(\theta_1, \ldots, \theta_r) = 0, \ \ldots, \ \Pi_r(\theta_1, \ldots, \theta_s) = 0, \quad r < s, \tag{5.2.4}$$

where the functions $\Pi_j(\theta_1, \cdots, \theta_s)$ ($j = 1, 2, \cdots, r$) are holomorphic and real on Ω. Furthermore let us assume that the functions $\Pi_1, \cdots, \Pi_r$ satisfy condition (Γ).

(Γ) *The holomorphy condition.* There exists an integer $N \ge 0$ such that the functions

$$[(\theta_1 + 1)\ldots(\theta_s + 1)]^{-N}\,\Pi_j(\theta_1, \ldots, \theta_s), \quad j = 1, 2, \ldots, r, \tag{5.2.5}$$

are holomorphic on the entire region $\mathcal{P}$, including its boundary.

Here the behavior of Π_j at infinitely distant points ($\theta_i = \infty$), $i = 1, 2, \cdots, s$, is significant. Condition (Γ) is not very restrictive and is satisfied for polynomials and other functions of similar asymptotic behavior. Thus the null hypothesis H_0 imposes r ($< s$) holomorphic relationships on the s natural parameters $\theta_1, \cdots, \theta_s$. From an analytical point of view these relationships generate a real analytic set (see Chapter I) contained in Ω. Let us denote it by Ω_0. By what was said at the end of the preceding section, we can confine ourselves to tests that depend only on sufficient statistics: $\phi(T_1, \cdots, T_s)$. Assuming that an alternative H_1 to H_0 is given in the Bayes formulation (as mentioned in the preceding section), we seek ϵ-optimum similar tests for the hypothesis H_0 that are defined in accordance with formula (5.1.6). Here we need to impose a number of conditions on the relations (5.2.4) determined by the null hypothesis H_0 and on the function $h(T_1, \cdots, T_s)$ defined in the region $\mathcal{T}: T_j > 0$ for $j = 1, 2, \cdots, s_1$. We define

Condition (A). The function $h(T_1, \cdots, T_s)$ is positive almost everywhere in $\mathcal{P}$.

Although this condition is satisfied in many cases, for example in many problems on the statistical analysis of normal vectors, it is still rather restrictive; but if it is not satisfied our investigations become quite complicated. We shall not go into this matter in the present book except in Chapter VII, which deals with unbiased statistics. One can easily see that the case in which the function $h(T_1, \cdots, T_s)$ vanishes outside an s_1-faced cone in s-dimensional Euclidean space but is nonzero almost everywhere inside that cone reduces to condition (A).

By applying a suitable nonsingular linear transformation to the sufficient statistics, we can then apply a linear transformation to the parameters, replacing them with new parameters, in such a way that condition (A) is satisfied.

§3. IDEALS OF PRECOTESTS

In line with the definition given in §1, we now define a precotest $\xi(T_1, \cdots, T_s)$ as any Lebesgue-measurable statistic such that the quantity

$$\dot{C}(\theta) \int_{\mathcal{T}} \cdots \int \xi(T_1, \ldots, T_s) \times \exp[-(\theta_1 T_1 + \ldots + \theta_s T_s)]\, h(T_1, \ldots, T_s)\, dT_1 \ldots dT_s \tag{5.3.1}$$

exists and vanishes for

$$(\theta_1, \cdots, \theta_s) \in \Omega_0. \tag{5.3.2}$$

We note that $C(\theta) \neq 0$ for $\theta \in \Omega$, since the quantity (5.3.1) must be equal to unity if $\xi(T_1, \cdots, T_s) \equiv 1$. Consider the set of all statistics $m(T_1, \cdots, T_s)$ for which the s-fold Laplace transform

$$L(m|\theta) = \int_{\mathcal{T}} \cdots \int m(T_1, \ldots, T_s) \times \exp[-(\theta_1 T_1 + \ldots + \theta_s T_s)]\, dT_1 \ldots dT_s \tag{5.3.3}$$

converges absolutely in our basic region $\mathcal{P}$. If we set $u(T_1, \cdots, T_s) = m(T_1, \cdots, T_s)/h(T_1, \cdots, T_s)$ for $(T_1, \cdots, T_s) \in \mathfrak{T}$, which we can do by virtue of condition (A), we find that

$$C(\theta)\, L(m\,|\,\theta) = E(u\,|\,\theta). \tag{5.3.4}$$

To find the precotests, we can find the statistics m for which

$$L(m\,|\,\theta) = 0 \text{ for } \theta \in \Omega_0. \tag{5.3.5}$$

Then $\xi = m/h$ is a precotest. Since $C(\theta) \neq 0$ for $\theta \in \Omega$, this procedure provides us with all the precotests.

The statistics $m(T_1, \cdots, T_s)$ for which the integral (5.3.3) converges absolutely constitute a linear space $\mathfrak{M}$. Here the region of scalar multipliers can be the set of all real numbers, or the set of all complex numbers if we are admitting complex-valued statistics. If $m_1, m_2 \in \mathfrak{M}$ then the convolution (cf. (1.1.2))

$$m_1 * m_2 = \int_0^{T_1} d\xi_1 \ldots \int_0^{T_{s_1}} d\xi_{s_1} \ldots \int_{-\infty}^{\infty} d\xi_s m_1(T_1 - \xi_1, \ldots, T_s - \xi_s) \times m_2(\xi_1, \ldots, \xi_s)\, d\xi_1 \ldots d\xi_s \qquad (5.3.6)$$

belongs to $\mathfrak{M}$ and

$$L(m_1 * m_2|\theta) = L(m_1|\theta)\, L(m_2, \theta) \qquad (5.3.7)$$

for $\theta \in R_1 \times \cdots \times R_s$ (see [16]).

Thus the elements of the space $\mathfrak{M}$ constitute a ring, with the convolution defined by formula (5.3.6) serving as the ring multiplication. The images $L(m|\theta)$ of the statistics m constitute a ring with respect to ordinary multiplication. We denote this ring by $L(\mathfrak{M})$.

If ξ is a precotest then the function $\xi \cdot h$ belongs to $\mathfrak{M}$ and

$$E(\xi|\theta) = L(\xi h|\theta) = 0 \quad \text{for} \quad \theta \in \Omega_0. \qquad (5.3.8)$$

Let m denote an arbitrary function belonging to $\mathfrak{M}$. From (5.3.8) and (5.3.7) we have

$$E\left(\frac{1}{h}\left[m * (\xi h)\right] | \theta\right) = L(m * (\xi h)|\theta) = 0 \quad \text{for} \quad \theta \in \Omega_0.$$

Thus if ξ is a precotest, then an arbitrary function of the form $[m^*(\xi h)]/h$ is also a precotest and functions of the form ξh constitute an ideal $I_{H_0} \subset \mathfrak{M}$. We shall call this ideal the *cotest ideal of the null hypothesis* or, more briefly, the *cotest ideal*. If $\xi h \in I_{H_0}$, then the expressions $E(\xi|\theta) = L(\xi h|\theta)$ constitute an ideal in the ring $L(\mathfrak{M})$. We shall call this ideal the *image* of the cotest ideal I_{H_0} and we shall denote it by $L(I_{H_0})$.

The investigation of similar tests under our assumptions reduces to study of the image of the cotest ideal $L(I_{H_0})$. If we construct a basis for the ideal $L(I_{H_0})$, we shall also have a basis for the cotest ideal and a description of all precotests ξ. Of these we then need to choose those cotests for which

$$-\alpha \leq \xi \leq 1 - \alpha. \qquad (5.3.9)$$

To choose an ϵ-optimum cotest ψ_ϵ, we impose, following (5.1.6), the condition

$$W(\psi_\varepsilon|B) = \int_{(\Omega \setminus \Omega_0)} E(\psi_\varepsilon|\theta)\, dB(\theta) \geqslant \sup_{\psi} \int_{(\Omega \setminus \Omega_0)} E(\psi|\theta)\, dB(\theta) - \varepsilon, \qquad (5.3.10)$$

where the supremum is over all cotests ψ_ϵ of level α subject to the condition

(5.3.9), in which we need to set $\xi = \psi_\epsilon$. The ϵ-optimum similar test is $\Phi_\epsilon = \psi_\epsilon + \alpha$. However, construction of a basis for the ideal $L(I_{H_0})$ in the ring $L(\mathfrak{M})$ is a difficult matter. Investigation of the ideals in rings of holomorphic functions has been carried out primarily for rings of ideals of all functions that are holomorphic on some compact polycylinder (see the articles of Cartan cited above, and also of K. Oka and other authors). In a recent article [29] Cartan investigated ideals in rings of all functions that are holomorphic in the region of real values of the arguments. We know virtually nothing about other types of rings of holomorphic functions, in particular about our basic ring $L(\mathfrak{M})$. Therefore, having in mind the application of Cartan's results (expounded in Chapter I, §3) to an investigation of the images of cotest ideals, let us look at an extension $\overline{L(I_{H_0})}$ of the ideal $L(I_{H_0})$ for the ring $\overline{L(\mathfrak{M})}$ of all functions that are holomorphic on $\widetilde{\Omega}$ (see Chapter I). We can describe a basis for this extension $\overline{L(I_{H_0})}$ under certain conditions in the same way as for an ideal in $\overline{L(\mathfrak{M})}$. By imposing further conditions on the function $h(T_1, \cdots, T_s)$, we can use our knowledge of the basis for the ideal $\overline{L(I_{H_0})}$ to find the ϵ-optimum cotests. The values assumed by the parameters $\theta_1, \cdots, \theta_s$ lie on the parallelepiped $\Omega \subset \Lambda$. To simplify the remainder of our discussion let us suppose that the parallelepiped Ω is of the form

$$\theta_j \geqslant \varepsilon_j, \ j=1, 2, \ldots, s_1; \ \varepsilon_j \leqslant \theta_j \leqslant E_j, \ j=s_1+1, \ldots, s, \qquad (5.3.11)$$

where the ϵ_j are small positive numbers and the E_j are given positive numbers less than the corresponding A_j.

The boundary of the region $\mathcal{P}$ contains points with infinite coordinates, in particular the point $(\infty, \cdots, \infty)$. We denote by $\partial^0 \mathcal{P}$ the set of such boundary points of the region $\mathcal{P}$. Functions of the form (5.3.3), i.e. elements of the ring $L(\mathfrak{M})$, will not in general be analytic at the points of $\partial^0 \mathcal{P}$. It turns out that for $s = 1$ analyticity of these elements at the points of $\partial^0 \mathcal{P}$ is equivalent to entireness of a certain order (see [10]) of the function $m(T_1)$ of the complex variable T_1.

The requirement that the precotests be holomorphic may restrict our choice of similar tests, and we shall not make this requirement. In such a case we need to reckon with the difficulty arising from the fact that we are unable to consider functions that are holomorphic on all $\mathcal{P}$ and must assume that they are holomorphic in the interior but not on the boundary of $\mathcal{P}$. One encounters considerable difficulties in studying ideals of functions that are holomorphic in an open region. Up to the present time such ideals have not been completely described even in the simplest cases (see Gleason [13]). What we have to do is impose stringent requirements in order to obtain a sufficiently complete description of the corresponding similar tests. With a view to applying the corollary to Theorem 1.3.1 (Cartan's theorem), we shall study first not the cotest ideal $L(l_{H_0})$ but its extension $\overline{L(l_{H_0})}$ in the ring of all functions that are holomorphic inside $\mathcal{P}$.

§4. APPLICATION OF CARTAN'S THEOREMS

The precotest ideal $L(l_{H_0})$ consists of all holomorphic functions in $L(\mathfrak{M})$ that vanish at all real points $(\theta_1, \cdots, \theta_s)$ of the region Ω satisfying (5.2.4). Our plan is to apply the theorems of Chapter I, §3 dealing with holomorphic functions $\overline{L(\mathfrak{M})}$ that vanish at all complex points of the analytic set (5.2.4) belonging to $\mathcal{P}$. Obviously the ideal that they form is only a portion of the ideal $\overline{L(l_{H_0})}$ in the general case. In view of this we need to admit a "complexification" of conditions (5.2.4) (see Cartan [29]). We shall do this in the elementary form indicated by Theorem 1.2.2.

Let us look at the relationships between the parameters in (5.2.4) when the parameters $(\theta_1, \cdots, \theta_s)$ are complex members of $\mathcal{P}$. We make an assumption completely analogous to the hypotheses of Theorem 1.2.2. We denote the analytic set (5.2.4) by $V_{\Pi_1 \cdots \Pi_r}$.

Condition (B). Inside $\mathcal{P}$ the analytic set $V_{\Pi_1 \ldots \Pi_r}$ can be decomposed into finitely many connected components $V^q_{\Pi_1 \ldots \Pi_s}$ each of which is of complex dimension $s - r$ and contains the connected set $R^q_{\Pi_1 \ldots \Pi_r} \subset \Omega_0$ of real points of real dimension $s - r$.

When the set $V_{\Pi_1 \ldots \Pi_r}$ has such a structure, the functions that are holomorphic in $\mathcal{P}$ and vanish in Ω_0 also vanish at all complex points inside $\mathcal{P}$ (by Theorem 1.2.2).

Thus the ideal of the functions $\overline{L(\mathfrak{M})}$ that vanish at all complex points of the analytic set (5.2.4) coincides with the original ideal $\overline{L(l_{H_0})}$ when condition (B) is satisfied.

It would also be desirable if the functions $\Pi_1, \cdots, \Pi_r$ defining the conditions connecting the parameters (5.2.4), which are holomorphic on the closed region $\mathcal{P}$, generate the ideal $L(l_{H_0})$ locally. To arrange for this, we impose on the relationships (5.2.4) the following condition.

Condition (C). The analytic set $\mathfrak{B}$ of points in the region $\mathcal{P}$ such that

$$\left.\begin{array}{c}\Pi_1(\theta_1, \ldots, \theta_s) = 0, \ \ldots, \ \Pi_r(\theta_1, \ldots, \theta_s) = 0, \\ \operatorname{rank} \left\| \begin{array}{ccc} \frac{\partial \Pi_1}{\partial \theta_1} & \ldots & \frac{\partial \Pi_1}{\partial \theta_s} \\ \cdot & \cdot\ \cdot\ \cdot & \cdot \\ \frac{\partial \Pi_r}{\partial \theta_s} & \ldots & \frac{\partial \Pi_r}{\partial \theta_s} \end{array} \right\| < r, \end{array}\right\} \tag{5.4.1}$$

has no points inside $\mathcal{P}$.

We note that condition (C) is not especially restrictive. The inequality regarding the rank of the matrix yields, in general, an analytic set of complex dimension not exceeding $s - r - 1$. In general the first r of conditions (5.4.1) make the set $\mathfrak{B}$ empty and not merely finite. Also, it is easy to exhibit many cases in which condition (C) is not satisfied. Thus, if the entire functions $\Pi_j(\theta_1, \cdots, \theta_s)$ are squares of other entire functions, then the set $\mathfrak{B}$ has, in general, complex dimension no less than $s - r$. Of course, in our present example it is not expedient to use relationships of the form (5.2.4). Without changing the problem we can replace them with the conditions

$$\Pi_j^{1/2} = 0 \qquad (j = 1, 2, \ldots, r).$$

Now we can prove

Theorem 5.4.1. *If conditions* (B) *and* (C) *are satisfied, the ideal* $\overline{L(I_{H_0})}$ *inside* $\mathcal{P}$ *is generated by the functions* $\Pi_1, \cdots, \Pi_r$. *Thus if* $F \in \overline{L(I_{H_0})}$, *then there exist functions* $G_1, \cdots, G_r$, *holomorphic inside* $\mathcal{P}$ *such that*

$$F = \Pi_1 G_1 + \cdots + \Pi_r G_r. \tag{5.4.2}$$

To prove this we note that, by virtue of condition (B), the ideal $\overline{L(I_{H_0})}$ coincides with the ideal of all functions that are holomorphic inside $\mathcal{P}$ and that vanish at all complex points in the analytic set $V_{\Pi_1 \cdots \Pi_r}$ inside $\mathcal{P}$. The representation (5.4.2) then follows from the corollary to Cartan's theorem (Theorem 1.3.2).

§5. THE BEHAVIOR OF SMOOTH PRECOTESTS

The results of the preceding section dealt with the ideal $\overline{L(I_{H_0})}$, which is an extension of the basic precotest ideal $L(I_{H_0})$. Theorem 5.4.2 yields a basis $\Pi_1, \cdots, \Pi_r$ for it inside $\mathcal{P}$. However this gives no immediate information regarding a basis for the ideal $L(I_{H_0})$ in the ring $L(\mathfrak{M})$; in fact it gives no information as to whether such a basis is finite or not. In this connection a simple example[1])

[1]) This example was communicated to the author by B. Z. Moroz.

from the theory of rings is instructive. Let $K[x, y, z]$ denote the ring of all polynomials in the three variables x, y, z over the field of real numbers. Let $K_1[(x), y, z]$ denote the ring of all polynomials in the variables y and z whose coefficients are rational functions over the same field. Then

$$K[x, y, z] \subset K[(x), y, z].$$

In the ring $K[x, y, z]$, consider the ideal I_0 of all polynomials without a constant term. This ring has the three-number basis $\{x, y, z\}$. An extension I in the ring $K_1[(x), y, z]$ is obviously the ring itself, which has the one-member basis $\{1\}$. Thus the basis of the original ideal is quite different from the basis of its extension. However, since we are interested not in the most general case of ideals in rings but in the examination of the behavior of the precotest ideals $L(I_{H_0})$, under certain conditions the basis of $\overline{L(I_{H_0})}$ gives us a great deal of information regarding the ideal $L(I_{H_0})$.

Let us look at smooth precotests $\xi(T_1, \cdots, T_s)$ in the precotest ideal I_{H_0}, the preimage of $L(I_{H_0})$. We shall say that a precotest ξ is smooth and that it belongs to the class $L^{(m)}$ (where m is a positive integer) if

$$E(\xi|\theta) = L(\xi h|\theta) = O\left(\frac{1}{(|\theta_1|+1)^m \cdots (|\theta_s|+1)^m}\right), \tag{5.5.1}$$

for the function $\xi(T_1, \cdots, T_s) h(T_1, \cdots, T_s)$ whenever at least one of the numbers $\theta_i \to \infty$. By Theorem 1.1.3 a smooth precotest $\xi(T_1, \cdots, T_s)$ has partial derivatives of the first m orders inclusive.

We now impose on the relationships (5.2.4) defined by the zero hypothesis H_0 other conditions in the region $\mathcal{P}$ that we shall need.

Condition (C′). Inside $\mathcal{P}$,

$$\left|\frac{\partial \Pi_i}{\partial \theta_{\alpha_j}}\right| \neq 0 \quad (i, j = 1, 2, \ldots, r). \tag{5.5.2}$$

for suitably chosen indices $\alpha_1, \cdots, \alpha_r$.

We recall that the functions $\Pi_j(\theta_1, \cdots, \theta_s)$, $j = 1, 2, \cdots, r$, are holomorphic on $\overline{\mathcal{P}}$. Suppose now that the function F belongs to the extended ideal $\overline{L(I_{H_0})}$ of the zero hypothesis H_0. By Theorem 5.4.1 we have inside $\mathcal{P}_{\eta_0}$ the representation

$$F = \Pi_1 G_1 + \cdots + \Pi_r G_r. \tag{5.5.3}$$

Keeping (5.2.5) in mind, we multiply both sides of this last equation by $[(\theta_1 + 1) \cdots (\theta_s + 1)]^{-N}$. We obtain

$$F_1 = F[(\theta_1+1)\dots(\theta_s+1)]^N = \Pi_1 G_1' + \dots + \Pi_r G_r', \tag{5.5.4}$$

where the functions $G_j' = [(\theta_1+1)\cdots(\theta_s+1)]^{-N} G_j$ are holomorphic inside $\mathcal{P}_{\eta_0}$. Suppose that F is bounded inside $\mathcal{P}$ by a constant M: $|F| \leq M$. If we define $M_1 = \sup_{\mathcal{P}} |(\theta_1+1)\cdots(\theta_s+1)|^{-N}$ we obtain $|F_1| \leq MM_1 = M_0$ on $\mathcal{P}$, where M_0 is a new constant. By virtue of condition (5.5.2), Theorem 1.2.7 is applicable, so that inside $\mathcal{P}$ we have the following bound for the functions G_j' $(j = 1, 2, \cdots, r)$:

$$|G_j'(\theta_1, \dots, \theta_s)| \leqslant \frac{K_1 M}{\delta^{K_2}} (|\theta_1| + |\theta_2| + \dots + |\theta_s| + 1)^{K_2},$$

where K_1 and K_2 are constants and δ is the distance from the point $(\theta_1, \cdots, \theta_s)$ to the boundary of the region $\mathcal{P}_{\eta_0}$. Since $G_j' = [(\theta_1+1)\cdots(\theta_s+1)]^{-N} G_j$, this inequality gives us the following bound for G_j:

$$|G_j(\theta_1, \dots, \theta_s)| \leqslant \frac{K_3 M}{\delta^{K_2}} (|\theta_1| + \dots + |\theta_s| + 1)^{K_2}, \tag{5.5.5}$$

where K_3 is a new constant. Here we remember that

$$|(\theta_1+1)\cdots(\theta_s+1)| \geq 1 \text{ in } \mathcal{P}.$$

Let us now suppose that $\xi(T_1, \cdots, T_s)$ is a smooth precotest of the class $L^{(m)}$, where $m \geq 1$. Then $F(\theta) = E(\xi|\theta)$ satisfies the relation (5.5.1).

If the function F belongs to the ideal $\overline{L(I_{H_0})}$, then so does the function $F_2 = [F(\theta_1, \cdots, \theta_s)]^m \in \overline{L(I_{H_0})}$, which is holomorphic inside $\mathcal{P}$. Also, $|F(\theta_1, \cdots, \theta_s)|^m < M$, where M is a constant. Therefore F_2 has a representation of the form (5.5.3):

$$F_2 = \Pi_1 G_1 + \cdots + \Pi_r G_r, \tag{5.5.6}$$

where the G_j satisfy an inequality of the form (5.5.5). If we multiply equation (5.5.6) by $[(\theta_1+1)\cdots(\theta_s+1)]^{-1}$ and set

$$G_j'' = [(\theta_1+1)\dots(\theta_s+1]^{-m} G_j \quad (j=1, \dots, s),$$

we obtain, by virtue of (5.5.5),

$$|G_j''| \leqslant \frac{K_2 M |(\theta_1+1)\dots(\theta_s+1)|^{-m}}{\delta^{K_3}}, \quad j=1, 2, \dots, r. \tag{5.5.7}$$

In the region $\mathcal{P}$ consider the Cartesian product $L_1 \times \cdots \times L_s$, where L_j is the vertical contour $[-i\infty + \eta/2, i\infty + \eta/2]$ with $0 < \eta < \eta_0$, where η_0 is a small number. Then on this Cartesian product we obtain the following bounds for the G_j'':

$$|G_j''(\theta_1, \ldots, \theta_s)| < \frac{K_\eta M}{[(|\theta_1|+1) \ldots (|\theta_s|+1)]^m},$$

where K_η is a constant depending on η.

Now let ξ denote a precotest of the smoothness class $L^{(m)}$, so that (5.5.1) holds inside $\mathcal{P}$. In view of this we obtain from (5.5.7)

$$E(\xi|\theta) = F(\theta) = \Pi_1 G_1'' + \ldots + \Pi_r G_r''. \tag{5.5.8}$$

If we now remember that (5.2.5) is holomorphic and if we multiply and divide every term on the right-hand side of (5.5.8) by $[(\theta_1 + 1) \cdots (\theta_s + 1)]^{N+2}$, we obtain inside $\mathcal{P}$

$$E(\xi|\theta) = P_1 A_1 + \ldots + P_r A_r, \tag{5.5.9}$$

where

$$P_j = \frac{\Pi_j}{(\theta_1 \ldots \theta_s)^{N+2}} (j = 1, 2, \ldots, s), \quad A_j = (\theta_1, \ldots, \theta_s)^{N+2} G_j''. \tag{5.5.10}$$

If $m \geq N + 3$, then by Theorem 1.1.3 the functions $A(\theta_1, \cdots, \theta_s)$ will be unilateral Laplace transforms of $A_j(\theta_1, \cdots, \theta_s) = L(H_j|\theta)$, where $H_j = H(T_1, \cdots, T_s)$ is a function that vanishes for $T_1 < 0, \cdots, T_{s_1} < 0$ and satisfies the relation

$$|H_j(T_1, \ldots, T_s)| = O(\exp \eta(|T_1| + \ldots + |T_s|)) \tag{5.5.11}$$

for arbitrary $\eta > 0$.

Furthermore, H has partial derivatives of at least the first $m - N$ orders. By Theorem 1.1.1 (the convolution theorem), if the precotest ξ is sufficiently smooth we have the representation

$$E(\xi|\theta) = R_1 * H_1 + \ldots + R_r * H_r, \tag{5.5.12}$$

where R_j and H_j $(j = 1, 2, \cdots, r)$ are the functions described above. Furthermore, we shall prove that every cotest can be replaced by a cotest of any desired degree of smoothness in such a way that the basic statistical properties change by an arbitrarily small amount.

§6. SMOOTHING OF PRECOTESTS

Let us generalize the concept of the gain function $W(\Phi | B)$ (see §1); in particular let us generalize formula (5.1.6) to precotests (including cotests) by setting

$$W(\xi|B) = \int_{(\Omega \setminus \Omega_0)} E(\xi|\theta)\, dB(\theta) \tag{5.6.1}$$

for a precotest ξ and a given Bayes distribution $B(\theta)$ on $\Omega\setminus\Omega_0$.

Let us look at the question of "smoothing" the precotests by replacing them with smooth precotests for which the gain function $W(\xi|B)$ will differ only slightly from the original one.

Let δ denote a number in the interval $(0, \frac{1}{2})$. Let $d_{\delta r}(x)$ denote a smooth function of a single variable that imitates the behavior of Dirac's function, has the form of a "smooth peak" of area 1 and has a given number $h \geq 1$ of derivatives. The function $d_{\delta r}(x)$ must vanish for $|x - \delta/2| \geq \delta^r$. Here $r \geq 2$ is an integer that will be fixed in what follows. Thus we have

$$\int_0^\delta d_{\delta r}(x)\,dx = 1; \qquad d_{\delta r}(x) = 0 \qquad \text{for} \qquad x \leqslant 0 \quad \text{or} \quad x \geqslant \delta.$$

Let us now define the function

$$D_\delta(T_1, \ldots, T_s) = d_{\delta r}(T_1) \ldots d_{\delta r}(T_s). \tag{5.6.2}$$

Let $\xi \in \mathfrak{R}$ denote a given precotest and define $\rho = \xi h$. Consider the convolution

$$\rho_\delta = D_\delta * \rho.$$

We have

$$\rho_\delta = D_\delta * \rho = \int_0^{T_1} d\xi_1 \ldots \int_0^{T_{s_1}} d\xi_{s_1} \int_{-\infty}^{\infty} d\xi_{s_1+1} \ldots$$

$$\ldots \int_{-\infty}^{\infty} d_s D_\delta(\xi_1, \ldots, \xi_s)\,\rho(T_1 - \xi_1, \ldots, T_s - \xi_s) \cdot d\xi_1 \ldots d\xi_s. \tag{5.6.3}$$

We shall call the function ξ_δ defined by

$$\xi_\delta(T_1, \ldots, T_s) = h^{-1}(D_\delta * \rho) = h^{-1}(D_\delta * \xi h) = h^{-1}\rho_\delta. \tag{5.6.4}$$

the smoothed precotest of ξ.

Theorem 5.6.1. *For given ϵ and sufficiently small $\delta > 0$,*

$$|W(\xi_\delta|B) - W(\xi|B)| \leqslant \varepsilon. \tag{5.6.5}$$

Proof. For $m \in \mathfrak{M}$ and $\theta \in \mathcal{P}$ we have

$$E(h^{-1}m\,|\,\theta) = L(m\,|\,\theta).$$

Therefore

$$E(\xi_\delta|\theta) = L(D_\delta * \rho|\theta) = L(D_\delta|\theta)\,L(\rho|\theta).$$

By the construction of the function D_δ and a well-known property of the Laplace transform for $\theta \in \Omega \setminus \Omega_0$ and given positive ϵ, we have

$$|L(D_\delta|\theta) - 1| \leqslant \varepsilon, \tag{5.6.6}$$

if $\delta > 0$ is sufficiently small (cf. condition (5.3.11)). Therefore

$$\begin{aligned} W(\xi_\delta|B) - W(\xi|B) &= W(\xi_\delta|B) - W(h^{-1}\rho|B) \\ &= \int_{(\Omega\setminus\Omega_0)} dB(\theta)(E(\xi_\delta|\theta) - E(h^{-1}\rho|\theta)) \\ &= \int_{(\Omega\setminus\Omega_0)} dB(\theta)(L(D_\delta * \rho|\theta) - L(\rho|\theta)) \\ &= \int_{(\Omega\setminus\Omega_0)} L(\rho|\theta)(L(D_\delta|\theta) - 1)\, dB(\theta). \end{aligned} \tag{5.6.7}$$

Since $\Omega \setminus \Omega_0$ is contained in the compact set Ω, in which all $\theta_j \geq \epsilon_j > 0$ (cf. (5.3.11)), we have

$$|L(\rho|\theta)| < C \quad \text{for} \quad \theta \in \Omega \setminus \Omega_0.$$

From this and (5.6.7) the conclusion of the theorem follows by virtue of (5.6.6).

Smoothing of the precotest ξ, i.e. its replacement with the precotest ξ_δ defined by formula (5.6.4), leads to improvement in the behavior of the function $E(\xi_\delta|\theta)$ introduced in §3 with increase in $|\theta_j|$, where j is one of the numbers $1, 2, \cdots, s$, in the region $\tilde{\Omega}$.

Theorem 5.6.2. *If the function $d_{\delta_r}(x)$ has $h \geq 1$ continuous derivatives, then in the region* $\tilde{\Omega}$

$$E(\xi_\delta|\theta) = O\left(\frac{1}{(|\theta_1|+1)^h \ldots (|\theta_s|+1)^h}\right). \tag{5.6.8}$$

To prove this we note that, by Theorems 1.1.4 and 5.6.1, we have, for $\theta \in \tilde{\Omega}$,

$$\begin{aligned} E(\xi_\delta|\theta) &= L(D_\delta|\theta)\, L(\rho|\theta) \\ &= O\left(\frac{1}{(|\theta_1|+1)^h \ldots (|\theta_s|+1)^h}\right) \cdot O(1), \end{aligned}$$

which proves (5.6.8).

For a sufficiently large value of h the precotest ξ_δ has, by virtue of what was said in §5, a representation of the form (5.5.12).

§7. FORMATION OF SMOOTH PRECOTESTS FROM A GIVEN ONE

We saw above that, under certain conditions of a rather general type, a smooth precotest ξ can be represented as a sum of convolutions

$$\xi = A_1 * H_1 + \dots + A_r * H_r, \tag{5.7.1}$$

where $A_1, \cdots, A_r$ are given functions corresponding to the connection conditions and $H_1, \cdots, H_r$ are smooth functions of a fairly arbitrary type. Conversely, every expression of the form (5.7.1) is a smooth precotest. For a precotest ξ to be a cotest of level α, it must satisfy the condition $-\alpha \le \xi \le 1-\alpha$.

If we began with the cotest ψ of level α and then smoothed it in accordance with formula (5.6.4), we would obtain the smoothed precotest

$$\psi_\delta = h^{-1}(D_\delta * \psi h). \tag{5.7.2}$$

For this to be a cotest of level α it is necessary and sufficient that

$$-\alpha h \leqslant D_\delta * \psi h \leqslant (1-\alpha) h. \tag{5.7.3}$$

Let us assume that the level α is neither 0 nor 1. In such a case, if we found a smoothing function D_δ that would lead not to the inequalities (5.7.3) but to the inequalities

$$(-\alpha - \eta) h \leqslant D_\delta * \psi h \leqslant (1 - \alpha + \eta) h, \tag{5.7.4}$$

where $\eta > 0$ is an arbitrarily small fixed number, then, by replacing D_δ with $(1-\eta')D_\delta$, where η' is sufficiently small, we would obtain a smoothed cotest ψ'_δ of level α. By formula (5.6.7) the Bayes gain $W(\psi'_\delta|B)$ would be arbitrarily close to $W(\psi|B)$. Such a situation is satisfactory, so that we can make not the requirement (5.7.3) but the requirement (5.7.4), which is equivalent to the inequalities

$$-\alpha - \eta \leqslant h^{-1}(D_\delta * \psi h) \leqslant 1 - \alpha + \eta. \tag{5.7.5}$$

We see that the zeros of the density function $h = h(T_1, \cdots, T_s)$ play an extremely important role here. To obtain a satisfactory description of ϵ-complete cotests of a given level we need to impose rather stringent restrictions on the behavior of the density h. However these restrictions are satisfied for an extremely large number of cases that we encounter in statistics. In the class of problems that we are considering, the function $h(T_1, \cdots, T_s)$ disappeared for $T_j < 0$, where j is one of the numbers $1, 2, \cdots, s_1$, so that we considered it in the region $\mathcal{T}$: $T_1 > 0, \cdots, T_{s_1} > 0$; $-\infty < T_j < \infty$ (for $j = s_1 + 1, \cdots, s$). Let us impose on the function h

Condition (N). The function $h(T_1, \cdots, T_s)$ does not vanish in the region $\mathcal{T}$. In that region it has continuous first partial derivatives. Furthermore, for every $\epsilon > 0$ the inequality

$$\left|\frac{\partial \ln h}{\partial T_1}\right| + \ldots + \left|\frac{\partial \ln h}{\partial T_s}\right| \leqslant \frac{A_1}{\varepsilon^a} + A_2, \tag{5.7.6}$$

where A_1, A_2, and a (< 1) are positive constants, holds in the region $\mathcal{T}_\epsilon$: $T_1 \geq \epsilon$, $T_2 \geq \epsilon, \cdots, T_{s_1} \geq \epsilon$.

This condition makes it possible to ensure that the inequalities (5.7.5) will be satisfied.

Theorem 5.7.1. *Suppose that condition* (5.7.6) *is satisfied. Then from a given cotest ψ of level α it is possible to construct an arbitrarily smooth cotest ψ'_δ for which the Bayes gain $W(\psi'_\delta | B)$ is arbitrarily close to the initial gain $W(\psi | B)$.*

To prove this theorem let us choose a smoothing function in accordance with formula (5.6.2). First we set

$$r = 2. \tag{5.7.7}$$

The parameter $\delta > 0$ will be determined later. By formula (5.7.2) we obtain, for a given cotest ψ of level α,

$$h^{-1}(D_\delta * \psi h) = \int_0^{T_1} d\xi_1 \ldots \int_0^{T_{s_1}} d\xi_{s_1} \cdot \int_{-\infty}^{\infty} d\xi_{s_1+1} \ldots \int_{-\infty}^{\infty} d\xi_s$$
$$\times d_{\delta r}(\xi_1) \ldots d_{\delta r}(\xi_s) \frac{h(T_1 - \xi_1, \ldots, T_s - \xi_s)}{h(T_1, \ldots, T_s)}$$
$$\times \psi(T_1 - \xi_1, \ldots, T_s - \xi_s). \tag{5.7.8}$$

By virtue of the construction of the function $d_{\delta r}(x)$, the integrand in (5.7.8) disappears for $T_j \leq \delta/4$ (where j is any one of the numbers $1, 2, \cdots, s_1$) because the inequalities $T_j \leq \delta/4$ and $\xi_j \leq T_j$ imply that $d_{\delta r}(\xi_j) = 0$ and the inequality $\xi_j > T_j$ implies that $h = 0$. Thus the integral (5.7.8) is over the region $T_j > \delta/4$ $(j = 1, 2, \cdots, s_1)$.

By (5.7.6), in this case we have, in accordance with Lagrange's theorem,

$$\ln h(T_1, \ldots, T_s) - \ln h(T_1 - \xi_1, \ldots, T_s - \xi_s)$$
$$= \xi_1 \left(\frac{\partial \ln h}{\partial T_1}\right) + \ldots + \xi_s \left(\frac{\partial \ln h}{\partial T_s}\right), \tag{5.7.9}$$

where the notation $(\partial \ln h / \partial T_j)$ means that the derivatives are evaluated at an intermediary point $(T_1 - \theta\xi_1, \cdots, T_s - \theta\xi_s)$, where $|\theta| \leq 1$. Furthermore, by virtue of the definition of the function $d_{\delta r}(x)$ and equation (5.7.7), $|\xi_j| \leq \delta/4$. Using (5.7.9) we get

$$|\ln h(T_1, \ldots, T_s) - \ln h(T_1 - \xi_1, \ldots, T_s - \xi_s)| \leqslant \frac{3}{4}\delta\left(\frac{4^a \cdot A_1}{\delta^a} + A_2\right) \leqslant \delta^{1-a}.$$

For sufficiently small δ we then see that for all points of the region of integration

$$\frac{h(T_1 - \xi_1, \ldots, T_s - \xi_s)}{h(T_1, \ldots, T_s)} = 1 + \xi_\delta, \tag{5.7.10}$$

where $\xi_\delta \to 0$ as $\delta \to 0$. Furthermore, for the cotest ψ we have the usual inequalities: $-\alpha \leq \psi \leq 1 - \alpha$. If we substitute (5.7.10) into (5.7.8) we obtain

$$-\alpha - \eta_\delta \leqslant h^{-1}(D_\delta * \psi h) \leqslant 1 - \alpha + \eta_\delta, \tag{5.7.11}$$

where $\eta_\delta \to 0$ as $\delta \to 0$. If we set $\psi_\delta = h^{-1}(D_\delta * \psi h)$ and choose δ sufficiently small, we obtain a cotest ψ'_δ of level α with the properties required by Theorem 5.7.1.

We note now that in many important cases, for example in problems of multivariant analysis, inequality (5.7.6) holds for $a = 1$. A theorem of the type of 5.7.1 holds in this case too, although the proof of it is different. We shall examine those cases in §8, where examples are given of testing of a linear hypothesis.

§8. FORMULATION OF THE FINAL RESULTS. EXAMPLES

We can now formulate the final results of this chapter.

Consider an exponential family defined by probability densities with respect to Lebesgue measure:

$$p_\theta(T_1, \ldots, T_s) = C(\theta)\exp -(\theta_1 T_1 + \ldots + \theta_s T_s)\, h(T_1, \ldots, T_s). \tag{5.8.1}$$

We impose the following conditions on $h(T_1, \cdots, T_s)$.

I. There exists a number s_1 in the interval $[0, s]$ such that $h(T_1, \cdots, T_s) = 0$ if at least one of the variables T_j for $j = 1, 2, \cdots, s_1$ is negative. This determines the carrier $\mathfrak{T}$ of the function $h(T_1, \cdots, T_s)$.

II. The function $h(T_1, \cdots, T_s)$ does not vanish at interior points of the region $\mathfrak{T}$ and it has continuous first derivatives at such points. Furthermore, in the region $\mathfrak{T}_\epsilon \subset \mathfrak{T}$ defined by the inequalities

$$T_1 \geqslant \varepsilon_1, \ldots, T_s \geqslant \varepsilon \tag{5.8.2}$$

for arbitrary $\epsilon > 0$, we have

$$\left|\frac{\partial \ln h}{\partial T_1}\right| + \ldots + \left|\frac{\partial \ln h}{\partial T_s}\right| = O\left(\frac{1}{\varepsilon^a} + 1\right), \qquad (5.8.3)$$

where $a < 1$ is a constant. In the case $a = 1$ a special examination is necessary (see the examples given below).

III. The integral $\int_{\mathcal{T}} \cdots \int p_\theta(T_1, \cdots, T_s)\, dT_1 \cdots dT_s$ converges absolutely for $\theta = (\theta_1, \cdots, \theta_s) \in \mathcal{P}$, where $\mathcal{P} = R_1 \times \cdots \times R_{s_1} \times S_{s_1+1} \times \cdots \times S_s$ is the Cartesian product of the s_1 right half-planes $\operatorname{Re}\theta_j > 0$ and the $(s - s_1)$ strips $0 < \operatorname{Re}\theta_j < A_j$.

The null hypothesis (H_0). For real points $\theta = (\theta_1, \cdots, \theta_s) \in \mathcal{P}$, we have r $(< s)$ relationships

$$\Pi_1(\theta_1, \ldots, \theta_s) = 0, \ \ldots, \ \Pi_r(\theta_1, \ldots, \theta_s) = 0. \qquad (5.8.4)$$

Here the functions $\Pi_1, \cdots, \Pi_r$ must be real for real $\theta_1, \cdots, \theta_s$. When we multiply by $1/[(\theta_1 + 1) \cdots (\theta_s + 1)]^N$, where N is a suitable integer, the functions $\Pi_1, \cdots, \Pi_r$ become holomorphic functions of $1/(\theta_1 + 1), \cdots, 1/(\theta_s + 1)$ on $\overline{\mathcal{P}}$. (The point $\theta_j = \infty$ is included.) The null hypothesis consists in satisfaction of conditions (5.8.4) on the compact set Ω of real points $(\theta_1, \cdots, \theta_s)$ defined by the inequalities

$$\varepsilon_j \leqslant \theta_j \leqslant E_j, \quad E_j < A_j \qquad (j = 1, 2, \ldots, s). \qquad (5.8.5)$$

We denote the set of corresponding points by Ω_0. Alternatives to H_0 are $(\theta_1, \cdots, \theta_s) \in \Omega \setminus \Omega_0$. The alternatives are equipped with the Bayes probability measure $B(\theta)$ defined on $\Omega \setminus \Omega_0$, which reduces them to a simple hypothesis.

For a test Φ that is similar with respect to H_0 we introduce the Bayes gain $W(\Phi | B) = \int_{(\Omega \setminus \Omega_0)} E(\Phi | \theta)\, dB(\theta)$. An analogous gain is introduced for the cotest.

Conditions on the relationships (complexification of H_0):

$(\mathrm{Y_I})$ *Equations* (5.8.4) *considered in a complex region* $\mathcal{P}$ *generate in that region an analytic set of points* $V_{\Pi_1, \cdots, \Pi_r}$ *that can be decomposed into a finite number of components (see Chapter* I, §2) *connected "along the strips"* $V^q_{\Pi_1, \cdots, \Pi_s}$, *each of which is of complex dimension* $s - r$ *and contains a connected set* $R^q_{\Pi_1, \cdots, \Pi_r}$ *of real points that is contained in* Ω_0 *and that has real dimension* $s - r$.

$(\mathrm{Y_{II}})$ *In the region* $\mathcal{P}$ *the relationships* (5.8.4) *admit choice of variables*

$\theta_{a_1}, \cdots, \theta_{a_r}$ such that the determinant $|\partial\Pi_i/\partial\theta_{a_j}|$, $i, j = 1, \cdots, r$, has no zeros inside $\mathcal{P}$.

Condition (Y_{II}) is somewhat restrictive. It follows from what was said in §3 of Chapter I that in the case $r = 1$ (hypothesis H_0 with one relationship) this condition can be replaced with the weaker condition

$$(Y'_{II}) \qquad \operatorname{rank}\left\|\frac{\partial\Pi_i}{\partial\theta_j}\right\| = r \quad \text{inside } \mathcal{P} \quad (i = 1, 2, \ldots, r, \quad j = 1, 2, \ldots, s).$$

It turns out that condition (Y_{II}) can be replaced with the condition (Y'_{II}) even when $r > 1$. However the proof of this is extremely complicated and we shall not stop to go through it.

Under the conditions indicated, it is possible to give a description of an ϵ-complete family of tests Φ that are similar with respect to the hypothesis H_0.

Theorem 5.8.1. *For given $\epsilon > 0$ and a given similar test Φ of level α there exists a cotest $\psi_\epsilon = \psi_\epsilon(T_1, \cdots, T_s)$ of level α such that $|W(\psi_\epsilon|B) - W(\psi|B)| \le \epsilon$, where $\psi = \Phi - \alpha$. Here $\psi_\epsilon(T_1, \cdots, T_s)$ has the prescribed number of partial derivatives and can be represented in the form*

$$\psi_\varepsilon(T_1, \ldots, T_s) = \frac{1}{h}(A_1 * H_1 + \ldots + A_r * H_r). \tag{5.8.6}$$

In this equation, the A_j are the preimages of the functions $\Pi_j/(\theta_1 \cdots \theta_s)^{N+1}$ under the unilateral Laplace transformation

$$L(A_j\,|\,\theta) = \frac{\Pi_j}{(\theta_1 \ldots \theta_{s+1})^{N+1}},$$

and $H_1, \cdots, H_r$ are functions possessing the prescribed number of partial derivatives and satisfying the relation

$$H_j(T_1, \ldots, T_s) = O(\exp\xi(|T_1| + \ldots + |T_s|)), \quad j = 1, 2, \ldots, r, \tag{5.8.7}$$

where ξ is a sufficiently small positive number. For $T_j < 0$, where j is one of the numbers $1, 2, \cdots, s_1$, the function $H_j(T_1, \cdots, T_s)$ vanishes.

Conversely, every expression of the form $\xi = 1/h(A_1 * H_1 + \cdots + A_r * H_r)$, where the H_j are continuous, which vanishes for $T_j < 0$ ($j = 1, 2, \cdots, s_1$) and satisfies condition (5.8.7), is a precotest. If $H_1, H_2, \cdots, H_r$ are chosen in such a way that

$$-\alpha \le \xi \le 1 - \alpha,$$

then ξ is a cotest of level α.

In view of this the search for ϵ-optimal similar tests in the problem posed leads to a variational problem with constraints: *Find continuous functions* $H_1, \cdots, H_r$ *satisfying conditions* (5.8.7) *such that setting*

$$\xi = \frac{1}{h}(A_1 * H_1 + \dots + A_r * H_r), \tag{5.8.8}$$

maximizes the Bayes gain $W(\xi|B)$, *where* ξ *satisfies the inequalities*

$$-\alpha \leq \xi \leq 1-\alpha. \tag{5.8.9}$$

In this form the question can be regarded as a problem of linear programming. There is a definite computational algorithm regarding this, but we will not stop for it here.

If $r = 1$ (the case of the single relationship $\Pi_1 = 0$) then (5.8.8) has only the single term

$$\xi = \frac{1}{h}(A_1 * H_1). \tag{5.8.10}$$

This corresponds to the principal ideal $\overline{I}_{H_0} = \{\Pi_1 G_1\}$ (cf. §3, Chapter I). Of all the conditions listed above the most restrictive is condition II that the function $h(T_1, \cdots, T_s)$ not vanish in the region $\mathcal{T}$. If this condition is not satisfied, all that follows from the above considerations is that every sufficiently smooth cotest ψ is of the form (5.8.6):

$$\psi = \frac{1}{h}(A_1 * H_1 + \dots + A_r * H_r),$$

where $H_1, \cdots, H_r$ are such that the expression in the parentheses vanishes wherever h vanishes. However we still do not know whether such a family of smooth cotests is ϵ-complete in the sense of the definition given above.

In the study of the local properties of similar tests we can pose the problems considered at the end of §1. In particular we can study tests that are unbiased with respect to the zero hypothesis H_0. Here we treat as alternatives for H_0: $\Pi_1 = 0, \cdots, \Pi_r = 0$ the hypotheses H_1: $\Pi_1 = \delta_1, \cdots, \Pi_r = \delta_r$, where not all the δ_i vanish. It should be borne in mind that the relationships $\Pi_1 = 0, \cdots, \Pi_r = 0$ have many equivalents and that, by shifting from a basis for the ideal $\overline{I}_{H_0}$ of given $\Pi_1, \cdots, \Pi_r$ to another basis, we can express H_0 in terms of the new basis; when we do this the alternatives H_1 obviously change.

Let us suppose that the alternatives are defined by means of the relationships (5.8.4) and that the conditions listed at the beginning of this section are satisfied. Then for sufficiently smooth cotests ψ we have, in accordance with what was shown above, the representation

$$E(\psi \mid \theta) = \Pi_1 G_1 + \ldots + \Pi_r G_r. \tag{5.8.11}$$

The assumption that the cotest is unbiased with respect to H_0 leads to the following equation on Ω_0:

$$\frac{\partial}{\partial \Pi_j} E(\psi \mid \theta) = 0 \quad (j = 1, 2, \ldots, r).$$

From (5.8.11) we see that the functions $G_1, \cdots, G_r$ must vanish for $\Pi_1 = \cdots \cdots = \Pi_r = 0$; in other words they must belong to the ideal $\bar{I}_{H_0}$. The functions G_j are holomorphic in $\mathcal{P}$. By what was shown above,

$$G_j = \Pi_1 G_{j1} + \ldots + \Pi_r G_{jr} \quad (j = 1, 2, \ldots, r),$$

where the G_{jl} $(l = 1, 2, \cdots, r)$ are functions of the same type as the G_j. Thus we obtain

$$E(\psi \mid \theta) = \sum_{i, j=1}^{r} G_{ij} \Pi_i \Pi_j. \tag{5.8.12}$$

From this we get

Theorem 5.8.2. *Under the conditions listed at the beginning of this section, every sufficiently smooth unbiased cotest* ψ *can be represented in the form*

$$\psi = \frac{1}{h} \sum_{i, j=1}^{r} H_{ij} * A_i * A_j, \tag{5.8.13}$$

where the A_i *are the same functions as in the hypotheses of Theorem* 5.8.1 *and the* H_{ij} *behave like the functions* H_j, $j = 1, 2, \cdots, r$, *defined in that theorem.*

Here smoothness of a cotest is to be understood in the sense of the definition given in §5, specifically in the sense of satisfaction of condition (5.5.1).

Let us turn to individual examples of the application of the results that we have obtained.

Let us first consider some examples in which the family of sufficient statistics is a unilateral exponential family. Such a situation is encountered in the problem of testing a linear hypothesis (see Example 1, §1, Chapter II). Let us consider the following special case, which is a generalization of the Behrens-Fisher problem. Suppose that we have a set of k normal samples $x_{j1}, x_{j2}, \cdots \cdots, x_{jn_j} \in N(a_j, \sigma_j^2)$, of sizes $n_1, n_2, \cdots, n_k$. Here sufficient statistics are

$$\bar{x}_j = \frac{1}{n_j}\sum_{l=1}^{n_l} x_{jl} \text{ and } s_j^2 = \frac{1}{n}\sum_{l=1}^{n_j}(x_{jl} - \bar{x}_j)^2.$$

By making translations if necessary, we may assume that the parameters vary on finite intervals:

$$\alpha_{1j} \leqslant a_j \leqslant \alpha_{2j};\quad \beta_{1j} \leqslant \sigma_j^2 \leqslant \beta_{2j};\quad \alpha_{ij} > 0,\quad \beta_{ij} > 0.$$

The hypothesis H_0 that

$$\lambda_1 a_1 + \ldots + \lambda_k a_k = 0,$$

where the λ_j are known quantities not all equal to 0, is being verified. To verify H_0, we use the statistic $X = \lambda_1 \bar{x}_1 + \cdots + \lambda_k \bar{x}_k$ and the statistics $V_1 = s_1^2, \cdots$ $\cdots, V_k = s_k^2$, so that we use only the subalgebra of the σ-algebra of sufficient statistics.

The statistics $V_1, V_2, \cdots, V_k$ are independent of X. Here

$$X \in N(0, \sigma^2), \text{ where } \sigma^2 = \frac{\lambda_1^2\sigma_1^2}{n_1} + \ldots + \frac{\lambda_k^2\sigma_k^2}{n_k}.$$

Furthermore, the statistics $V_1, \cdots, V_k$ have the distribution (see formula (3.3.4))

$$p_k(v_k) = \left(\frac{n_k}{2}\right)^{\frac{n_k-1}{2}} \frac{1}{\Gamma\left(\frac{n_k-1}{2}\right)} \sigma_k^{-n_k+1} \cdot v_k^{\frac{n_k}{2}-\frac{3}{2}} \exp - \left(\frac{n_k v_k}{2\sigma_k^2}\right).$$

From this we see that the family of distributions defined by the random vector $(X, V_1, \cdots, V_k)$ has sufficient statistics $(X^2, V_1, \cdots, V_k)$ and that it is a unilateral exponential family. Let us set

$$\theta_1 = \frac{1}{2\sigma^2} = \frac{1}{2\left(\frac{\lambda_1^2\sigma_1^2}{n_1} + \ldots + \frac{\lambda_k^2\sigma_k^2}{n_k}\right)};$$

$$\theta_2 = \frac{1}{2\sigma_1^2}, \ \ldots, \ \theta_{k+1} = \frac{1}{2\sigma_k^2};$$

$$p_j = \frac{1}{\theta_j} \quad (j = 1, 2, \ldots, k+1).$$

If this family is defined in terms of the natural parameters, the null hypothesis H_0 consists in

$$p_1 - \frac{\lambda_1^2}{n_1} p_2 - \frac{\lambda_2^2}{n_2} p_3 - \ldots - \frac{\lambda_k^2}{n_k} p_{k+1} = 0.$$

Here the conditions listed above are easily verified.

On the basis of what was said above, from a given Bayes distribution $B(\theta)$ on an alternative hypothesis $\Omega \setminus \Omega_0$ for the construction of an ϵ-optimal sufficiently smooth cotest $\psi_\epsilon(x, v_1, \cdots, v_k)$, where $x = X^2$ and $v_1, \cdots, v_k$ are the values of the statistics $V_1, \cdots, V_k$, we can use the expression

$$\psi_\varepsilon(x, v_1, \ldots, v_k)$$
$$=[F_0(x, v_1, \ldots, v_k) * H(x, v_1, \ldots, v_k)]\, x^{1/2} v_1^{-m_1} \ldots v_k^{-m_k}.$$

Here $F_0 = x - \lambda_1^2 v_1/n_1 - \cdots - \lambda_k^2 v_k/n_k$ and $H(x, v_1, \cdots, v_k)$ is a smooth function such that

$$H(x, v_1, \ldots, v_k) = O(\exp \varepsilon(x + v_1 + \ldots + v_k))$$
for arbitrary $\varepsilon > 0$.

The cotest ψ_ϵ must satisfy the conditions

$$-\alpha \leqslant x^{1/2} v_1^{-m_1} \ldots v_k^{-m_k} F_0 * H \leqslant 1 - \alpha,$$

where α is a level. Under such restrictions, we pose for H the variational problem of maximizing

$$W(\psi_\epsilon | B) = E(\psi_\epsilon | \theta)\, dB(\theta). \qquad (5.8.14)$$

In the class of smooth functions H this problem can be solved "with accuracy up to ϵ"; i.e., for arbitrary ϵ one can find $H = H_\epsilon$ for which the cotest ψ_ϵ is ϵ-optimal in the sense indicated by inequality (5.1.6).

It is natural to ask whether it is possible to construct from a given cotest ψ of level α a sufficiently smooth cotest ψ_δ for which the Bayes gain $W(\psi_\delta | B)$ is arbitrarily close to $W(\psi | B)$, so that to find an ϵ-optimal cotest we need only consider smooth cotests. The answer to this question is affirmative. To show that this is the case we need to modify somewhat the proof of Theorem 5.7.1.

We set

$$T_1 = V_1,\ T_2 = V_2,\ \ldots,\ T_k = V_k,\ T_{k+1} = x,\ s = k + 1;$$
$$h(T_1, \ldots, T_s) = T_1^{m_1} \ldots T_k^{m_k} T_{k+1}^{-\frac{1}{2}}.$$

Let us set up a formula of the type (5.7.8) taking, for small given δ, the functions $d_\delta^{(j)}(\xi_j)$, $j = 1, 2, \cdots, k$, in the form of "smooth peaks" of area 1 with carriers $[\delta/2 - \delta^2, \delta/2 + \delta^2]$. For $j = k + 1$, we put the function $d_\delta^{(k+1)}(\xi_{k+1})$ in the form of a "smooth peak" of area 1 with carrier $[-\delta/2 - \delta^2, -\delta/2 + \delta^2]$. Define $D_\delta = d_\delta^{(1)} \cdots d_\delta^{(k+1)}$. A formula of the form (5.7.8) is written in the following form:

$$h^{-1}(D_\delta * \psi h) = \int_0^{T_1} d\xi_1 \ldots \int_0^{T_k} d\xi_k \int_{-\infty}^{+\infty} d\xi_{k+1} d_\delta^{(1)}(\xi_1) \cdot \ldots$$

$$\ldots \cdot d_\delta^{(k)}(\xi_k) d_\delta^{(k+1)}(\xi_{k+1}) \left(\frac{T_1 - \xi_1}{T_1}\right)^{m_1} \cdot \ldots \cdot \left(\frac{T_k - \xi_k}{T_k}\right)^{m_k}$$

$$\times \left(\frac{T_{k+1} - \xi_{k+1}}{T_{k+1}}\right)^{-\frac{1}{2}} \psi(T_1 - \xi_1, \ldots, T_s - \xi_s).$$

By the construction of the functions $d_\delta^{(1)}(\xi_1), \cdots, d_\delta^{(k+1)}(\xi_{k+1})$ the product

$$P = \left(\frac{T_1 - \xi_1}{T_1}\right)^{m_1} \ldots \left(\frac{T_k - \xi_k}{T_k}\right)^{m_k} \left(\frac{T_{k+1} - \xi_{k+1}}{T_{k+1}}\right)^{-\frac{1}{2}}$$

satisfies the inequality $0 \leq P \leq 1$ in our region of integration, so that the expression $\psi_\delta = h^{-1}(D_\delta * \psi h)$ provides a cotest if ψ is a cotest. The level of the cotest ψ_δ is equal to the level of the cotest ψ. Their powers differ by an arbitrarily small amount for sufficiently small δ. This answers our question.

If we set $k = 2$ in the above statement of the problem of testing a linear hypothesis, then without loss of generality we can take the condition of connectedness in the form $a_1 - a_2 = 0$. Thus we obtain the Behrens-Fisher problem. Here $F_0 = x - v_1/n_1 - v_2/n_2$, so that the formulas have a simple form. (In [56] the letters x, u, and v denote variables proportional to the quantities for which we have been using them, so that the linear form F_0 has a slightly different appearance.)

Let us calculate the Bayes gain (5.8.14). We set $x = X^2$, $\psi_\epsilon(x, v_1, v_2) = \psi_\epsilon(X^2, v_1, v_2)$; $\Phi_\epsilon = \psi_\epsilon + \alpha$. The joint distribution of X, v_1, and v_2 has probability density

$$p(X, v_1, v_2)$$

$$= \frac{1}{\sqrt{2\pi}\,\sigma} \prod_{l=1}^{2} \left[\left(\frac{n_l}{2}\right)^{\frac{n_l - 1}{2}} \frac{1}{\Gamma\left(\frac{n_l - 1}{2}\right)} \sigma_l^{-n_l+1} v_l^{\frac{n_l - 3}{2}} \right]$$

$$\times \exp\left[-\frac{1}{2}\left(\frac{(X-\delta)^2}{\sigma^2} + \frac{n_1 v_1}{\sigma_1^2} + \frac{n_2 v_2}{\sigma_2^2}\right)\right]; \quad \sigma^2 = \frac{\sigma_1^2}{n_1} + \frac{\sigma_2^2}{n_2}.$$

For given $\sigma_1, \sigma_2, a_1 - a_2 = \delta$, the power function is

$$E(\Phi_\varepsilon(X^2, v_1, v_2) \mid \sigma_1, \sigma_2, \delta)$$

$$= \int_{-\infty}^{\infty} dX \int_0^{\infty} dv_1 \int_0^{\infty} dv_2 \Phi_\varepsilon(X^2, v_1, v_2). \tag{5.8.15}$$

By virtue of the evenness of our similar test with respect to X, we see that the test is unbiased: for arbitrary σ_1 and σ_2 we have

$$\left.\begin{aligned}\frac{\partial}{\partial\delta}E(\Phi_\varepsilon(X^2, v_1, v_2)\,|\,\sigma_1, \sigma_2, \delta) = 0 \quad \text{for } \delta = 0,\\ \frac{\partial^2}{\partial\delta^2}E(\Phi_\varepsilon(X^2, v_1, v_2)\,|\,\sigma_1, \sigma_2, \delta) > 0 \quad \text{for } \delta = 0.\end{aligned}\right\} \tag{5.8.16}$$

These relationships mean that the similar test Φ_ϵ is unbiased. If the Bayes distribution $B(\theta)$ on the alternative $\Omega\setminus\Omega_0$ is concentrated on intervals of a straight line $\sigma_1 = \sigma_1^{(e)}$, $\sigma_1 = \sigma_1^{(0)}$, $-\infty < \delta < \infty$, $\delta \neq 0$, then we obtain the locally most powerful similar test if we maximize the functional (5.8.16) (see §1). Thus for the similar test Φ_ϵ we have a variational problem, that of maximizing the left-hand member of (5.8.16) with the constraints

$$-\alpha \leqslant x^{1/2} v_1^{-m_1} v_2^{-m_2} F_0 * H_\varepsilon \leqslant 1-\alpha. \tag{5.8.17}$$

This problem is amenable to numerical solution by the methods of linear programming.

Let us turn to the general problem of testing a linear hypothesis with unknown weights.

Let $X_1, \cdots, X_n$ denote independently distributed normal variables such that $X_j \in N(\alpha_j, \gamma_j^2)$. Here, the a_j are *a priori* subject to $n-s$ "initial" linear relationships. Specifically, if

$$\alpha = \begin{Vmatrix} \alpha_1 \\ \cdot \\ \cdot \\ \cdot \\ \alpha_n \end{Vmatrix},$$

then there exists a constant matrix $C = C_{n-s,n}$ such that $C\alpha = 0$ and rank $C = n-s$.

Thus $\alpha_1, \cdots, \alpha_n$ can be expressed as a linear combination of the parameters $b_1, \cdots, b_s$.

Suppose we are verifying the hypothesis H_0 that, in addition to these relationships, there are still $r\,(<s)$ relationships (which may be thought of as imposed on $b_1, \cdots, b_s$). Specifically, we introduce a matrix $F = F_{rs}$ of rank $r < s$ such that

$$Fb = 0, \quad \left(b = \begin{Vmatrix} b_1 \\ \cdot \\ \cdot \\ \cdot \\ b_s \end{Vmatrix}\right).$$

One can show that the theory expounded earlier, except possibly for the conditions on $h(T_1, \cdots, T_s)$, is applicable here.

Let us look at an individual case of a linear hypothesis, the general Behrens-Fisher problem. Here $n = n_1 + n_2$ is partitioned into two samples $x_1, \cdots, x_{n_1}$ and $y_1, \cdots, y_{n_2}$ of sizes n_1 and n_2. The initial relationships are of the forms $\alpha_1 = \alpha_2 = \cdots = \alpha_{n_1} = a_1$ and $\alpha_{n_1+1} = \cdots = \alpha_n = a_2$. We also introduce the condition $\gamma_1 = \gamma_2 = \cdots = \gamma_{n_1} = \sigma_1$; $\gamma_{n_1+1} = \cdots = \gamma_n = \sigma_2$. Sufficient statistics are $X_1 = \bar{x}$, $X_2 = \bar{y}$, $V_1 = s_1^2$, $V_2 = s_2^2$. By formula (3.3.14) the probability density is

$$p(x_1, x_2, v_1, v_2) = C_n \sigma_1^{-n_1} \sigma_2^{-n_2} \exp\left(-\frac{n_1 a_1^2}{2\sigma_1^2} - \frac{n_2 a_2^2}{2\sigma_2^2}\right) \cdot v_1^{\frac{n_1-3}{2}} v_2^{\frac{n_2-3}{2}}$$

$$\times \exp\left[-\frac{n_1}{2\sigma_1^2}(v_1 + x_1^2) - \frac{n_2}{2\sigma_2^2}(v_2 + x_2^2) + \frac{n_1 a_1}{\sigma_1^2} x_1 + \frac{n a_2}{\sigma_2^2} x_2\right].$$

Just as in §3 of Chapter III, we introduce the natural parameters

$$\frac{n_1}{2\sigma_1^2} = \lambda_1, \quad \frac{n_2}{2\sigma_2^2} = \lambda_2, \quad \frac{n_1 a_1}{\sigma_1^2} = \mu_1, \quad \frac{n_2 a_2}{\sigma_2^2} = \mu_2.$$

The hypothesis H_0 imposes one relationship (see Chapter III):

$$\frac{\mu_1}{\lambda_1} - \frac{\mu_2}{\lambda_2} = 0. \tag{5.8.18}$$

Let us consider H_0 in the region $\Omega_0 : \epsilon_j \leq \theta_j \leq E_j$, $j = 1, 2, 3, 4$, where θ_j is one of the numbers $\lambda_1, \lambda_2, \mu_1, \mu_2$ (if necessary the parameters can be displaced). In the present situation all the conditions of the theory expounded above are satisfied except for the conditions on the density h.

If we divide equation (5.8.18) by $\lambda_1 \lambda_2 \mu_1^2 \mu_2^2$ we obtain the equivalent relationship in Ω_0:

$$\frac{1}{\lambda_1^2 \lambda_2 \mu_1 \mu_2^2} - \frac{1}{\lambda_1 \lambda_2^2 \mu_1^2 \mu_2} = 0. \tag{5.8.19}$$

If we now set $X_1 = T_1$, $X_2 = T_2$, $V_1 + X_1^2 = T_3$, and $V_2 + X_2^2 = T_4$, we obtain, corresponding to the left-hand member of (5.8.19), its preimage under the unilateral Laplace transformation (see §1 of Chapter I):

$$A(T) = \begin{cases} T_1 T_4 - T_2 T_3 & \text{if } T_i > 0 \text{ for } i = 1, 2, 3, 4, \\ 0 & \text{otherwise.} \end{cases}$$

The general form of sufficiently smooth cotests $\psi(T)$ is[1)]

$$\psi(T)=\frac{1}{h}(A*H), \tag{5.8.20}$$

where H is a function of the type described in the preceding section, and

$$h(T)=V_1^{\frac{n_1-3}{2}}V_2^{\frac{n_2-3}{2}}=(T_3-T_1^2)^{\frac{n_1-3}{2}}\cdot(T_4-T_2^2)^{\frac{n_2-3}{2}}$$

for $T_1\leq\sqrt{T_3}$ and $T_2\leq\sqrt{T_4}$; $h(T)=0$ otherwise. For the expression (5.8.20) to be a cotest it is necessary that H be chosen in such a way that $A*H$ vanishes for $T_1>\sqrt{T_3}$ or $T_2>\sqrt{T_4}$. The fact that such a choice of the functions H is possible will be shown in §3 of Chapter IX. In view of the complexity of the question of constructing a sufficiently broad family of such functions, we shall not stop to do this here.

---

1) *Added in proof.* V. P. Palamodov has shown that all cotests have an analogous form: *Testing of a multidimensional polynomial hypothesis*, Dokl. Akad. Nauk SSSR 172 (1967), 291–293 = Soviet Math. Dokl. 8 (1967), 95–97. *Editor's note.* See also the Supplement to the present book, by Palamodov and Kagan.

CHAPTER VI

WIJSMAN'S *D*-METHOD

§1. THE *D*-METHOD AND THE CONDITIONS UNDER WHICH IT CAN BE APPLIED

Let us again look at exponential families, where the probability density with respect to Lebesgue measure can be expressed by formula (5.2.3) without the corresponding conditions on the T_j $(j = 1, \cdots, s)$. As was stated in §§2 and 8 of Chapter V, we can under rather general conditions construct a precotest ideal and similar tests if we can exhibit a cone in the space E_s of sufficient statistics $T_1, \cdots, T_s$ inside which $h(T_1, \cdots, T_s)$ does not vanish and, after a suitable linear transformation, satisfies certain conditions. The relationships between the parameters $\theta_1, \cdots, \theta_s$ defining the null hypothesis H_0 are assumed to be holomorphic of the type (5.2.4). Of course this does not give a sufficiently complete description of the similar tests.

If we cannot exhibit a cone of the type indicated we cannot construct similar tests by this method.

In 1958, in an interesting article [11], Wijsman presented a method enabling us to construct similar tests under weaker assumptions regarding $h(T_1, \cdots, T_s)$ even when the null hypothesis H_0 reduces to a single polynomial relationship among $\theta_1, \cdots, \theta_s$. However only a part of the entire family of similar tests can be obtained in this way, and it is only in rare cases that we can determine their optimality or ϵ-optimality. Nonetheless, Wijsman's method is extremely elegant and it discloses interesting relationships between statistics and the theory of partial differential equations. Therefore we shall stop to examine this method in the present chapter. As stated above, suppose that an exponential family is defined by the probability density with respect to Lebesgue measure:

$$p_\theta(T_1, \ldots, T_s) = C(\theta)\exp-(\theta_1 T_1 + \ldots + \theta_s T_s)h(T_1, \ldots, T_s) \quad (6.1.1)$$

and suppose that $h(T_1, \cdots, T_s)$ satisfies the following condition.

Condition (W). There exists a closed s-dimensional cube C on which

$$h(T_1, \ldots, T_s) \geqslant \varepsilon_0 > 0. \tag{6.1.2}$$

The zero hypothesis H_0 is assumed to be defined by a single polynomial relationship

$$H_0 : P(\theta_1, \ldots, \theta_s) = 0, \tag{6.1.3}$$

where P is a polynomial of degree $d \geq 1$.

To construct a similar test $\Phi = \Phi(T_1, \cdots, T_s)$ of level α, let us construct the corresponding cotest $\psi = \psi(T_1, \cdots, T_s)$. To do this we construct a bounded function $F(T_1, \cdots, T_s) = F(T)$ which vanishes outside the cube C and for which the Laplace transform $L(F|\theta)$ vanishes when condition (6.1.3) is satisfied. By virtue of condition (6.1.2) and the boundedness of F, the function $\psi(T)$ defined by

$$\left.\begin{aligned} \psi(T) &= \varepsilon \frac{F(T)}{h(T)} && \text{for } T \in C, \\ \psi(T) &= 0 && \text{for } T \notin C, \end{aligned}\right\} \tag{6.1.4}$$

is a cotest for sufficiently small ϵ. Thus the problem reduces to construction of this function $F(T)$. To construct this function we take an entire function $G(T) = G(T_1, \cdots, T_s)$ which possesses all partial derivatives of the first d orders inside C, which vanishes outside C, and for which all the derivatives of the first $(d-1)$ orders are continuous on the boundary of C. An example of such a function is the one defined as follows. Suppose that the cube C is defined by the inequalities

$$a_j \leqslant T_j \leqslant a_j + l, \quad j = 1, 2, \ldots, s.$$

Then define

$$G(T) = \prod_{j=1}^{s} (T_j - a_j)^d (a_j + l - T_j)^d$$

for $T \in C$ and $G(T) = 0$ for $T \notin C$. The function $G(T)$ is the desired one.

Consider the differential operator

$$D = P\left(\frac{\partial}{\partial T_1}, \ldots, \frac{\partial}{\partial T_s}\right). \tag{6.1.5}$$

By virtue of the properties of $G(T)$ (see §1 of Chapter I) we have

$$L(DG|\theta) = P(\theta) L(G|\theta). \tag{6.1.6}$$

From this it is obvious that the function $\xi(T) = DG(T)/h(T)$ is a precotest. Since $\epsilon\xi(T)$ is bounded, it is a cotest for sufficiently small ϵ. If the space of sufficient statistics $T_1, \cdots, T_s$ also contains several cubes or parallelepipeds in which $\inf h(T)$ is positive, we can construct cotests for them by the method described and set up linear combinations of them from which we can also obtain cotests. Following Wijsman we shall call this method of constructing similar tests the "D-method".

§2. EXAMPLES OF APPLICATION OF THE D-METHOD

Example 1. Construction of certain similar tests for the Behrens-Fisher problem.

Here, in the notation of Chapter III, §3, we obtain the equation

$$P(\theta)=\theta_1\theta_4-\theta_2\theta_3=0,$$

for defining the null hypothesis H_0, so that here we need to set

$$D=\frac{\partial}{\partial T_2}\frac{\partial}{\partial T_3}-\frac{\partial}{\partial T_1}\frac{\partial}{\partial T_4}.$$

For the cube C we may take any cube in which the function $h(T)$ defined by formula (3.3.14) is bounded below by a positive number.

Of course the cotests constructed in this way constitute only a part of the entire family of cotests.

Example 2. Verification of a hypothesis regarding a standardized mean.

Here we succeed, to a sufficient degree, in getting a complete description of the similar tests. Suppose that we make a repeated sample from a normal set $N(a, \sigma^2)$. Let us verify the hypothesis H_0: $a/\sigma=\gamma_0$ that the standardized mean has a given value γ_0. We wish to get as complete as possible a description of the similar tests H_0. Let us make an orthogonal transformation of the chosen variables, taking the new variables $X_0, X_1, \cdots, X_n$ in accordance with the formulas (see Chapter III, §3)

$$X_0=\bar{x}\sqrt{n+1};\ X_1=\left(1-\frac{1}{n+1}\right)^{-\frac{1}{2}}(x_1-\bar{x}),\ \ldots$$

$$\ldots,\ X_n=\left(1-\frac{1}{n+1}\right)^{-\frac{1}{2}}(x_n-\bar{x}).$$

These variables are normal and independent, and

$$X_0\in N(a\sqrt{n+1},\ \sigma^2),\qquad X_j\in N(0,\ \sigma^2),\qquad j=1,\ 2,\ \ldots,\ n.$$

If we set $a\sqrt{n+1}=\mu$ and $r_0=\gamma_0\sqrt{n+1}$, we reduce the problem to verification of the hypothesis H_0: $\mu/\sigma=r_0$ for the sample $(X_0, X_1, \cdots, X_n)$. For this sample the probability density has the form

$$\frac{1}{(2\pi)^{\frac{n+1}{2}}\sigma^n}\exp\left\{-\frac{1}{2\sigma^2}\left[(X_0-\mu^2)+\sum_{j=1}^{n}X_j^2\right]\right\}$$

$$=C(\mu,\ \sigma)\exp\left[-\frac{1}{2\sigma^2}\sum_{j=0}^{n}X_j^2+\frac{r_0}{\sigma}X_0\right]. \qquad (6.2.1)$$

Sufficient statistics are $T_1(x)=\Sigma_{j=0}^n X_j^2$ and $T_2(x)=X_0$. Furthermore, we set

$$\theta_1=\frac{1}{2\sigma^2},\quad \theta_2=-\frac{r_0}{\sigma}.$$

Under the hypothesis H_0 we have $\theta_2^2-2r_0^2\theta_1=0$, so that

$$P(\theta_1,\ \theta_2)=\theta_2^2-2r_0^2\theta_1 \qquad (6.2.2)$$

and

$$D=\frac{\partial^2}{\partial T_2^2}-2r_0^2\frac{\partial}{\partial T_1}. \qquad (6.2.3)$$

We return to the function $h(T_1,\ T_2)$. The statistics T_2 and $V=T_1-T_2^2$ are independent. Also, $T_2\in N(\mu,\ \sigma^2)$ and $V=\sigma^2\chi_n^2$, so that the corresponding probability densities $p(T_2)$ and $q(V)$ are equal to

$$p(T_2)=\frac{1}{\sqrt{2\pi}\sigma}\exp-\frac{(X_0-\mu)^2}{2\sigma^2}$$

and

$$q(V)=\frac{1}{2^{\frac{n}{2}}\Gamma\left(\frac{n}{2}\right)\sigma^n}V^{\frac{n}{2}-1}\exp-\left(\frac{V}{2\sigma^2}\right).$$

Furthermore, for $T_1\geq T_2^2$,

$$\frac{\partial(T_2,\ V)}{\partial(T_2,\ T_1)}=\begin{vmatrix}1 & 0\\ -T_2 & 1\end{vmatrix}=1,$$

so that the common probability density for T_1 and T_2 is of the form

$$p(T_1,\ T_2)=C_1(\mu,\ \sigma)\exp-(\theta_1T_1+\theta_2T_2)\,h(T_1,\ T_2),$$

where $C_1(\mu,\ \sigma)$ depends only on the parameters μ and σ and where

$$h(T_1,\ T_2)=\begin{cases}(T_1-T_2^2)^{\frac{n}{2}-1} & \text{for } T_1\geq T_2^2\\ 0 & \text{for } T_1<T_2^2.\end{cases}$$

Thus for suitably chosen $G(T_1,\ T_2)$ we can, on the basis of what has been said above, obtain the cotest

$$\psi(T_1, T_2) = (T_1 - T_2^2)^{1-\frac{n}{2}} \left(\frac{\partial^2}{\partial T_2^2} - 2r_0^2 \frac{\partial}{\partial T_1}\right) G(T_1, T_2) \tag{6.2.4}$$

if

$$T_1 \geq T_2^2. \tag{6.2.5}$$

If inequality (6.2.5) is not satisfied, then $\psi(T_1, T_2) = 0$. We note that, by Chapter III, §3, the statistic $T_2/\sqrt{T_1}$ has a distribution depending only on a/σ, so that tests of the form $\Phi(T_2/\sqrt{T_1})$, where $\Phi \in (0,1)$ is an arbitrary measurable function, are similar. From these tests we can construct the cotests $\psi(T_2/\sqrt{T_1})$. For a suitable choice of $G(T_1, T_2)$ formula (6.2.4) yields cotests that do not reduce to the type indicated.

Let us now discuss Wijsman's reasoning with regard to the construction of similar tests of this problem by the method described. We need several tedious analytical lemmas, which we use without proof, referring the reader to Wijsman's article [11].

We can extend somewhat the general idea of the D-method. We shall show that for a given level α every cotest can be expressed by formula (6.2.4) for suitable $G(T_1, T_2)$ satisfying certain additional conditions. We do not require that $G(T_1, T_2)$ vanish whenever $h(T_1, T_2)$ vanishes. It is important only that equation (6.1.6) be satisfied, and in fact even this requirement can be somewhat weakened.

Equation (6.2.4) can be written in the form

$$\left(\frac{\partial}{\partial T_1} - \frac{1}{2r_0^2} \frac{\partial^2}{\partial T_2^2}\right) G = \frac{\sqrt{2\pi}}{r_0} \Phi, \tag{6.2.6}$$

where

$$\Phi(T) = \frac{1}{\sqrt{8\pi}\, r_0} h(T) \psi(T). \tag{6.2.7}$$

Here $\psi(T)$ is a cotest of level α.

Equation (6.2.6) can be regarded as the one-dimensional heat-flow equation. Here T_1 should be interpreted as time and T_2 as the space coordinate, G as the temperature, and $(\sqrt{2\pi}/r_0)\,\Phi$ as a heat source (either positive or negative) that is variable in space and time. If we were dealing with an ordinary heat-flow problem we could write its solution with the aid of the corresponding Green's function in the form

$$G(T_1, T_2) = \int_{-\infty}^{\infty} dT_2' \int_0^{T_1} dT_1 \Phi(T_1', T_2')(T_1 - T_1')^{-\frac{1}{2}} \times \exp\left[-\frac{r_0^2}{2}\frac{(T_2 - T_2')^2}{T_1 - T_1'}\right]. \tag{6.2.8}$$

Since $\Phi(T') = 0$ whenever $T_2'^2 > T_1'$, we can integrate over the interval $T_2'^2 \leq T_1' \leq T_1$.

We need to investigate the question as to whether the formal solution (6.2.8) of equation (6.2.6) actually yields the cotest that we need. To do this, let us set $\Phi(T) = \psi(T) + \alpha$ and examine the power function

$$\beta(r, \sigma) = E(\Phi(T_1, T_2)|r, \sigma).$$

Suppose that ψ satisfies (6.2.4). Then

$$\beta(r, \sigma) = \alpha + C(r, \sigma)\int\int \exp\left(-\frac{T_1}{2\sigma^2} + \frac{r}{\sigma}T_2\right) \times\left(\frac{\partial^2}{\partial T_2^2} - 2r_0^2\frac{\partial}{\partial T_1}\right)G(T_1, T_2)\,dT_1\,dT_2, \tag{6.2.9}$$

where the integration is over the region $0 \leq T_1 < \infty$, $-\infty < T_2 < \infty$ and where $C(r, \sigma)$ is a normalizing constant.

A detailed investigation of this integration leads to the formula (see [11], pp. 1037–1038 and 1042–1045)

$$\beta(r, \sigma) = \alpha + C(r, \sigma)\frac{r^2 - r_0^2}{\sigma^2} \lim_{A\to\infty}\lim_{B\to\infty}\int_0^A dT_1 \int_{-\infty}^B G(T_1, T_2) \times \exp\left[-\frac{T_1}{2\sigma^2} + \frac{r}{\sigma}T_2\right]dT_2. \tag{6.2.10}$$

Here we take the limit first as $B \to \infty$, then as $A \to \infty$. From equation (6.2.10) we see that under the hypothesis H_0: $r = r_0$ the power function is independent of the nuisance parameter σ, which means that the test Φ_1 is similar.

To describe all similar tests let us consider the class L of functions $G(T_1, T_2)$ defined in the right (T_1, T_2)-half-plane $(T_1 > 0)$ that satisfy the following four conditions.

I. $DG(T_1, T_2) = 0$ for $T_2^2 > T_1$.

II. $-\alpha \leqslant (T_1 - T_2^2)^{1-\frac{n}{2}} DG(T_1, T_2) \leqslant 1 - \alpha$, if $T_2^2 \leqslant T_1$.

III. $G(0, T_2) = 0$; $G(T_1, T_2) \to 0$ as $T_2 \to -\infty$ for arbitrary prescribed values of T_1.

IV. The integrals

$$\int_{-\infty}^{B} G(A, T_2) \exp\left[-\frac{A}{2\sigma^2}+\frac{r}{\sigma} T_2\right] dT_2,$$

$$\int_{0}^{A} G(T_1, B) \exp\left[-\frac{T_1}{2\sigma^2}+\frac{r}{\sigma} B\right] dT_1,$$

$$\int_{0}^{A} \frac{\partial G(T_1, B)}{\partial T_2} \exp\left[-\frac{T_1}{2\sigma^2}+\frac{r}{\sigma} B\right] dT_1$$

converge to 0 as first B and then A approaches ∞. Then (see [11], pp. 1037–1038 and 1042–1045) there is a one-to-one correspondence between these functions $G(T_1, T_2) \in L$ and the cotests ψ for the level α in accordance with formulas (6.2.6) and (6.2.8). Here conditions IV are difficult to verify. They are always satisfied, however, if

$$G(T_1, T_2) \equiv 0 \quad \text{for} \quad T_2 > \sqrt{T_1}.$$

CHAPTER VII

UNBIASED ESTIMATES

§1. UNBIASED ESTIMATES FOR INCOMPLETE EXPONENTIAL FAMILIES DEPENDING ON SUFFICIENT STATISTICS

The question of the relationship between the theory of sufficient statistics and unbiased estimates was considered in Chapter II, §6. Theorems 2.6.1 and 2.6.2 tell us that for a broad class of questions we can confine ourselves to unbiased tests depending on sufficient statistics.

Let us consider an exponential family of the form (5.2.3) under the assumptions listed in §2 of Chapter V. Specifically, we assume that the function $h(T_1, \cdots, T_s)$ does not vanish in the region $\mathcal{J}$: $T_j < 0$, $j = 1, 2, \cdots, s$, that $s_1 \leq s$, and that the integral of the function (5.2.3) converges absolutely in the Cartesian product $\mathcal{P}$ of the s_1 half-planes R_j: $\operatorname{Re}\theta_j > 0$ $(j = 1, 2, \cdots, s_1)$ and the $s - s_1$ strips S_j: $0 < \operatorname{Re}\theta_j \leq A_j$. We denote the region of real values $(\theta_1, \cdots, \theta_s) \in \mathcal{P}$ by $\mathcal{P}_R$. If we impose no additional constraints on the parameters $\theta_1, \cdots, \theta_s$, then the family of sufficient statistics $T_1, \cdots, T_s$ is complete (see §2, Chapter IV). Let $\xi(T_1, \cdots, T_s)$ denote a statistic satisfying the condition

$$\xi(T_1, \ldots, T_s) = O(\exp \varepsilon(|T_1| + \ldots + |T_s|)) \tag{7.1.1}$$

for every $\epsilon > 0$. Then for all real values of $\theta = (\theta_1, \cdots, \theta_s) \in \mathcal{P}_R$ the mathematical expectation

$$E(\xi \mid \theta) = f(\theta_1, \ldots, \theta_s) \tag{7.1.2}$$

and the variance

$$D(\xi|\theta) = E((\xi^2|\theta) - f(\theta_1, \ldots, \theta_s))^2 \tag{7.1.3}$$

exist.

The statistic ξ is an unbiased estimate of the function $f(\theta_1, \cdots, \theta_s)$ in the region $\mathcal{P}_R$. If we impose no additional constraints on the parameters $(\theta_1, \cdots, \theta_s)$, then by Theorem 4.2.2 the statistic ξ is the unique unbiased estimate of $f(\theta_1, \cdots, \theta_s)$. Every other unbiased estimate $f(\theta_1, \cdots, \theta_s)$ coincides with it with probability 1. On the other hand, if there are additional constraints on the parameters then equation (7.1.2) may hold only under the assumption of these relationships. We then do not have completeness and there may be many unbiased estimates.

Let ξ_1 and ξ_2 denote two unbiased estimates of $f(\theta_1, \cdots, \theta_s)$ for $(\theta_1, \cdots, \theta_s) \in \Omega_0$, where $\Omega_0 \subset \mathcal{P}_R$. Then $\chi = \xi_1 - \xi_2$ is an unbiased estimate of zero. We shall denote this concept simply by UEZ. If ξ is an unbiased estimate of $f(\theta_1, \cdots, \theta_s)$, then all other unbiased estimates of f are of the form $\xi + \chi$, where χ is a UEZ. To describe all unbiased estimates of $f(\theta_1, \cdots, \theta_s)$, it will be sufficient to find one of them and all the UEZ's χ.

If $f(\theta_1, \cdots, \theta_s)$ is a function defined on $\mathcal{P}_R$, then the question of existence of an unbiased estimate of it in $\mathcal{P}_R$, that is, of an equation of the form (7.1.2), is, for the given families, a special question in the theory of the multiple Laplace transformation. The function $f(\theta_1, \cdots, \theta_s)$ must always be holomorphic in $\mathcal{P}$. It is sufficient that it be holomorphic in a neighborhood of the point $(\theta_1 = \infty, \cdots, \theta_s = \infty)$. Then there exists a smooth function $\rho = \rho(T_1, \cdots, T_s)$ that vanishes if one of the numbers $T_j < 0$ $(j = 1, 2, \cdots, s)$ and that satisfies the equation $L(\rho|\theta) = f(\theta_1, \cdots, \theta_s)$. Since by hypothesis $h(T_1, \cdots, T_s)$ does not vanish in $\mathcal{T}$, it follows that $\xi = \rho/h$ is an unbiased estimate of $f(\theta_1, \cdots, \theta_s)$ in $\mathcal{P}_R$. Furthermore, it is the only unbiased estimate since there are no relationships between the parameters.

Let us now describe the set of UEZ's for incomplete exponential families, i.e. when we have relationships of the form

$$\Pi_1(\theta_1, \ldots, \theta_s) = 0, \ \ldots, \ \Pi_r(\theta_1, \ldots, \theta_s) = 0, \ r < s. \qquad (7.1.4)$$

We impose on these relationships the same conditions as in §8 of Chapter V. The functions $\Pi_1, \cdots, \Pi_r$ must be real for real $(\theta_1, \cdots, \theta_s)$ and there must exist an integer $N \geq 0$ such that the functions $\Pi_j/[(\theta_1 + 1) \cdots (\theta_s + 1)]^N$ are holomorphic functions of $1/(\theta_1 + 1), \cdots, 1/(\theta_s + 1)$, on $\overline{\mathcal{P}}$ for $j = 1, \cdots, r$. Furthermore, just as in §8 Chapter V, the complexification conditions (Y_I) and the conditions (Y_{II}) are imposed on them.

So far the only condition that we have imposed on the function $h(T_1, \cdots, T_s)$ is condition III of §8, Chapter V, i.e. the requirement that the integral of (5.2.3) be absolutely convergent in $\mathcal{P}$.

Let us now investigate the UEZ ξ under the assumption that (7.1.1) holds. Under this assumption $E(\xi|\theta)$ exists and is a holomorphic function of $(\theta_1, \cdots, \theta_s)$ in $\mathcal{P}$. Furthermore, under relations (7.1.4) $E(\xi|\theta) = 0$, so that ξ plays the role of a precotest and the reasoning of Chapter V can be applied to $E(\xi|\theta)$. If ξ is a sufficiently smooth unbiased estimate, we obtain the representation

$$\xi \cdot h = A_1 * H_1 + \dots + A_r * H_r, \tag{7.1.5}$$

where the functions A_j and H_j $(j = 1, 2, \cdots, r)$ are the same as those introduced in §8 of Chapter V. The degree of smoothness required for a representation of the form (7.1.5) to exist is also indicated by the reasoning of Chapter V.

Conversely, if $H_1, \cdots, H_r$ are such functions and if the sum of the convolutions of the right-hand side of (7.15) vanishes whenever h vanishes, then

$$\xi = \frac{1}{h}(A_1 * H_1 + \dots + A_r * H_r) \tag{7.1.6}$$

will be a UEZ in $\mathcal{P}_R$.

In particular, if the function h does not vanish in $\mathcal{T}$, then formula (7.1.5) yields a UEZ in $\mathcal{P}_R$ for arbitrary $H_1, \cdots, H_r$ subject to the conditions listed in §8 of Chapter V.

§2. ON THE BEHAVIOR OF THE VARIANCE OF UNBIASED ESTIMATES [1)]

In this section we consider the same exponential families and the same unbiased estimates of zero as in the preceding one. Let $\xi = \xi(T_1, \cdots, T_s)$ denote an unbiased estimate of the function $f(\theta_1, \cdots, \theta_s)$. Suppose that ξ possesses a variance $D(\xi|\theta)$ in the region $\mathcal{P}_R$. Let Ω_0 denote a compact set of real values of the parameters in $\mathcal{P}_R$. We shall say that the unbiased estimate ξ is *admissible* on the compact set Ω_0 if there does not exist a UEZ χ such that $D(\xi + \chi|\theta) \leq D(\xi|\theta)$ for all $\theta \in \Omega_0$ with equality holding for at least one value of θ. Otherwise ξ is said to be *inadmissible* on Ω_0. An estimate ξ is said to be the *best*

1) This section was written by the author in collaboration with A. M. Kagan.

estimate on Ω_0 if for an arbitrary UEZ χ $D(\xi+\chi|\theta)\geq D(\xi|\theta)$. Furthermore, if χ is a UEZ, then for an arbitrary constant γ the statistic $\gamma\cdot\chi$ is a UEZ and

$$D(\xi+\gamma\chi\mid\theta)=E(\xi-f(\theta)+\gamma\chi)^2$$
$$=E(\xi-f(\theta))^2+2\gamma E(\xi\chi)+\gamma^2E\chi^2.$$

Or, singling out the dependence on θ,

$$D(\xi+\gamma\chi\mid\theta)=D(\xi\mid\theta)+2\gamma E(\xi\chi\mid\theta)+\gamma^2E(\chi^2\mid\theta). \tag{7.2.1}$$

Here the behavior of the covariance $E(\xi\chi|\theta)$ is extremely important. If

$$E(\xi\chi\mid\theta)\neq 0, \tag{7.2.2}$$

at the point θ, then the third term on the right-hand side of equation (7.2.1) will in many cases be less than the second term for sufficiently small γ and the estimate ξ will not be the best. If $E(\xi\chi|\theta)$ retains its sign for all $\theta\in\Omega_0$, then ξ will be an inadmissible estimate on Ω_0. On the other hand, if inequality (7.2.2) is satisfied at a given point $\theta\in\Omega_0$ then the unbiased estimate cannot be the best.

If χ is a UEZ satisfying the conditions of the preceding section, then we have the representation (7.1.6). If h does not vanish in $\mathcal{T}$ we can choose rather freely the functions $H_1,\cdots,H_r$. In this case

$$E(\chi\mid\theta)=\Pi_1G_1+\ldots+\Pi_rG_r, \tag{7.2.3}$$

where $\sigma_1,\cdots,\sigma_r$ are functions that are holomorphic in $\mathcal{P}$ and that are subject to the conditions described earlier. Let us write equation (7.2.3) in greater detail:

$$\int_{\mathcal{T}}\cdots\int\chi h\exp[-(\theta_1T_1+\ldots+\theta_sT_s)]\,dT_1\ldots dT_s$$
$$=\frac{1}{C(\theta)}(\Pi_1G_1+\ldots+\Pi_sG_s).$$

Here, $1/C(\theta)=\int\cdots\int_{\mathcal{T}}\exp[-(\theta_1T_1+\cdots+\theta_sT_s)]dT_1\cdots dT_s$ is a function that is holomorphic in $\mathcal{P}$ and positive for $\theta\in\mathcal{P}_R$. Thus

$$\int_{\mathcal{T}}\cdots\int\chi h\exp[-(\theta_1T_1+\ldots+\theta_sT_s)]\,dT_1\ldots dT_s\in\bar{I}_{H_0}. \tag{7.2.4}$$

Let $L=P(\partial/\partial\theta_1,\cdots,\partial/\partial\theta_s)$ denote a linear differential operator that is a polynomial in $\partial/\partial\theta_1,\cdots,\partial/\partial\theta_s$ with constant coefficients. If we apply this operator to the left-hand member of (7.2.4), we get the expression

$$\int_{\mathcal{T}}\cdots\int\chi Qh\exp[-(\theta_1T_1+\ldots+\theta_sT_s)]\,dT_1\ldots dT_s, \tag{7.2.5}$$

where Q is a polynomial that is easily determined from the polynomial P. Conversely, for each polynomial Q we can choose the corresponding polynomial P. If we treat $Q = Q(T_1, \cdots, T_s)$ as an unbiased estimate of $E(Q|\theta)$ for $\theta \in \Omega_0$, we can use (7.2.5) to judge the quality of that estimate.

In particular, the question as to whether the estimate Q is admissible or not reduces to the question of whether the ideal $\bar{I}_{H_0}$ contains a function $f \in \bar{I}_{H_0}$ such that Lf is negative for $\theta \in \Omega_0$. Here L is an operator such that $P(L)$ corresponds to the polynomial Q.

To find the polynomial statistic $Q(T_1, \cdots, T_s)$, which would be the best unbiased estimate of its mathematical expectation on the compact set Ω_0, we need to find differential operators L such that

$$L \cdot \bar{I}_{H_0} \subset \bar{I}_{H_0}. \tag{7.2.6}$$

Obviously such operators constitute a ring admitting multiplication by arbitrary complex numbers.

Let us give some examples of the application of the methods described to determine the inadmissability of certain polynomial estimates, for example the method of sample moments.[1] Consider a one-dimensional exponential family with a single parameter and with density (with respect to Lebesgue measure) of the form

$$p(x \mid \theta) = C(\theta)\exp(\theta\varphi_1(x) + \ldots + \theta^s\varphi_s(x))\exp l_0(x), \tag{7.2.7}$$

where $\phi_j(x)$ and $l_0(x)$ are continuous functions and $x \in [a, b]$ for $a = -\infty$ or $b = +\infty$. We have $l_0(x) \to -\infty$ as $x \to -\infty$; $l_0(x) \to -\infty$ as $x \to \infty$; $|\phi_j(x)/l_0(x)| \to 0$ as $x \to \pm\infty$; and $\int_{-\infty}^{\infty} \exp((1-\epsilon)l_0(x))\,dx < \infty$ for some $\epsilon > 0$.

Consider a repeated sample $(x_1, \cdots, x_n)$ out of a set with probability density (7.2.7). This sample corresponds to an exponential family with probability density

$$p(x_1, \ldots, x_n \mid \theta) = \\ = C_1(\theta)\exp(\theta_1 T_1 + \ldots + \theta_s T_s) R(T_1, \ldots, T_s). \tag{7.2.8}$$

(Earlier we used notation with the minus sign in the argument of the exponential.) Here $T_j = \Sigma_{i=1}^n \phi_j(x_i)$ $(j = 1, 2, \cdots, s)$; $h = h(T_1, \cdots, T_s)$ is the corresponding

---

1) These examples were given by A. M. Kagan.

probability density. The parameters $\theta_1, \cdots, \theta_s$ are related by

$$\Pi_1 = \theta_2 - \theta_1^2 = 0, \ \Pi_2 = \theta_3 - \theta_1^3 = 0, \ \ldots, \ \Pi_{s-1} = \theta_s - \theta_1^s = 0, \quad (7.2.9)$$

which define an analytic set Π. Here $r = s - 1$. The statistics $T_1, \cdots, T_s$ are sufficient statistics. Let us show that an arbitrary polynomial $Q(T_2, \cdots, T_s) \neq$ const is an inadmissible estimate of the function $E(Q|\theta)$ on an arbitrary compact set Θ of values of the parameter θ.

In particular, if $\phi_j(x) = x^j$ $(j = 1, 2, \cdots, s)$, we may assert that the sample moments $a_k = (1/n)\Sigma_{i=1}^n x_i^k$ for $k = 2, 3, \cdots, s$ are inadmissible estimates of the general initial moments α_k (for $k = 2, \cdots, s$) on an arbitrary compact set Θ.

As a second example consider a set with densities $p(x|\theta) = C_0 \exp[-(x-\theta)^{2k}]$, where k is an integer ≥ 2. For a repeated sample $(x_1, \cdots, x_n)$ the sums $T_1 = \Sigma_{i=1}^n x_i^{2k-1}$, $T_2 = \Sigma_{i=1}^n x_i^{2k-2}, \cdots, T_{2k-1} = \Sigma_{i=1}^n x_i = n\bar{x}$ are sufficient statistics. We may assert, in particular, that the sample moments $\bar{x}, a_2, \cdots, a_{2n-2}$ are inadmissible estimates of the corresponding general moments.

By virtue of the conditions imposed on the functions $\phi_j(x)$, the quantity $E(Q|\theta)$ exists for an arbitrary polynomial $Q(T)$ and for arbitrary values of $\theta_1, \cdots, \theta_s$. Suppose that a polynomial $Q = Q(T_2, \cdots, T_s)$ depends only on the $s - 1$ sufficient statistics indicated. This polynomial is an unbiased estimate of the function $E(Q|\theta) = f(\theta)$. We shall be interested in the properties of this estimate under conditions (7.2.9) on an arbitrary compact set of values of θ_1. We have

$$C_1(\theta) \int_{\mathcal{T}} \cdots \int Q(T_2, \ldots, T_s) h(T) \times \exp(\theta_1 T_1 + \ldots + \theta_s T_s) \, dT_1 \ldots dT_s = f(\theta), \quad (7.2.10)$$

where $\mathcal{T}$ is the space of values of $(T_1, \cdots, T_s)$. By virtue of the conditions on $h(x)$ in the space $\mathcal{T}$, we can choose a bounded closed cube R: $T_{j1} \leq T_j \leq T_{j2}$ $(j = 1, 2, \cdots, s)$ in which

$$h(T) \geqslant \varepsilon_0 > 0.$$

Let $m \geq 1$ denote the degree of the polynomial Q and let $a_{k_2 \cdots k_s} T_2^{k_2} \cdots T_s^{k_s}$ denote one of its mth-degree terms $(a_{k_2 \cdots k_s} \neq 0$ and $k_2 + \cdots + k_s = m)$. We shall now apply considerations of the nature of Wijsman's D-method (see Chapter

VI).

Let $\psi_0(T_1, \cdots, T_s)$ denote a function satisfying the following conditions.

(1) $\psi_0(T_1, \cdots, T_s) > 0$ for $(T_1, \cdots, T_s)$ inside R.

(2) $\psi_0(T_1, \cdots, T_s) = 0$ for $T \notin R$.

(3) $\psi_0(T_1, \cdots, T_s)$ is everywhere continuous and has no fewer than $2k_2 + \cdots + sk_s$ partial derivatives with respect to each argument. Here all derivatives vanish on the boundary of R.

We set

$$\int \cdots \int \psi_0(T_1, \ldots, T_s) \exp(\theta_1 T_1 + \ldots + \theta_s T_s)\, dT_1 \ldots \ldots dT_s = V(\theta).$$

Consider the polynomial

$$w(\theta) = (\theta_2 - \theta_1^2)^{k_2} \ldots (\theta_s - \theta_1^s)^{k_s} = \sum_{l_1, \ldots, l_s} w_{l_1, \ldots, l_s} \theta_1^{l_1} \ldots \theta_s^{l_s}$$

of degree $(2k_2 + \cdots + sk_s)$.

We set

$$\psi(T_1, \ldots, T_s) = W\psi_0(T_1, \ldots, T_s),$$

where the operator W is given by

$$W = \sum w_{l_1, \ldots, l_s} \frac{\partial^{l_1 + \cdots + l_s}}{\partial \theta_1^{l_1} \ldots \partial \theta_s^{l_s}}.$$

By virtue of condition (3), W is applicable to the function $\psi_0(T_1, \cdots, T_s)$.

A familiar property of the Laplace transformation (see §1, Chapter I) leads to the relation

$$\int \cdots \int \psi(T_1, \ldots, T_s) \exp(\theta_1 T_1 + \ldots + \theta_s T_s)\, dT_1 \ldots dT_s = w(\theta) V(\theta) = (\theta_2 - \theta_1^2)^{k_2} \ldots (\theta_s - \theta_1^s)^{k_s} V(\theta).$$

Now define

$$\chi(T_1, \ldots, T_s) = \begin{cases} 0, & T \notin R, \\ \dfrac{\psi(T_1, \ldots, T_s)}{h(T_1, \ldots, T_s)}, & T \in R. \end{cases}$$

We have

$$E(\chi \mid \theta) = C_1(\theta) \int \ldots$$
$$\ldots \int \psi(T_1, \ldots, T_s) \exp(\theta_1 T_1 + \ldots + \theta_s T_s)\, dT_1 \ldots dT_s$$
$$= C_1(\theta)(\theta_2 - \theta_1^2)^{k_2} - (\theta_s - \theta_1^s)^{k_s} V(\theta).$$

Therefore χ is a UEZ under the relations (7.2.9), i.e. for $\theta \in \Pi$.

Furthermore,

$$E(\chi Q \mid \theta) = C_1(\theta) \int \ldots \int \psi(T_1, \ldots, T_s)$$
$$\times Q(T_2, \ldots, T_s) \exp(\theta_1 T_1 + \ldots + \theta_s T_s)\, dT_1 \ldots dT_s$$
$$= C_1(\theta) L\left((\theta_2 - \theta_1^2)^{k_2} \ldots (\theta_s - \theta_1^s)^{k_s} V(\theta)\right),$$

where

$$L = \sum_{l_2, \ldots, l_s} a_{l_1, \ldots, l_s} \frac{\partial^{l_2 + \ldots + l_s}}{\partial \theta_2^{l_2} \ldots \partial \theta_s^{l_s}}.$$

Obviously, under conditions (7.2.9),

$$L\left((\theta_2 - \theta_1^2)^{k_2} \ldots (\theta_s - \theta_1^s)^{k_s} V(\theta)\right)$$
$$= a_{k_2, \ldots, k_s} k_2! \ldots k_s! V(\theta), \quad \theta \in \Pi. \tag{7.2.11}$$

If we assume that $a_{k_2, \ldots, k_s} > 0$, this equation yields

$$E(\chi Q \mid \theta) > 0, \quad \theta \in \Pi. \tag{7.2.12}$$

Now, let $\tilde{\Theta}$ denote an arbitrary compact subset of Ω_0 and let $\tilde{\Pi}$ denote the corresponding compact subset of Π; i.e.

$$\tilde{\Pi} = \{\theta_1, \ \theta \in \Pi, \ \theta_1 \in \tilde{\Theta}\}.$$

Obviously, for $\theta \in \widetilde{\Pi}$,

$$E(\chi Q \mid \theta) \geqslant c_0 > 0 \qquad (7.2.13)$$

and

$$E(\chi^2 \mid \theta) \leqslant C_0 < \infty. \qquad (7.2.14)$$

Consider the statistic

$$\tilde{Q}(T_1, \ldots, T_s) = Q(T_2, \ldots, T_s) - \gamma\chi(T_1, \ldots, T_s). \qquad (7.2.15)$$

We have

$$E(\tilde{Q} \mid \theta) = C_1(\theta) \int_{\mathcal{T}} \ldots \int \tilde{Q}(T) h(T) \times \exp(\theta T_1 + \ldots + \theta^s T_s)\, dT_1 \ldots dT_s$$

and

$$E(\tilde{Q}^2 \mid \theta) = E(Q^2 \mid \theta) - 2\gamma E(\chi Q \mid \theta) + \gamma^2 E(\chi^2 \mid \theta). \qquad (7.2.16)$$

It is obvious from (7.2.13), (7.2.14), and (7.2.16) that γ can be chosen in such a way that, for $\theta \in \widetilde{\Theta}$,

$$-2\gamma E(\chi Q \mid \theta) + \gamma^2 E(\chi^2 \mid \theta) < 0, \qquad (7.2.17)$$

which proves the inadmissibility of the estimate $Q(T_2, \cdots, T_s)$ on an arbitrary compact set $\widetilde{\Theta}$. If we consider the polynomials depending on all sufficient statistics as estimates, this inadmissibility may obviously fail to occur.

§3. A THEOREM OF S. R. RAO ON THE INADMISSIBILITY OF CERTAIN ESTIMATES [1)]

Densities with "shifting parameter" of the form

$$p(x \mid \theta) = C_0 \exp P(x - \theta), \qquad (7.3.1)$$

where $P(x)$ is a polynomial of even degree with negative leading coefficient, constitute a special case of densities of the form (7.2.7). When a repeated sample is made the sample mean $\bar{x}$ is obviously an unbiased estimate of θ. From Kagan's example at the end of the preceding section one can easily see that $\bar{x}$ is and inadmissible estimate of θ if the degree of the polynomial exceeds 2. (If the degree is equal to 2, i.e. if the sample is a normal one, $\bar{x}$ is, as we know, the best estimate of θ in the sense described earlier.)

However, inasmuch as this result has to do with families with "shifting parameter," it follows from a more general result proved by Rao [62] in 1952

1) This section was written by the author in collaboration with A. M. Kagan.

dealing with such families.

Consider a family of distributions on E_n with probability density (with respect to Lebesgue measure)

$$f(x_1 - \theta, \ldots, x_n - \theta) \tag{7.3.2}$$

and a class $\mathfrak{A}$ of unbiased estimates A of the parameter θ that satisfy the condition

$$A(x_1 + c, \ldots, x_n + c) = c + A(x_1, \ldots, x_n). \tag{7.3.3}$$

Obviously the class $\mathfrak{A}$ coincides with the set of statistics of the form $\bar{x} + f(x_i - x_j)$, where $f(x_i - x_j)$ is a statistic depending only on the observed differences and satisfying the equation $E(f|0) = 0$.

Theorem 7.3.1. *Every estimate $A \in \mathfrak{A}$ is inadmissible in the class of unbiased estimates θ for the region of all values of θ except when*

$$E(A|x_2 - x_1, \ldots, x_n - x_1, \theta) = \theta. \tag{7.3.4}$$

To prove this we note that

$$\begin{aligned} &E(A(x_1, \ldots, x_n)|x_2 - x_1, \ldots, x_n - x_1, \theta) \\ &= E(A(x_1 + \theta, \ldots, x_n + \theta)|x_2 - x_1, \ldots, x_n - x_1, 0) \\ &= \theta + E(A(x_1, \ldots, x_n)|x_2 - x_1, \ldots, x_n - x_1, 0). \end{aligned} \tag{7.3.5}$$

We set

$$\chi = E(A(x_1, \ldots, x_n)|x_2 - x_1, \ldots, x_n - x_1, 0). \tag{7.3.6}$$

Obviously χ is a UEZ. Let us now consider the estimate $B(x_1, \cdots, x_n) = A - \chi$. We have

$$E(B|\theta) = E(A|\theta) - E(\chi|\theta) = \theta.$$

Furthermore,

$$\begin{aligned} D(A|\theta) &= E[(A - \theta)^2|\theta] = E[(B - \theta + \chi)^2] = \\ &= E[(B - \theta)^2|\theta] + E(\chi^2|\theta) + 2E(\chi(B - \theta)|\theta). \end{aligned}$$

But

$$E[\chi(B - \theta)|\theta] = E[(E\chi(B - \theta)|x_2 - x_1, \ldots, x_n - x_1, \theta)].$$

The statistic χ depends by definition only on $x_2 - x_1, \cdots, x_n - x_1$. Because of this and equations (7.3.5) and (7.3.6),

$$E(\chi(B-\theta)|x_2-x_1, \ldots, x_n-x_1, \theta)$$
$$=\chi[E[(B-\theta)|x_2-x_1, \ldots, x_n-x_1, \theta]]=0.$$

Therefore $E[\chi(B-\theta)|\theta]=0$ and $D(A|\theta)=D(B)+E(\chi^2|\theta)>D(B)$ except when $\chi=0$ with probability 1.

Let us look at those cases when this theorem of Rao gives inadmissibility of a classical estimate $\bar{x}$ for the parameter θ. For the statistic $A=\bar{x}$, let us find the "correction" χ defined by (7.3.6). To do this we need to calculate the conditional probability density for $\bar{x}$ with fixed $x_2-x_1, \cdots, x_n-x_1$, and $\theta=0$. We make the change of variables

$$\xi_1=\bar{x}=\frac{1}{n}(x_1+\ldots+x_n),$$
$$\xi_2=x_2-x_1,$$
$$\cdots\cdots\cdots$$
$$\xi_n=x_n-x_1.$$

The determinant of the transformation is

$$\begin{vmatrix} \frac{1}{n} & \frac{1}{n} & \frac{1}{n} & \cdots & \frac{1}{n} \\ -1 & 1 & 0 & \cdots & 0 \\ -1 & 0 & 1 & \cdots & 0 \\ \cdot & \cdot & \cdot & \cdots & \cdot \\ -1 & 0 & 0 & \cdots & 1 \end{vmatrix}=1,$$

which is easily verified by adding all columns of the determinant to the first column. If we set $\bar{\xi}=(\xi_2+\cdots+\xi_n)/n$, we find, after some simple calculation, that

$$\begin{aligned}\chi&=E(\bar{x}|x_2-x_1, \ldots, x_n-x_1, 0)\\ &=E(\xi_1|\xi_2, \ldots, \xi_n, 0)\\ &=\bar{\xi}+E(x_1|\xi_2, \ldots, \xi_n, 0)\\ &=\bar{x}-\frac{\int_{-\infty}^{\infty}\xi f(x_1-\xi, \ldots, x_n-\xi)\,d\xi}{\int_{-\infty}^{\infty}f(x_1-\xi, \ldots, x_n-\xi)\,d\xi}.\end{aligned} \tag{7.3.7}$$

Let us now consider the question as to when the "correction" χ defined by formula (7.3.7) is nontrivial, i.e. when $\chi = E(\bar{x}|x_2 - x_1, \cdots, x_n - x_1, 0) = 0$. We confine ourselves to the case of a repeated sample of size n, that is, to the case in which

$$f(x_1-\theta,\ x_2-\theta,\ \dots,\ x_n-\theta)=f(x_1-\theta)\ \dots\ f(x_n-\theta).$$

Theorem 7.3.2.[1] *For the case of a repeated sample of size* $n \geq 3$, *if* Ex_j *exists, the equation*

$$\chi=E(\bar{x}\,|\,x_2-x_1,\ \dots,\ x_n-x_1;\ 0)=0 \tag{7.3.8}$$

is satisfied only for a normal sample. If $n = 2$ *this equation will be satisfied if the distribution of* x_j *is symmetric about* 0.

To prove this we note that (7.3.8) implies immediately that for $n \geq 3$,

$$E(\bar{x}\,|\,x_2-x_1,\ x_3-x_1;\ 0)=0. \tag{7.3.9}$$

We introduce the characteristic function $\phi(t) = Ee^{itx_j}$, which is independent of j since the sample is repeated. For arbitrary real t_1 and t_2 it follows from (7.3.9) that

$$E(\bar{x}\exp[it_1(x_1-x_2)+it_2(x_1-x_3)])=0. \tag{7.3.10}$$

Since $Ex_j = 0$ exists, we have $\phi'(t) = iE(x_j e^{itx_j})$ (see, for example, [33]). Therefore we obtain from (7.3.7) that

$$\begin{aligned} &E(x_1+x_2+x_3)\exp i[(t_1+t_2)x_1-t_1x_2-t_2x_3] \\ &=-i\varphi'(t_1+t_2)\varphi(-t_1)\varphi(-t_2)-i\varphi(t_1+t_2)\varphi'(-t_1)\varphi(-t_2) \\ &\qquad -i\varphi(t_1+t_2)\varphi'(-t_2)\varphi(-t_1)=0. \end{aligned}$$

In a sufficiently small neighborhood of the point (0,0) we have $|t_1| \leq \epsilon$ and $|t_2| \leq \epsilon$, and $\phi(t) \neq 0$. We may write

$$\frac{\varphi'(t_1+t_2)}{\varphi(t_1+t_2)}+\frac{\varphi'(-t_1)}{\varphi(-t_1)}+\frac{\varphi'(-t_2)}{\varphi(-t_2)}=0. \tag{7.3.11}$$

In such a neighborhood of 0 we obtain $\phi'(t)/\phi(t) = \psi(t)$. Let us show that

1) See the article by A. M. Kagan, Ju. V. Linnik, and S. R. Rao, *On a characterization of the normal law based on a property of the sample average*, Sankhyā, Ser, A, 27 (1965), 405–406.

$\psi(t) = at$. From (7.3.8) we have $\psi(t_1 + t_2) = -\psi(-t_1) - \psi(-t_2)$. Furthermore, $\psi(0) = 0$. If we set $t_1 = -t_2$ we find that $\psi(t_1) = -\psi(-t_1)$; that is, the function $\psi(t)$ is odd. Thus $\psi(t_1 + t_2) = \psi(t_1) + \psi(t_2)$ in this neighborhood of zero.

It then follows from the continuity of $\psi(t)$ that $\psi(t)$ is a linear function, so that $\phi(t) = \exp(at^2/2 + C)$. Since $\phi(t)$ is the characteristic function, we have $c = 0$ and $a \leq 0$, so that x_j is normal.

If $n = 2$ then

$$E[(x_1 + x_2)(\exp it(x_1 - x_2)] = 0,$$

and hence $\phi'(t)/\phi(t) = -\phi'(-t)/\phi(-t)$ in some neighborhood of 0, and $\phi(t)$ is an even function in that neighborhood. On the other hand, if $\phi(t)$ is everywhere an even function, the equation is satisfied. Thus for $n = 2$ condition (7.2.5) is satisfied for all symmetric x_j.

CHAPTER VIII

ANALYTICAL METHODS OF STUDYING UNRANDOMIZED TESTS. APPLICATION TO THE BEHRENS-FISHER PROBLEM

§ 1. QUESTIONS OF EXISTENCE OF UNRANDOMIZED SIMILAR TESTS FOR INCOMPLETE EXPONENTIAL FAMILIES

In this chapter we shall look at the construction of ϵ-complete families of randomized similar tests for exponential families under holomorphic relationships of the form (5.2.4). These tests depend only on sufficient statistics, which, by what was said in § 1 of Chapter V, does not violate their ϵ-completeness. Furthermore, as was shown in the same section, when we have a randomized test depending only on sufficient statistics, we can, generally speaking, construct from it an unrandomized test that is measurable with respect to the entire σ-algebra of the sample but not with respect to the σ-subalgrbra of sufficient statistics. This test possesses the same power function as the original one (and thus is equivalent to it in this respect).

Thus by leaving the σ-algebra of sufficient statistics we are enabled to construct an ϵ-complete family of unrandomized tests for the cases considered above. In particular, on the basis of the discussion in Chapter V, this can be done for the Behrens-Fisher problem.

However, if we wish to investigate unrandomized similar tests for exponential families that depend only on sufficient statistics under null hypotheses of the form (5.2.4), we encounter considerable difficulties even in such a simple case of an exponential family as that to which the Behrens-Fisher problem leads. Below, we shall investigate the Behrens-Fisher problem in this respect.

Consider an exponential family of the form

$$p_\theta(T_1, \ldots, T_s) = (C(\theta) \exp[-(\theta_1 T_1 + \ldots + \theta_s T_s)] h(T_1, \ldots, T_s). \qquad (8.1.1)$$

Following [42], let us consider unrandomized similar tests $\phi(T_1, \cdots, T_s)$ for the hypothesis H_0 defined with the aid of $r\ (< s)$ polynomial relations of the form

$$P_1(\theta) = 0, \ \ldots, \ P_r(\theta) = 0. \qquad (8.1.2)$$

where the polynomials $P_j(\theta)$, $j = 1, 2, \cdots, r$, are assumed to be homogeneous of degree ≥ 1.

The test $\phi(T_1, \cdots, T_s)$ is characterized by its critical zone, for which it is the indicator function. Of course, we need to require of this test only that it be measurable with respect to the σ-algebra of sufficient statistics. Furthermore, we must require satisfactory behavior of its power function. Finally, to apply the test, it is desirable that the boundary of the critical zone have as simple a form as possible, for example that it consist of finitely many smooth pieces.

The problem of describing unrandomized similar tests for exponential families that depend only on sufficient statistics is still far from solved. However, investigations that have been pursued in this direction (which will be discussed to some degree in what follows) show that such tests, if they exist, apparently have critical zones whose boundaries are extremely complicated. In general the function itself or its derivatives of low orders have discontinuities on such boundaries. Roughly speaking, we should not expect the existence of tests of such a form with sufficiently smooth boundaries of the critical zone. Fairly detailed investigations in this connection have been made only for the Behrens-Fisher problem by the method of analytic continuation with respect to the parameters.

Let us look briefly at the application of such a method to exponential families (8.1.1) and the null hypothesis of the form (8.1.2). Let $\mathcal{T} = (T_1, \cdots, T_s)$ denote the space of values of the sufficient statistics. This space is contained in E_s. Let us require that the integral

$$\int_{\mathcal{T}} \cdots \int |\exp[-(\theta_1 T_1 + \ldots + \theta_s T_s)]| \, h(T_1, \ldots, T_s) \, dT_1 \ldots dT_s$$

converge in the Cartesian product of the half-spaces $\operatorname{Re}\theta_j > 0$ $(j = 1, 2, \cdots, s)$. If $\phi(T_1, \cdots, T_s)$ is an unrandomized similar test of level α with similar zone Z, then

$$\int_Z \cdots \int \exp[-(\theta_1 T_1 + \ldots + \theta_s T_s)] \times h(T_1, \ldots, T_s) \, dT_1 \ldots dT_s = \frac{C_0(\alpha)}{C(\theta)} \qquad (8.1.3)$$

for all real values of $\theta_1, \cdots, \theta_s$ that satisfy conditions (8.1.2). Here $C_0(\alpha)$ is a positive constant depending only on the level α and the critical zone Z. The relationships (8.1.2) constitute an algebraic set in the projective space. Let $(\theta_1, \cdots, \theta_s)$

denote a point in this algebraic set. For arbitrary $\omega > 0$ the point $(\theta_1/\omega, \cdots, \theta_s/\omega)$ also belongs to that set.

We set $\theta_j = \omega\vartheta_j$ for $j = 1, 2, \cdots, s$. Let us multiply both sides of equation (8.1.3) by $\omega^b e^{-a\omega}$, where b is an integer ≥ 1 and a is a positive number. If we formally integrate both sides of the equation with respect to ω from 0 to ∞, we obtain the formal relationship

$$\int \cdots \int_Z \frac{h(T_1, \ldots, T_s)\, dT_1 \ldots dT_s}{(a + \vartheta_1 T_1 + \ldots + \vartheta_s T_s)^{b+1}} = C_0(a) f(\vartheta, a, b). \tag{8.1.4}$$

Here $f(\vartheta, a, b)$ is the result of integrating the quantity

$$\frac{\omega^b e^{-a\omega}}{C(\theta)} = \frac{\omega^b e^{-a\omega}}{C(\vartheta\omega)}$$

with respect to ω from 0 to ∞.

Let us suppose now that both sides of (8.1.4) are meaningful for real $\vartheta_j > 0$ $(j = 1, 2, \cdots, s)$. A sufficient condition for this is absolute convergence of the integral on the left side of (8.1.4) for $\vartheta_j > 0$. Then the left-hand member of (8.1.4) must coincide with the right-hand member if the relations

$$P_1(\vartheta) = 0, \ \ldots, \ P_r(\vartheta) = 0 \tag{8.1.5}$$

are satisfied. For real parameters ϑ these relations can be extended to a complex region. Let us suppose that such an extension is possible on both sides of equation (8.1.4). Then this equation itself will remain valid for the corresponding complex values of $\vartheta_1, \cdots, \vartheta_s$ in the corresponding region $\overline{\Omega}$. Here the form of the function $f(\vartheta, a, b)$ is determined by the exponential family (8.1.1), and investigation of the possible similar tests leads to investigation of the zones Z satisfying the identity (8.1.4) when the relations (8.1.5) are satisfied. For $Z = \mathfrak{I}$ the identity (8.1.4) obviously holds with $C_0(a) = C_0(1)$; this is a trivial unrandomized similar test. For other Z the zone has a nontrivial boundary.

For a given a consider the expression in the denominator of the left-hand member of (8.1.4)

$$D(a, \vartheta, T) = a + \vartheta_1 T_1 + \ldots + \vartheta_1 T_s \tag{8.1.6}$$

with relations (8.1.5) holding.

In the study of equations of the type (8.1.4), a special role is played by the real zeros of the expression (8.1.6) when the relationships mentioned hold, i.e. the geometric loci in the space E_s that are defined by the equations

$$D(a, \vartheta, T) = 0; \quad P_1(\vartheta) = 0, \ \ldots, \ P_s(\vartheta) = 0. \tag{8.1.7}$$

Here we admit arbitrary complex values for ϑ that yield real values for $T_1, \cdots, T_s$. We shall call such real zeros "critical surfaces" or "critics" of the exponential family (they depend on a number a fixed in advance).

The values $\vartheta^{(0)} = (\vartheta_1^{(0)}, \cdots, \vartheta_s^{(0)})$ generated by the critics lie outside the region $\overline{\Omega}$ in which equation (8.1.4) is valid, since the left-hand member of (8.1.4) becomes meaningless for such values. However in many cases it is possible to approximate such values without leaving the region $\overline{\Omega}$. Here the denominator of the fraction on the left side of (8.1.4) approximates 0. In studying the asymptotic behavior of the left and right sides of this equation as we let $\vartheta \in \overline{\Omega}$ approach $\vartheta^{(0)}$, we can frequently obtain somewhat remarkable infomation regarding the behavior of the similar zones Z. Here the structure of their boundaries is of particular importance. Under certain conditions one can use this procedure to show that there ate no similar critical zones with sufficiently smooth boundaries.

We note also that to carry out an investigation of this type it is essential to have a relation of the form (8.1.4), where the denominator in the integrand vanishes at the critics of the exponential family. The origin of such a relationship is of no significance. We obtained it by an integral transfomation, but it can also be obtained by other approaches, especially when we are considering tests that are measurable with respect to some (incomplete) subalgebra of the algebra of sufficient statistics. We shall encounter such examples when we study the Behrens-Fisher problem.

§2. STATEMENT OF THE PROBLEM OF AN UNRANDOMIZED HOMOGENEOUS SIMILAR TEST IN THE BEHRENS-FISHER PROBLEM

For the Behrens-Fisher problem (in the usual notation) we have four sufficient statistics: $\overline{x}$, $\overline{y}$, s_1, and s_2. We define a homogeneous similar unrandomized test as a similar test whose critical zone is of the form

$$G\left(\frac{\overline{x}-\overline{y}}{s_2},\ \frac{s_1}{s_2}\right) \geqslant 0,$$

where G is a Lebesgue-measurable function. We shall see in §1 of Chapter X that such a test always exists for an arbitrary level α in the case of samples of sizes n_1 and n_2 of unlike parity.

From an analytical point of view, however, the boundary of this critical zone is extremely complicated. We shall see that the problem centers around this difficulty.

Since we shall be dealing with the random vector $((\bar{x}-\bar{y})/s_2,\ s_1/s_2)$, we need to derive its distribution. Let us set $\xi=(\bar{x}-\bar{y})/s_2$ and $\eta=s_1/s_2$. Let us assume also that $n_1 \geq 2$ and $n_2 \geq 2$. We define

$$X_1=(\bar{x}-\bar{y})\left(\frac{n_1\sigma_2^2+n_2\sigma_1^2}{n_1n_2}\right)^{1/2}.$$

This quantity is independent of s_1^2 and s_2^2 (see Chapter IV). Here $E(X_1)=0$ and $D(X_1)=1$, so that $X_1 \in N(0, 1)$.

Let us also define $u=n_1s_1^2/\sigma_1^2$ and $v=n_2s_2^2/\sigma_2^2$. These quantities are independent and they have χ^2 distributions with (n_1-1) and (n_2-1) degrees of freedom respectively. Finally, we define $u_1=\sqrt{u}$ and $v_1=\sqrt{v}$. The quantities X_1, u_1 and v_1 are independent and have probability densities

$$p(X_1)=\frac{1}{\sqrt{2\pi}}\exp\left(-\frac{x_1^2}{2}\right),$$

$$q(u_1)=\frac{1}{2^{\frac{n_1-3}{2}}\Gamma\left(\frac{n_1-1}{2}\right)}u_1^{n_1-2}\exp\left(-\frac{u_1^2}{2}\right),\quad u_1>0,$$

$$r(v_1)=\frac{1}{2^{\frac{n_2-3}{2}}\Gamma\left(\frac{n_2-1}{2}\right)}v_1^{n_2-2}\exp\left(-\frac{v_1^2}{2}\right),\quad v_1>0.$$

Define

$$\tilde{x}=\frac{X_1}{v_1},\quad \tilde{y}=\frac{u_1}{v_1},\quad \tilde{z}=v_1.$$

Then $X_1=\tilde{x}\tilde{z}$, $u_1=\tilde{y}\tilde{z}$, $v_1=\tilde{z}$, and

$$\frac{\partial(X_1,\ u_1,\ v_1)}{\partial(\tilde{x},\ \tilde{y},\ \tilde{z})}=\begin{vmatrix}\tilde{z} & 0 & 0\\ 0 & \tilde{z} & 0\\ \tilde{x} & \tilde{y} & 1\end{vmatrix}=\tilde{z}^2.$$

Thus for arbitrary x_0 and arbitrary positive y_0 the common density is

$$\rho(\tilde{x},\ \tilde{y})=K_{n_1n_2}\int_0^\infty d\tilde{z}\ \tilde{y}^{n_1-2}\,\tilde{z}^{n_1+n_2-2}\exp\left[-\frac{\tilde{z}^2}{2}(1+\tilde{x}^2+\tilde{y}^2)\right],$$

where

$$K_{n_1n_2}=\frac{1}{\sqrt{\pi}\,2^{\frac{n_1+n_2-5}{2}}\Gamma\left(\frac{n_1-1}{2}\right)\Gamma\left(\frac{n_2-1}{2}\right)}$$

Integrating with respect to $\tilde{z}$, we find that

$$\int_0^\infty d\tilde{z} \cdot \tilde{z}^{n_1+n_2-2} \exp\left[-\frac{\tilde{z}^2}{2}(1+\tilde{x}^2+\tilde{y}^2)\right] = \frac{K'_{n_1n_2}}{(1+\tilde{x}^2+\tilde{y}^2)^{\frac{n_1+n_2-1}{2}}},$$

where

$$K'_{n_1n_2} = 2^{\frac{n_1+n_2-3}{2}} \Gamma\left(\frac{n_1+n_2-1}{2}\right).$$

From this we get

$$\rho(\tilde{x}, \tilde{y}) = \tilde{C}_{n_1n_2} \frac{\tilde{y}^{n_1-2}}{(\tilde{x}^2+\tilde{y}^2+1)^{\frac{n_1+n_2-1}{2}}}, \tag{8.2.1}$$

where

$$\tilde{C}_{n_1n_2} = K'_{n_1n_2} K_{n_1n_2} = \frac{2\Gamma\left(\frac{n_1+n_2-1}{2}\right)}{\Gamma\left(\frac{n_1-1}{2}\right)\Gamma\left(\frac{n_2-1}{2}\right)}.$$

Let us now define $\theta = n_2\sigma_1^2/n_1\sigma_2^2$. After some simple manipulations we obtain

$$\tilde{x} = \frac{\xi}{\sqrt{1+\theta}}, \quad \tilde{y} = \frac{\eta}{\sqrt{\theta}}.$$

Let $\phi(\xi, \eta)$ denote the indicator function of the critical zone $G(\xi, \eta) \geq 0$. Then

$$E(\varphi \mid \theta) = \int_\Omega\int \varphi(\tilde{x}\sqrt{1+\theta}, \tilde{y}\sqrt{\theta})\rho(\tilde{x}, \tilde{y})\, d\tilde{x}\, d\tilde{y} = \alpha, \tag{8.2.2}$$

where α is the level of the test and Ω is the upper half-plane

$$-\infty < \tilde{x} < \infty, \quad \tilde{y} \geqslant 0.$$

Define

$$N = \frac{n_1+n_2-1}{2}.$$

Let us return to the variables ξ and η. Then (8.2.2) may be written in the form

$$\int_\Omega\int \varphi(\xi, \eta) \frac{\eta^{n_1-2}\, d\xi\, d\eta}{(\theta^2+\theta(1+\xi^2+\eta^2)+\eta^2)^N}$$
$$= C(n_1, n_2)\alpha\theta^{-\frac{n_2}{2}}(1+\theta)^{-N+\frac{1}{2}}, \tag{8.2.3}$$

where $C(n_1, n_2) > 0$. This is the basic integral relationship for a homogeneous unrandomized similar test.

Following the exposition of §1, let us investigate the critical curves for the family of measures appearing in formula (8.2.3). We shall simply refer to these as "critics". They are geometric loci obtained by setting the expression

$$\theta^2+\theta(1+\xi^2+\eta^2)+\eta^2,$$

equal to 0, i.e. they correspond to values of $\theta = -D \leq 0$. For $D = 0$ this is the abscissa $\eta = 0$. For $0 < D < 1$, we obtain the family of confocal hyperbolas

$$\frac{\eta^2}{D}-\frac{\xi^2}{1-D}=1 \tag{8.2.4}$$

with common focus (0, 1), the limiting position of these hyperbolas being the line $\eta = 0$.

For $D = 1$ we have $\xi = 0$, i.e. the equation for the ordinate. It is natural to consider that portion of the ordinate corresponding to $\eta \geq 1$ (twice covered) as belonging to our family of hyperbolas. Thus we obtain the family of confocal hyperbolas

$$A(\xi, \eta)=D, \tag{8.2.5}$$

where $0 \leq D \leq 1$. For $D = 0$ and $D = 1$ the hyperbolas degenerate into segments of straight lines. For $D > 1$, instead of hyperbolas, we have ellipses

$$\frac{\xi^2}{D-1}+\frac{\eta^2}{D}=1 \tag{8.2.5'}$$

with common focus (0, 1). For $D = 1$, they degenerate into the twice covered segment $\xi = 0$, $0 \leq \eta \leq 1$. We write the family of ellipses in the form

$$B(\xi, \eta)=D, \quad D \geqslant 1. \tag{8.2.6}$$

When we study the distribution of the critics with respect to the boundary of the test zone and apply the principle of analytic continuation with respect to θ, we discover a number of properties of the unrandomized test $\phi(\xi, \eta)$ (see the articles by Linnik [42], [44] and Šalaevskiĭ [77]). (Šalaevskiĭ applied the Laplace transformation to get these results. He obtained simpler and more general results by this procedure for continuous families of tests.)

We shall say that a continuous family of unrandomized similar tests with critical zones $Z_C = \{G(\xi, \eta) \geq C\}$ is defined if a continuous function $G(\xi, \eta)$ is constructed such that, for arbitrary given C and an arbitrary value of the parameter θ, the quantity $p_C = P[(\xi, \eta) \in Z_C]$ is independent of θ, so that the distribution of $G(\xi, \eta)$ is independent of θ. Trivial continuous families are generated by the functions $G(\xi, \eta) = \text{const}$.

Theorem 8.2.1 (Šalaevskiĭ). *There are no nontrivial continuous families of unrandomized similar tests for the Behrens-Fisher problem.*

The proof of this interesting theorem is rather laborious, and we omit it.

However, if we drop the requirement that the function $G(\xi, \eta)$ be continuous this theorem is no longer valid, as we shall show in Chapter X.

§ 3. HOMOGENEOUS FISHER-WELCH-WALD TESTS

In a well-known article [9] Wald examined unrandomized tests that satisfy four axioms regarding the critical region Z and that are quite natural from a statistical point of view. These axioms are as follows:

(I) The critical region Z lies in the space of the sufficient statistics $\bar{x}, \bar{y}, s_1^2, s_2^2$; that is, if a sample $(x_1, \cdots, x_{n_1}; y_1, \cdots, y_{n_2})$ belongs to Z, then the sample $(x'_1, \cdots, x'_{n_1}; y'_1, \cdots, y'_{n_2})$ also belongs to Z if $\bar{x}' = \bar{x}$, $\bar{y}' = \bar{y}$, $s'^2_1 = s_1^2$, and $s'^2_2 = s_2^2$.

(II) If the sample $(x_1, \cdots, x_{n_1}; y_1, \cdots, y_{n_2})$ belongs to Z, then for an arbitrary number c the sample $(x_1 + c, \cdots, x_{n_1} + c; y_1 + c, \cdots, y_{n_2} + c)$ also belongs to Z.

(III) If the sample $(x_1, x_2, \cdots, x_{n_1}; y_1, \cdots, y_{n_2})$ belongs to Z then for arbitrary nonzero k the sample $(kx_1, \cdots, kx_{n_1}; ky_1, \cdots, ky_{n_2})$ also belongs to Z.

(IV) Let $(x_1, \cdots, x_{n_1}; y_1, \cdots, y_{n_2})$ denote a sample belonging to Z. Let $(x'_1, \cdots, x'_{n_1}; y'_1, \cdots, y'_{n_2})$ denote another sample such that $|\bar{y}' - \bar{x}'| > |\bar{y} - \bar{x}|$, $s'^2_1 = s_1^2$, and $s'^2_2 = s_2^2$. Then the sample $(x'_1, \cdots, x'_{n_1}; y'_1, \cdots, y'_{n_2})$ also belongs to Z.

It follows from these four axioms that the critical region of a test must be of the form

$$\frac{|\bar{x} - \bar{y}|}{s_2} \geqslant \varphi\left(\frac{s_1}{s_2}\right), \tag{8.3.1}$$

where ϕ is a single-valued measurable function.

Proof. It follows from axiom (I) that the region Z is determined only by $\bar{x}, \bar{y}, s_1^2$, and s_2^2. It follows from (II) that if the point $(\bar{x}, \bar{y}, s_1^2, s_2^2)$ belongs to Z, then the point $(\bar{x} - \bar{y}, 0, s_1^2, s_2^2)$ also belongs to Z, so that Z is determined by the variables $\bar{x} - \bar{y}$, s_1^2, and s_2^2. From (III) we conclude, by setting $k = -1$, that Z is symmetric in the sense that it is determined only by the values of $|\bar{x} - \bar{y}|$, s_1, and s_2.

Let us set $k = 1/s_2$. We see that, to determine Z, all we need are the values of $|\bar{x} - \bar{y}|/s_2$ and s_1/s_2, so that the zone Z can be represented in the half-plane: $\xi = |\bar{x} - \bar{y}|/s_2$, $\eta = s_1/s_2$, $\eta \geq 0$, and so that it is symmetric about the η-axis. We can consider only the half of Z lying in the quadrant Ω_1: $\xi \geq 0$, $\eta \geq 0$.

Let us show now that every straight line $\eta = \eta_0$ $(\eta_0 > 0)$ intersects the boundary of Z in the quadrant Ω_1 at no more than one point. Suppose that this is not the case, i.e. that there are two points (ξ_1, η_0) and (ξ_2, η_0) belonging to Z such that $\xi_2 > \xi_1 \geq 0$. Then there exist two points (ξ_1, η_0) and (ξ_3, η_0) such that $\xi_1 < \xi_3 < \xi_2$ and $(\xi_3, \eta_0) \notin Z$. Suppose that corresponding to the point (ξ_1, η_0) are the sufficient statistics $\bar{x}, \bar{y}, s_1, s_2$ and $\bar{x}', \bar{y}', s'_1$, and s'_2. Here $s_1/s_2 = s'_1/s'_2$ and $(\bar{x}' - \bar{y}')/s'_2 > (\bar{x} - \bar{y})/s_2$. Let us set $k = s'_1 s_1^{-1}$ and take a point to which the sufficient statistics $k\bar{x}, k\bar{y}, ks_1$ and ks_2 correspond. This point also belongs to Z. Furthermore $ks_2 = s'_2$ and $(\bar{x}' - \bar{y}')/s'_2 > (k\bar{x} - k\bar{y})/ks_2$. According the axiom (IV) the point (ξ_3, η_0) must also belong to Z, which is impossible. Thus the critical zone Z must be of the form

$$\frac{|\bar{x}-\bar{y}|}{s_2} \geqslant \varphi\left(\frac{s_1}{s_2}\right),$$

where ϕ is a measurable single-valued function, which however may assume infinite values. If we exclude this possibility and suppose that the function ϕ in (8.3.1) assumes only finite values, we obtain a test, which we may call the *Fisher-Welch-Wald* test in honor of the authors who first investigated it.

Sometimes this test is written in the form

$$\frac{|\bar{x}-\bar{y}|}{\sqrt{s_1^2+s_2^2}} \geqslant \varphi_1\left(\frac{s_1}{s_2}\right),$$

which we obtain from the preceding form by setting $\phi_1 = \phi \cdot (1 + s_1^2/s_2^2)^{1/2}$. A fairly detailed bibliography regarding this test and others like it that were discovered up to 1955 can be found in the survey by Brény [6].

In his article [9] Wald considered approximate similar tests of the form (8.3.1), where ϕ is a rational function, and he raised the question of constructing precisely a similar test for which the function ϕ is analytic for $\eta = s_1/s_2 \geq 0$. We shall see later that such a construction is impossible—that a similar test for which the function ϕ is analytic does not exist. We shall also prove a stronger assertion of this type.

Let us return to the coordinates ξ and η. In the quadrant Ω_1: $\xi \geq 0$, $\eta \geq 0$, the test (8.3.1) becomes

$$\xi \geqslant \varphi(\eta), \tag{8.3.2}$$

so that the critical zone Z is a symmetric doubly-connected region $\xi \geq \phi(\eta)$ or $\xi \leq -\phi(\eta)$. Following articles[1] by the author [45] and I. L. Romanovskiĭ [63],

1) These two articles contain an omission, which is corrected in [49].

we now prove

Theorem 8.3.1. *Suppose that* $\phi(\eta)$ *is continuous for* $0 \leq \eta \leq 1$ *and that it has a finite derivative for* $0 \leq \eta < 1$, *where* $\phi'(0)$ *means the right-hand derivative. Suppose that* $\phi(\eta)$ *satisfies a Lipschitz condition with exponent* 1 *for* $1 \leq \eta \leq 2(M+1)$, *where* $M = \sup_{0 \leq \eta \leq 1} \phi(\eta)$. *Then the test* (8.3.2) *cannot be similar if the size of the test* $n_2 \geq 4$.

The proof is based on the method of analytic continuation with respect to a parameter, as described in § 1. Several lemmas are necessary for it. In what follows, the symbols l_i, c_i, and K_i denote positive constants and the symbols ϵ_i, η_i, and ξ_i denote small positive constants.

Lemma 8.3.1. *Suppose that* $\xi_0 = \phi(0)$. *Then*

$$c_2 \alpha \leqslant \xi_0 \leqslant c_1 \alpha. \tag{8.3.3}$$

To prove this we use formula (8.2.3). By this fomula we have

$$I = \int_0^\infty d\eta \int_0^{\varphi(\eta)} \frac{\eta^{n_1-2}\, d\xi\, d\eta}{(\theta\xi^2 + (1+\theta)\eta^2 + \theta + \theta^2)^N}$$
$$= C(n_1, n_2)\, \alpha\theta^{-\frac{n_2}{2}} (1+\theta)^{-N+\frac{1}{2}}. \tag{8.3.4}$$

By the conditions of the theorem $\phi(\eta)$ is continuous in a neighborhood of 0. Hence for given small $\epsilon > 0$ there exists a $\delta > 0$ such that the double inequality $0 \leq \eta \leq \delta$ implies the inequality $|\phi(\eta) - \xi_0| \leq \epsilon$. Let us suppose that ϵ and δ are both less than 1.

Consider the following regions in the quadrant Ω_1:

$$Z_\delta\colon\ 0 \leqslant \xi \leqslant \varphi(\eta), \qquad 0 \leqslant \eta \leqslant \delta;$$
$$Z'_\delta\colon\ 0 \leqslant \xi < \infty, \qquad \delta \leqslant \eta < \infty.$$

Let us find an estimate for

$$\iint_{Z'_\delta} \frac{\eta^{n_1-2}\, d\xi\, d\eta}{(\theta\xi^2 + (1+\theta)\eta^2 + \theta + \theta^2)^N} = I'_\delta. \tag{8.3.5}$$

We define $\rho(\eta, \theta) = (1+\theta)\eta^2 + \theta + \theta^2$. We find that

$$\int_0^\infty \frac{d\xi}{(\theta\xi^2 + \rho(\eta, \theta))^N} = C_N \theta^{-\frac{1}{2}} (\rho(\eta, \theta))^{-N+\frac{1}{2}},$$

where $C_N > 0$ is a constant. Furthermore, since $n_2 \geq 2$,

$$\int_{\delta}^{\infty} \frac{\eta^{n_1-2}\,d\eta}{(\rho(\eta,\theta))^{N-\frac{1}{2}}} = (1+\theta)^{-N+\frac{1}{2}} \int_{\delta}^{\infty} \frac{\eta^{n_1-2}\,d\eta}{(\eta^2+\theta)^{N-\frac{1}{2}}}$$
$$= (1+\theta)^{-N+\frac{1}{2}}\theta^{-\frac{n_2}{2}} \int_{\frac{\delta}{\sqrt{\theta}}}^{\infty} \frac{du}{(u^2+1)^{\frac{n_2}{2}}}. \qquad (8.3.6)$$

If θ is chosen in such a way that $\delta/\sqrt{\theta} = K$, where K is sufficiently large constant, then (8.3.6) has the estimate

$$(1+\theta)^{-N+\frac{1}{2}}\theta^{-\frac{n_2}{2}}\varepsilon_K O(1),$$

where $\epsilon_K \to 0$ as $K \to \infty$. Thus

$$I'_\delta = (1+\theta)^{-N+\frac{1}{2}}\theta^{-\frac{n_2}{2}}\varepsilon_K O(1) \quad \text{for} \quad \theta = \frac{\delta^2}{K^2}. \qquad (8.3.7)$$

Now let us look at the integral $\int\int_{Z_\delta} = I_\delta = I - I'_\delta$. Obviously

$$I_\delta \leqslant (\xi_0+\varepsilon)\,\theta^{-N}\delta^{n_1-1} = (\xi_0+\varepsilon)\,\theta^{-\frac{n_2}{2}}K^{n_1-1}$$

For sufficiently large K we obtain from (8.3.7) and (8.3.4)

$$K^{n_1-1}\theta^{-\frac{n_2}{2}}(\xi_0+\varepsilon) \geqslant \frac{1}{2}C(n_1, n_2)\,\alpha\theta^{-\frac{n_2}{2}},$$

and hence

$$\xi_0 \geqslant c_3\alpha - \varepsilon.$$

If we take $\epsilon \leq c_3\alpha/2$, we get

$$\xi_0 \geqslant c_2\alpha. \qquad (8.3.8)$$

To get an upper bound for ξ_0 let us find a lower bound for I_δ. We have

$$I_\delta \geqslant \frac{1}{K_1(K)}(\xi_0 - \varepsilon)\,\theta^{-\frac{n_2}{2}},$$

where $K_1(K)$ depends only on K. For sufficiently large K,

$$2C(n_1, n_2)\,\alpha\theta^{-\frac{n_2}{2}} \geqslant \frac{1}{K_1(K)}(\xi_0 - \varepsilon)\,\theta^{-\frac{n_2}{2}},$$

so that

$$(\xi_0 - \varepsilon) \leqslant K_2\alpha, \quad \xi_0 \leqslant K_2\alpha + \varepsilon < (K_2+1)\,\alpha,$$

if ϵ is chosen less than α. This completes the proof of Lemma 8.3.1.

Let us now look at the operation of "upsetting" tests. Suppose that we are given the test $|\xi| \geq \phi(\eta)$. We define

$$\varphi_1(\eta) = \frac{\varphi(\eta)}{\sqrt{1+\eta^2}},$$

so that

$$\varphi(\eta) = \varphi_1(\eta)\sqrt{1+\eta^2}.$$

Then the test takes the form

$$|\xi| \geqslant \varphi_1(\eta)\sqrt{1+\eta^2}$$

or

$$\frac{|\bar{x}-\bar{y}|}{s_2} \geqslant \varphi_1\left(\frac{s_1}{s_2}\right)\sqrt{1+\frac{s_1^2}{s_2^2}},$$

which may be written in the form

$$\frac{|\bar{x}-\bar{y}|}{\sqrt{s_1^2+s_2^2}} \geqslant \varphi_1\left(\frac{s_1}{s_2}\right),$$

or in the form

$$\frac{|\bar{y}-\bar{x}|}{\sqrt{s_2^2+s_1^2}} \geqslant \varphi_1\left(\left(\frac{s_2}{s_1}\right)^{-1}\right)$$

or

$$\frac{|\bar{y}-\bar{x}|}{s_1} \geqslant \varphi_1\left(\left(\frac{s_2}{s_1}\right)^{-1}\right)\sqrt{1+\left(\frac{s_2}{s_1}\right)^2}.$$

This last expression yields the critical zone of the test in which the role of the sample $(x_1, \cdots, x_{n_1})$ is played by $(y_1, \cdots, y_{n_2})$ and vice versa. Thus, corresponding to the test $|\xi| \geq \phi_1(\eta)\sqrt{1+\eta^2}$ is the test $|\xi| \geq \phi_1(1/\eta)\sqrt{1+\eta^2}$ of the same level. Here we need to switch n_1 and n_2 in formula (8.3.4).

Consider the projective transformation of the plane $-\infty < \xi < \infty$, $-\infty < \eta < \infty$ defined by $\xi' = \xi/\eta$, $\eta' = 1/\eta$. This is an involution of the plane with fixed axis $\eta = 1$ and fixed center $(0, -1)$.

Each quadrant of the upper half-plane is mapped by this transformation into itself. In the upper right quadrant the test boundary $\xi = \phi_1(\eta)\sqrt{1+\eta^2}$ is mapped into $\xi = \phi_1(1/\eta)\sqrt{1+\eta^2}$. From what was said above, this yields a test boundary of a similar test of the same level. In formula (8.3.4) we need to replace n_1 with n_2, n_2 with n_1, and $\phi(\eta)$ with $\phi_1(1/\eta)\sqrt{1+\eta^2}$.

Let D denote a number in the interval $(0, 1)$ and let the expression $A(\xi, \eta) = D$ define a nongenerate critic-hyperbola. The equation for this hyperbola is

$$\frac{\eta^2}{D} - \frac{\xi^2}{1-D} = 1.$$

The involution described above maps the hyperbola into the ellipse

$$\frac{\xi^2}{D^{-1}-1}+\frac{\eta^2}{D^{-1}}=1.$$

Here $D^{-1}>1$ and this ellipse is the critic $B(\xi, \eta)=D^{-1}$. This means that every nondegenerate critic-hyperbola is mapped into a critic-ellipse. Since out transformation is an involution, the converse is also true. Furthermore, if the test boundary $\xi=\phi(\eta)$ is tangent at any point to a nondegenerate critic-hyperbola (resp. nondegenerate critic-ellipse), then the "upset" similar test with test boundary

$$\xi=\sqrt{1+\eta^2}\,\varphi_1\left(\frac{1}{\eta}\right) \quad \left(\text{where } \varphi_1(\eta)=\frac{\varphi(\eta)}{\sqrt{1+\eta^2}}\right)$$

is tangent to a nondegenerate critic-ellipse (resp. nondegenerate critic-hyperbola) and conversely. If a point lies inside a quadrant (i.e. if it has positive abscissa and ordinate) then its image under our involution lies inside a quadrant.

§ 4. LEMMAS ON TANGENCY OF A TEST BOUNDARY TO A CRITIC[1)]

Lemma 8.4.1. *Suppose that a test boundary $\xi=\phi(\eta)$ is tangent to a nondegenerate critic at a point (ξ_0, η_0), so that ϕ has a finite derivative $\phi'(\eta_0)$ at $\eta=\eta_0$ and the tangent line to the boundary coincides with the tangent line to the critic. Suppose that the critic intersects the $(\xi=0)$-axis at a point $(0, \eta'_0)$ and that it satisfies a Lipschitz condition with exponent 1 in a neighborhood of the point $\eta'_0\phi(\eta)$. Then the test $|\xi|\geq\phi(\eta)$ cannot be a similar test.*

The proof of this lemma is rather complicated. We introduce the family of critics $A(\xi, \eta)$ and $B(\xi, \eta)$ and we write the basic expression (8.2.3) in the form

$$\iint\limits_{Z_1}\frac{\eta^{n_1-2}\,d\xi\,d\eta}{(\theta^2+\theta(1+\xi^2+\eta^2)+\eta^2)^N}=C_{n_1n_2}\frac{\alpha}{2}\,\theta^{-\frac{n_2}{2}}(1+\theta)^{-N+\frac{1}{2}}. \quad (8.4.1)$$

The basic equation (8.4.1) must be satisfied for all $\theta>0$.

Let us construct the anayltic continuation of the left and right sides of equation (8.4.1) onto the plane Λ of complex values of the parameter $\theta=\tau+i\zeta$. On the right side of (8.4.1) the points $\theta=0$ and $\theta=-1$ are singular points. Since the right-hand member is a function of θ, it can be extended to the plane Λ with a cut along the negative half of the real axis from 0 to $-\infty$. The integral on the left side of (8.4.1) converges absolutely and unifomly in every closed bounded region of the plane Λ with a cut along the negative path of the real axis. (All the zeros of the denominator in the integrand are real and negative.) The left-hand

1) This section was written by the author in colloboration with I. L. Romanovskiĭ.

member of (8.4.1) is a single-valued analytic function of θ in the region Λ and it too can be extended to the region of complex values of the parameter.

Since the left and right sides of (8.4.1) must coincide for positive values of θ, they must coincide for the entire region Λ. Therefore a test that is similar for positive values of θ is also similar for all values of $\theta \in \Lambda$.

By choosing the appropriate branches of the factors and using the concept of critics that we have introduced, one can represent the denominator in the left-hand member of (8.4.1) as the product

$$(\theta^2+\theta(1+\xi^2+\eta^2)+\eta^2)^N=(\theta+B(\xi,\ \eta))^N(\theta+A(\xi,\ \eta))^N.$$

Thus the fundamental equation takes the form

$$\int\int_{Z_1}\frac{\eta^{n_1-2}d\xi\, d\eta}{(\theta+A(\xi,\ \eta))^N(\theta+B(\xi,\ \eta))^N}$$

$$=C_{n_1n_2}\frac{\alpha}{2}\,\theta^{-\frac{n_2}{2}}(1+\theta)^{-N+\frac{1}{2}}. \qquad (8.4.2)$$

By considering the values of θ of the form $\theta=-D_0+i\zeta$, where ζ is a small real number, we shall obtain a contradiction, thus proving Lemma 8.4.1.

We may assume that the test boundary is tangent to the critic-hyperbola $A(\xi,\ \eta)=D_0$, $0<D_0<1$, because if it is tangent to the critic-ellipse $B(\xi,\ \eta)=D_1$, $D_1>1$, it will be tangent ot the critic-hyperbola for the "upset" test and the analytical properties of the lemma are maintained.

Let us divide the region of integration Z_1 into several parts. First of all, we denote by $\Pi_{\epsilon 0}$ the infinite half-horseshoe-shaped region containing the critic $A(\xi,\ \eta)=D_0$ bounded by the ordinate and the two confocal hyperbolas

$$A(\xi,\ \eta)=D_0+\varepsilon_0 \text{ and } A(\xi,\ \eta)=D_0-\varepsilon_0$$

(where ϵ_0 is a sufficiently small positive number such that $D_0+\epsilon_0<1$ and $D_0-\epsilon_0>0$). We then denote by $\Pi^{(0)}_{\epsilon 0}$ the finite subregion of this region bounded by the ordinate and the line $\eta=\eta_1$, where η_1 is such that the distance from the ordinate to the point of tangency $(\xi_0,\ \eta_0)$ is, for, example, several times as great as the distance along this line from the ordinate to the half-horseshoe. Thus the region $\Pi_{\epsilon 0}\backslash\Pi^{(0)}_{\epsilon 0}$ contains the point of tangency to the critic and the test boundary. It is for just this region that we shall obtain a contradiction in the behavior of the left and right sides of the basic equation as $\zeta\to 0$. The integral in the left-hand member of the basic equation (8.4.2) taken over this region approaches $-\infty$, whereas the right-hand member remains bounded.

In what follows we shall take $\theta = -D_0 + i\zeta$, where $\zeta < \zeta_1 \ll \epsilon_0$. We denote by $\widetilde{\theta}$ that region of values of θ that is defined by the conditions listed.

In the region $\Omega_1 \backslash \Pi_{\epsilon 0}$ for $\psi(\xi, \eta) \neq 0$,

$$|\theta + A(\xi, \eta)| \geqslant \varepsilon_0$$

and

$$|\theta + B(\xi, \eta)| \geqslant \epsilon_1 > 0.$$

Therefore the denominator of the integrand in (8.4.2) differs from 0 by some constant and its behavior plays no significant role.

In the region $\Pi_{\epsilon 0}$, for $\theta \in \widetilde{\theta}$,

$$|\theta + A(\xi, \eta)| \leqslant |-D_0 + i\zeta + D_0 + \varepsilon_0| \leqslant \sqrt{\varepsilon_0 + \zeta^2} < 2\varepsilon_0 \quad (8.4.3)$$

and

$$|B(\xi, \eta) - A(\xi, \eta)| \geqslant 1 - A(\xi, \eta) \geqslant \varepsilon_2 > 0. \quad (8.4.4)$$

Thus what is important is the behavior of the basic equation in the region $\Pi_{\epsilon 0}$. Let us denote by $I(\theta)$ the integral in the left-hand member of the basic equation and let us denote the same integral over the subregion $\Pi_{\epsilon 0}$ by $I_0(\theta)$. Let us study the integral $I_0(\theta)$ in greater detail. The quantity $(\theta + B)^{-N}$ can be rewritten

$$(B - A)^{-N}\left(1 + \frac{\theta + A}{B - A}\right)^{-N}. \quad (8.4.5)$$

Let us choose ϵ_0 in such a way that the series that we shall be considering converge; for example let us take $\epsilon_0 < \epsilon_2/10$. Since inequality (8.44) holds and since for $\theta > 0$ we are considering that branch of the function $(\theta + B)^{-N}$ with positive values, we can expand (8.4.5) according to the binomial formula.

Let us set

$$M = \begin{cases} N + \frac{1}{2} & \text{if } n_1 \text{ and } n_2 \text{ are of like parity,} \\ N & \text{if } n_1 \text{ and } n_2 \text{ are of opposite parity.} \end{cases}$$

Let $\psi(\xi, \eta)$ denote the indicator function of the zone z_1. Then

$$I_0(\theta) = \sum_{k=0}^{M} E_k \iint_{\Pi_{\varepsilon_0}} \psi(\xi, \eta) \frac{\eta^{n_1 - 2}\, d\xi\, d\eta}{(B - A)^{N+K}(\theta + A)^{N-K}}$$

$$+ \sum_{k=M+1}^{\infty} E_k \iint_{\Pi_{\varepsilon_0}} \psi(\xi, \eta) \frac{\eta^{n_1 - 2}\, d\xi\, d\eta}{(B - A)^{N+k}(\theta + A)^{N-k}}$$

$$= \sum_{k=0}^{M} E_k \iint_{\Pi_{\varepsilon_0}} \psi(\xi, \eta) \frac{\eta^{n_1 - 2}\, d\xi\, d\eta}{(B - A)^{N+k}(\theta + A)^{N-k}} +$$

$$+\iint_{\Pi_{\varepsilon_0}} \psi(\xi, \eta)\,\eta^{n_1-2} E_{M+1}\left(\frac{\theta+A}{B-A}\right) d\xi\, d\eta,$$

where the E_k are the coefficients in the expansion, $E_0 = 1$, the function $E_{M+1}((\theta + A)/(B - A))$ is a regular function of ξ and η in the region $\Pi_{\epsilon 0}$ and the last term is a bounded analytic function. As $B \to \infty$ in $\Pi_{\epsilon 0}$, this function has by (8.4.5) the estimate $O(B^{-N})$, and all the integrals converge absolutely in $\Pi_{\epsilon 0}$.

This expansion shows that quantities of different orders are "mixed" in the integral $I_0(\theta)$, where in the region $\Pi_{\epsilon 0}$ $|\theta + A| < 2\epsilon_0$ for $\theta \in \widetilde{\theta}$ the maximum order will be provided by the integral with maximum degree $(\theta + A)$; that is, the fundamental term is the integral

$$\iint_{\Pi_{\varepsilon_0}} \psi(\xi, \eta) \frac{\eta^{n_1-2}\, d\xi\, d\eta}{(B-A)^N (\theta+A)^N}. \tag{8.4.6}$$

To single out this fundamental quantity, we apply the following integro-differential operator to the integral (8.4.6).

Let $L_0^{(-k)}$ denote the operator for k-fold integration along the line segment connecting the points θ and θ_1 (where $\theta \in \widetilde{\theta}$ and $\theta_1 = 1 + i$). This path lies entirely in the region Λ. Let us apply the operator $L_\theta^{(-M_1)}$ to the integral $I(\theta)$, where

$$M_1 = \begin{cases} N - \tfrac{1}{2} \text{ if } n_1 \text{ and } n_2 \text{ are of like parity,} \\ N - 2 \text{ if } n_1 \text{ and } n_2 \text{ are of opposite parity.} \end{cases}$$

If n_1 and n_2 are of opposite parity, application of the operator $L_\theta^{(-M_1)}$ yields

$$L_\theta^{(-M_1)}(I_0(\theta))$$

$$= \frac{1}{(-N+1)(-N+2)\dots(-N+M_1)} \iint_{\Pi_{\varepsilon_0}} \psi(\xi, \eta) \frac{\eta^{n_1-2}\, d\xi\, d\eta}{(B-A)^N (\theta+A)^2} +$$

$$+ \frac{1}{(-N+2)\dots(-N+1+M_1)} \iint_{\Pi_{\varepsilon_0}} \psi(\xi, \eta) \frac{\eta^{n_1-2}\, d\xi\, d\eta}{(B-A)^{N+1} (\theta+A)} +$$

$$+ \frac{1}{(-N+3)\dots(-N+2+M_1)} \iint_{\Pi_{\varepsilon_0}} \psi(\xi, \eta) \frac{\eta^{n_2-2}\, d\xi\, d\eta}{(B-A)^{N+2}} + u_1(\theta),$$

where $u_1(\theta)$ is a function that is analytic and bounded and has analytic bounded first and second derivatives for $\theta \in \widetilde{\theta}$. If we apply this same operator to the integral $I_1(\theta)$, we again obtain an analytic bounded function with analytic bounded first and second derivatives. Application of the operator $L_\theta^{(-M_1)}$ in the case of samples of a single parity yields

$$L_\theta^{(-M_1)}(I(\theta)) = C_1' \iint_{\Pi_{\varepsilon_0}} \psi(\xi, \eta) \frac{\eta^{n_1-2}\, d\xi\, d\eta}{(B-A)^N (\theta+A)^{1/2}}$$

$$+ C_2' \iint_{\Pi_{\varepsilon_0}} \psi(\xi, \eta) \frac{\eta^{n_1-2} (\theta+A)^{1/2}\, d\xi\, d\eta}{(B-A)^{N+1}}$$

$$+ C_3' \iint_{\Pi_{\varepsilon_0}} \psi(\xi, \eta) \frac{\eta^{n_1-2} (\theta+A)^{3/2}\, d\xi\, d\eta}{(B-A)^{N+2}} + v'(\theta). \tag{8.4.7}$$

All these integrals converge absolutely. The function $v'(\theta)$ is also an analytic bounded function with analytic bounded first and second derivatives.

In the case of samples of like parity, to avoid fractional exponents we can apply a different operator to (8.4.7). Let us multiply this equation by $(\theta - \theta')^{-1/2} d\theta$, where θ varies over the line segment from $\theta = \theta' \in \tilde{\theta}$ to $\theta = \theta_1 = 1 + i$, and then let us integrate the result over that line segment: we obtain

$$\int_{\theta'}^{\theta_1} \frac{d\theta}{(\theta+A)^{1/2} (\theta-\theta')^{1/2}}$$

$$= \ln\left[2(\theta_1 + A)^{1/2} (\theta_1 - \theta')^{1/2} + 2\theta_1 + (A - \theta')\right] - \ln(A + \theta'),$$

$$\int_{\theta'}^{\theta_1} \frac{(\theta+A)^{1/2}\, d\theta}{(\theta-\theta')^{1/2}} = (\theta_1 - \theta')^{1/2} (\theta_1 - A)^{1/2}$$

$$+ (A+\theta') \ln\left[(\theta_1 - \theta')^{1/2} + \theta_1^{1/2} - A\right] - (A+\theta') \ln (A+\theta')^{1/2},$$

$$\int_{\theta'}^{\theta_1} \frac{(\theta+A)^{3/2}}{(\theta-\theta')^{1/2}}\, d\theta = \left[\frac{\theta_1 + A}{8} + \frac{3}{16}(\theta' + A)\right] \cdot 2 (\theta_1 - \theta')^{1/2}$$

$$\times (\theta_1 + A)^{1/2} + \frac{8}{3}(A+\theta')^2 \ln\left[2(\theta_1 + A)^{1/2} + 2(\theta_1 - \theta')^{1/2}\right]$$

$$- \frac{3}{8}(A+\theta') \ln\left[2(\theta' + A)^{1/2}\right].$$

Now let us twice differentiate the results obtained with respect to θ' for $\theta' \in \tilde{\theta}$. We denote by $\tilde{L}_\theta$ the operator producing the transformation that we have just performed. Then we can easily show that

$$\tilde{L}_\theta L_\theta^{(-M_1)}(I(\theta)) = C_1'' \iint_{\Pi_{\varepsilon_0}} \psi(\xi, \eta) \frac{\eta^{n_1-2}\, d\xi\, d\eta}{(B-A)^N (\theta+A)^2} +$$

$$+C_2'' \iint_{\Pi_{\varepsilon_0}} \psi(\xi, \eta) \frac{\eta^{n_1-2}\, d\xi\, d\eta}{(B-A)^{N+1}(\theta+A)}$$

$$+C_3'' \iint_{\Pi_{\varepsilon_0}} \psi(\xi, \eta) \frac{\eta^{n_1-2} \ln(\theta+A)\, d\xi\, d\eta}{(B-A)^{N+2}} + v(\theta), \qquad (8.4.8)$$

where c''_1, c''_2, and c''_3 are real constants, $c'_1 \neq 0$, and the function $v(\theta)$ is analytic and bounded for $\theta \in \tilde{\theta}$. We choose that branch of $\ln(\theta + A)$ that has real values for $\theta > 0$. Again, all the integrals converge absolutely.

The expressions (8.4.7) and (8.4.8) differ only in the third term. Therefore we may write (combining the two cases)

$$L_\theta(I(\theta)) = C_1 \iint_{\Pi_{\varepsilon_0}} \psi(\xi, \eta) \frac{\eta^{n_1-2}\, d\xi\, d\eta}{(B-A)^N (\theta+A)^2}$$

$$+C_2 \iint_{\Pi_{\varepsilon_0}} \psi(\xi, \eta) \frac{\eta^{n_1-2}\, d\xi\, d\eta}{(B-A)^{N+1}(\theta+A)}$$

$$+C_3 \iint_{\Pi_{\varepsilon_0}} \psi(\xi, \eta) \frac{\eta^{n_1-2} C(\theta)\, d\xi\, d\eta}{(B-A)^{N+2}} + u(\theta), \qquad (8.4.9)$$

where L_θ is either $L_\theta^{(-M\,1)}$ or $\tilde{L}_\theta(L_\theta^{(-M\,1)})$ and

$$C(\theta) = \begin{cases} 1 \text{ if } n_1 \text{ and } n_2 \text{ are of like parity,} \\ \ln(A+\theta) \text{ if } n_1 \text{ and } n_2 \text{ are of opposite parity.} \end{cases}$$

For the coefficients and the last terms we have used the conventional notation. Let us now show that the imaginary part of the first integral in (8.4.9) approaches ∞ as $\zeta \to 0$. We denote by $\Pi_{\epsilon 0}^{(1)} = \Pi_{\epsilon 0} \setminus \Pi_{\epsilon 0}^{0}$ that portion of the half-horseshoe bounded below by the horizontal straight line $\eta = \eta_1$. This region contains the point of tangency of the critic and the test boundary. Consider the integral over this region

$$\iint_{\Pi_{\varepsilon_0}^{(1)}} \psi(\xi, \eta) \frac{\eta^{n_1-2}\, d\xi\, d\eta}{(\theta+A)^2 (B-A)^N}, \qquad (8.4.10)$$

where $\theta \in \tilde{\theta}$; that is, $\theta = -D_0 + i\zeta$. In the region $\Pi_{\epsilon 0}^{(1)}$ let us replace the variables ξ and η with a and η just as in [45]. We set $a = A - D_0$. Here the Jacobian $\partial(a, \eta)/\partial(\xi, \eta) < 0$ in the region $\Pi_{\epsilon 0}^{(1)}$. This Jacobian is an analytic function which, when multiplied by $(B-A)^{-N}$, decreases at least as fast as $1/\eta^2$ in the region $\Pi_{\epsilon 0}^{(1)}$ as $\eta \to \infty$.

Using the new variables, we rewrite the integral (8.4.10) in the form

$$-\iint\limits_{\Pi_{\varepsilon_0}^{(1)}} \psi(a,\ \eta)\frac{F(a,\ \eta)}{(a+i\zeta)^2}\,da\,d\eta, \tag{8.4.11}$$

where

$$F(a,\ \eta)>0.$$

Since

$$\frac{\eta^2}{D_0+a}-\frac{\xi^2}{(1-(D_0+a)}=1,$$

it follows that a is a single-valued analytic function of ξ and η for given D_0. Here $a(\xi,\ \eta)$ is a strictly decreasing function of ξ for fixed η and a strictly increasing function of η for fixed ξ. For the new variables, instead of the characteristic function $\psi(\xi,\ \eta)$ we write the corresponding limits of integration. For given η we define

$$s_0(\eta)=\min(\varepsilon_0,\ a(\varphi(\eta),\ \eta))$$

and

$$F(a,\ \eta)=F_0(0,\ \eta)+aF_1(a,\ \eta).$$

Then the integral in (8.4.11) is equal to

$$-\int\limits_{\eta_1}^{\infty}d\eta\int\limits_{-\varepsilon_0}^{s_0(\eta)}\frac{F_0(0,\ \eta)+aF_1(a,\ \eta)}{(a+i\zeta)^2}\,da.$$

Here $F_0(0,\ \eta)=O(1/\eta^2)$ and $aF_1(a,\ \eta)=O(1/\eta^2)$ as $\eta\to\infty$.

We estimate the imaginary part of this integral:

$$\begin{aligned}-\operatorname{Im}\int\limits_{\eta_1}^{\infty}d\eta\int\limits_{-\varepsilon_0}^{s_0(\eta)}\frac{F_0(0,\ \eta)+aF_1(a,\ \eta)}{(a+i\zeta)^2}\,da &=-\int\limits_{\eta_1}^{\infty}d\eta\int\limits_{-\varepsilon_0}^{s_0(\eta)}-\frac{2a^2\zeta F_1(a,\ \eta)}{(a^2+\zeta^2)^2}\,da\\ &\quad+\int\limits_{\eta_1}^{\infty}d\eta\int\limits_{-\varepsilon_0}^{s_0(\eta)}\frac{2a\zeta F_0(0,\ \eta)\,da}{(a^2+\zeta^2)^2}.\end{aligned}$$

As $\zeta\to 0$ the integral

$$\int\limits_{\eta_1}^{\infty}d\eta\int\limits_{-\varepsilon_0}^{s_0(\eta)}\frac{2a^2\zeta F_1(a,\ \eta)\,da}{(a^2+\zeta^2)^2}$$

behaves like $O(1)$. The imaginary part of the second integral is equal to

$$-\int_{\eta_1}^{\infty} \frac{2\zeta F_0(0, \eta)}{(s_0(\eta))^2+\zeta^2} d\eta + H, \qquad (8.4.12)$$

where H is a bounded number. Here we should keep in mind the fact that

$$F_0(0, \eta) \geqslant \varepsilon_3 > 0.$$

for all η in a neighborhood of η_0.

The denominator of the integrand approaches 0 as $\zeta \to 0$ when $s_0(\eta) = 0$, as is the case at common points of the critic $A(\xi, \eta) = D_0$ and the boundary of the test. These common points may be points of either intersection or tangency. In either case the half-horseshoe $\Pi_{\epsilon 0}^{(1)}$ has at least one point of tangency, namely the point (ξ_0, η_0).

Consider the region $\Pi_{\epsilon 0}^{(1)}$ containing the point of tangency. As $\eta \to \eta_0$,

$$s_0(\eta) = o|\eta - \eta_0|.$$

Let K denote a large given number. For sufficiently small ζ and for $|\eta - \eta_0| \leq K\zeta$ we have the inequality $|s_0(\eta)| \leq \zeta$, thanks to the existence (assumed in the lemma) of a finite derivative of $\phi(\eta)$, and we can estimate the integral in (8.4.12) as follows:

$$2\zeta \int_{\eta_1}^{\infty} \frac{F_0(0, \eta)\, d\eta}{(s_0(\eta))^2+\zeta^2} > 2\zeta \int_{-K\zeta}^{+K\zeta} \frac{\varepsilon}{2\zeta^2} d\eta = 2\varepsilon_3 K,$$

where $\epsilon_3 > 0$. For sufficiently large K, we see that the integral (8.4.12) can be made arbitrarily large. Thus

$$\operatorname{Im} \int_{\eta_1'}^{\eta'} d\eta \int_{-\varepsilon_0}^{s_0(\eta)} \frac{F(a, \eta)}{(a+i\zeta)^2} da \underset{\zeta \to 0}{\to} -\infty.$$

Now let us look at the other two integrals in (8.4.9). We can get bounds for their imaginary parts as $\zeta \to 0$ just as we did for the first integral, this time replacing $(a+i\zeta)^{-2}$ with $(a+i\zeta)^{-1}$ and $\ln(a+i\zeta)$ or 1 depending on whether the difference $n_1 - n_2$ is odd or oven. Obviously, $\operatorname{Im} 1/(a+i\zeta) = -\zeta/(a^2+\zeta^2)$ and $\operatorname{Im} \ln(a+i\zeta) = o(1)$. As $\zeta \to 0$, we see that the imaginary parts of the second and third integrals in (8.4.9) are bounded. The function $u(\theta)$ and its imaginary part are also bounded for $\theta \in \tilde{\theta}$ as $\zeta \to 0$. We now need to investigate the integral over the region $\Pi_{\epsilon 0}^{(0)}$.

Let us consider that portion of the half-horseshoe $\Pi_{\epsilon 0}^{(0)}$ adjacent to the ordinate and bounded by the line $\xi = \epsilon_4$. The behavior of the integral in the left-hand member of (8.4.2) over the region $\Pi_{\epsilon 0}^{(0)}$ depends on the behavior of the test boundary $\xi = \phi(\eta)$ at the point at which the critic $A(\xi, \eta) = D_0$ leaves the ordinate, i.e. at the point

$(0, \sqrt{D_0})$. Two cases are possible.

Case 1: $\phi(\sqrt{D_0}) \neq 0$. We denote by $\widehat{\Pi}^{(0)}_{\epsilon 0}$ the intersection of $\Pi^{(0)}_{\epsilon 0}$ and the strip $0 \le \xi \le \epsilon_4$. Here the indicator function $\psi(\xi, \eta) = 0$ and the integration may be taken over the region $\Pi^{(0)}_{\epsilon 0}$. In this case $\partial(a, \eta)/\partial(\xi, \eta)$ neither vanishes nor becomes infinite and it is an analytic function. The integration is carried out just as in the region $\Pi^{(1)}_{\epsilon 0}$, but η varies from $-\epsilon_0$ to

$$s_0(\eta) = \min(\varepsilon_0,\ a(\varphi(\eta), \eta),\ a(\varepsilon_4, \eta)).$$

Case 2: $\phi(\sqrt{D_0}) = 0$. Then by the hypothesis of the lemma the function $\phi(\eta)$ satisfies a Lipschitz condition. Let us fix ξ in $(0, \epsilon_4)$ and consider the set of roots of the equation $\xi = \phi(\eta)$. Because of the continuity of the function $\phi(\eta)$, the set of these roots is closed. Among them there are two roots $s_+(\eta)$ and $s_-(\eta)$ such that $s_+(\eta) > \sqrt{D_0}$ and $s_-(\eta) < \sqrt{D_0}$ which out of all such roots are closest to the number $\sqrt{D_0}$. Since by the hypothesis of the lemma the function $\phi(\eta)$ satisfies a Lipschitz condition, we have

$$|s_+(\eta) - \sqrt{D_0}| \geqslant \varepsilon_5 \xi,$$
$$|s_-(\eta) - \sqrt{D_0}| \geqslant \varepsilon_5 \xi,$$

where ϵ_5 is a positive constant. Integration over the region $\Pi^{(0)}_{\epsilon 0} \setminus \widehat{\Pi}^{(0)}_{\epsilon 0}$ leads to the same results as before. In the region $\widehat{\Pi}^{(0)}_{\epsilon 0}$ the function $\psi(\xi, \eta)$ is nonzero and the Jacobian $\partial(a, \eta)/\partial(\xi, \eta)$ becomes infinite (the tangents at the vertices of the critic-hyperbolas are parallel to the η-axis). Therefore we must use a different approach. Let us replace the variables (ξ, η) with the variables (ξ, a), where $a = A - D_0$. Here the Jacobian $\partial(\xi, a)/\partial(\xi, \eta) > 0$ is an analytic function that neither vanishes nor becomes infinite. Just as in the region $\Pi^{(1)}_{\epsilon 0}$ we consider the integral

$$\iint_{\widehat{\Pi}^{(0)}_{\varepsilon_0}} \psi(\xi, \eta) \frac{\eta^{n_1 - 2}\, d\xi\, d\eta}{(B - A)^N (\theta + A)^2} = \iint_{\widehat{\Pi}^{(0)}_{\varepsilon_0}} \frac{\nu(\xi, a)\, da\, d\eta}{(a + i\zeta)^2},$$

where $\nu(\xi, a)$ is a regular function in the region $\widehat{\Pi}^{(0)}_{\epsilon 0}$ and $\nu(\xi, a) > 0$. For fixed ξ the limits of integration with respect to a depend on the roots of the equation $\xi = \phi(\eta)$. Just as in the region $\Pi^{(1)}_{\epsilon 0}$, we may write

$$\nu(\xi, a) = \nu_0(\xi, 0) + a\nu_1(\xi, a),$$

where $\nu_1(\xi, a)$ is a function that is regular in $\widehat{\Pi}^{(0)}_{\epsilon 0}$. What is significant is the integral

$$\iint\limits_{\Pi^{(\widehat{0})}_{\varepsilon_0}} \frac{\nu_0(\xi, a)\, da\, d\xi}{(a+i\zeta)^2}.$$

The imaginary part of this integral is of the form

$$\int\limits_0^{\varepsilon_4} -2\zeta\nu_0(\xi, 0)\left\{\sum_{m=1}^{T_1}\left[\frac{1}{(a_{1m}(\xi))^2+\zeta^2}-\frac{1}{(a_{2m}(\xi))^2+\zeta^2}\right]\right.$$
$$\left.-\sum_{m=1}^{T_2}\left[\frac{1}{(\bar{a}_{1m}(\xi))^2+\zeta^2}-\frac{1}{(\bar{a}_{2m}(\xi))^2+\zeta^2}\right]\right\}d\xi, \qquad (8.4.13)$$

where $a_{1m}(\xi)$ and $a_{2m}(\xi)$ correspond to those roots of the equation $\phi(\eta) = \xi$ that exceed $\sqrt{D_0}$, and $\bar{a}_{1m}(\xi)$ and $\bar{a}_{2m}(\xi)$ correspond to those roots that are less than $\sqrt{D_0}$. The upper limits of summation T_1 and T_2 may be infinite. Also, $a_{1m}(\xi) < a_{2m}(\xi)$ and $\bar{a}_{1m}(\xi) < \bar{a}_{2m}(\xi)$. Since the function $\phi(\eta)$ satisfies a Lipschitz condition in the interval in question, we have

$$|a_{1m}(\xi)| \geqslant \varepsilon_5\xi,$$
$$|\bar{a}_{1m}(\xi)| \geqslant \varepsilon_5\xi.$$

Each of the summations in (8.4.13) consists of positive quantities. These summations do not exceed

$$\frac{1}{(a_{1m}(\xi))^2+\zeta^2} \quad\text{and}\quad \frac{1}{(\bar{a}_{1m}(\xi))^2+\zeta^2},$$

respectively, where $|a_{1m}|$ and $|\bar{a}_{1m}|$ are minimal. The absolute value of the integrand in (8.4.13) does not exceed

$$2\zeta\nu_0(\xi, 0)\frac{2\, d\xi}{\varepsilon_5^2\xi^2+\zeta^2}. \qquad (8.4.14)$$

If we integrate (8.4.14) we obtain a quantity that remains bounded as $\zeta \to 0$. The remaining integrals, analogous to those in (8.4.10) that are taken over the region $\Pi^{(\widehat{0})}_{\epsilon 0}$, are investigated in the same way and they have a bounded imaginary part as $\zeta \to 0$.

Thus for $\theta = -D_0 + i\zeta$, where $\zeta > 0$, we have $L_\theta(I_0(\theta)) \to \infty$ as $\zeta \to 0$, where $I_0(\theta)$ is the basic integral and ∞ is the infinite point of the complex plane. Let us return to equation (8.4.2). If we consider the expression

$$L_\theta\left(C_{n_1n_2}\frac{a}{2}\,\theta^{-\frac{n_2}{2}}(1+\theta)^{-N+\frac{1}{2}}\right),$$

we see that it is bounded for $\theta = -D_0 + i\zeta$, which leads to a contradiction.

Thus, the test curve and the critic-hyperbola $A(\xi, \eta) = D$, $0 < D < 1$, cannot be tangent at a finite distance.

It follows that the test boundary cannot under such conditions be tangent to the critic-ellipse

$$B(\xi, \eta) = D \quad \text{for} \quad D > 1.$$

To see this note that, in accordance with what was said in §3, such tangency would lead to tangency of the test boundary of the "upset" test with the critic-hyperbola $A(\xi, \eta) = D$, and this, as we have shown, is impossible. This completes the proof of the lemma.

Lemma 8.4.2. *If the test boundary $\xi = \phi(\eta)$ is continuous on* [0, 1], *then it has a zero on* (0, 1].

In fact, this follows at once from the theorem on "null-regular tests" which is proved in §1 of Chapter IX; the proof of this theorem is completely independent of §§3 and 4 of Chapter VIII. By Lemma 8.3.1 we know moreover that $\xi_0 = \phi(0) \neq 0$.

Lemma 8.4.3.[1] *Let $\xi = \phi(\eta)$ denote a function that is continuous on* [0, 1] *and that has a finite derivative everywhere on* [0, 1), *with the right-hand derivative meant at $\eta = 0$. Suppose that the size of the sample is $n_2 \geq 4$. Then the curve $\xi = \phi(\eta)$ is tangent to a nondegenerate critic-ellipse at a point (ξ_0, η_0), $0 < \eta_0 < 1$.*

To prove this lemma we need only show that the (one-sided) derivative $\phi'(0)$ is positive at the point $\eta = 0$. This is true because $\phi(0) = \xi_0 > 0$ in accordance with Lemma 8.3.1 and, if $\phi'(0)$ is greater than 0, the curve $\xi = \phi(\eta)$ issuing from the point $(\xi_0, 0)$ has a right-hand tangent.

By Lemma 8.4.2 the curve $\xi = \phi(\eta)$ then reaches the point $(0, \gamma)$, where $\gamma \in (0, 1]$ is a zero of the function $\phi(\eta)$; i.e. it must "turn to the left". It is obvious from geometric considerations that it is tangent to a suitably chosen critic-ellipse $B(\xi, \eta) = D$, $D > 1$, that encircles the curve $\xi = \phi(\eta)$ for $0 \leq \eta \leq 1$. If $M = \sup_{0 \leq \eta \leq 1} \phi(\eta)$ we can find an upper bound for the ordinate of the point of intersection $(0, \eta'_0)$ of the critic $B(\xi, \eta) = D$ with the η-axis.

Indeed, the equation of our critic is

$$\frac{\xi^2}{D-1} + \frac{\eta^2}{D} = 1.$$

1) This lemma was proved under somewhat more stringent hypotheses by Šalaevskiĭ (see [49]).

From this one can easily see that

$$\eta_0' \leqslant \frac{3}{2}(M+1). \qquad (8.4.15)$$

It remains to show that

$$\xi_0' = \varphi'(0) > 0. \qquad (8.4.16)$$

The derivation of this inequality is rather complicated.

We write the zone of acceptance of the null hypothesis H_0: $a_1 = a_2$ (complement to the critical zone) in the form

$$|\xi| \leqslant \varphi(\eta) \quad \text{or} \quad \left|\frac{\bar{x}-\bar{y}}{s_2}\right| \leqslant \varphi\left(\frac{s_1^2}{s_2^2}\right).$$

In accordance with what was shown in § 3, we can write this inequality in the form

$$|X_1| \leqslant \frac{\sqrt{\chi'^2_{n_2-1}}}{\sqrt{1+\theta}}\,\varphi\left(\theta\,\frac{\chi^2_{n_1-1}}{\chi'^2_{n_2-1}}\right), \qquad (8.4.17)$$

where $X_1 \in N(0, 1)$, $\chi^2_{n_1-1}$ and $\chi'^2_{n_2-1}$ are independent random variables, two of which have a χ^2 distribution with the number of degrees of freedom indicated in the subscripts, and $\theta = n_2\sigma_1^2/n_1\sigma_2^2$.

We define

$$G(u) = \frac{2}{\sqrt{2\pi}}\int_0^u e^{-\frac{t^2}{2}}\,dt = P(|X_1| \leqslant u), \text{ and } g(u) = \frac{2}{\sqrt{2\pi}}e^{-\frac{u^2}{2}}.$$

If $\alpha \in (0, 1)$ is the level of our similar test, then

$$EG\left[\frac{\sqrt{\chi'^2_{n_2-1}}}{\sqrt{1+\theta}}\,\varphi\left(\theta\,\frac{\chi^2_{n_1-1}}{\chi'^2_{n_2-1}}\right)\right] = 1-\alpha, \quad 0 < \theta < \infty.$$

If we set $\chi^2_{n_1-1} = X$ and $\chi'^2_{n_2-1} = Y$, we have

$$EG\left[\frac{\sqrt{Y}}{\sqrt{1+\theta}}\,\varphi\left(\theta\frac{X}{Y}\right)\right] = 1-\alpha, \quad 0 < \theta < \infty. \qquad (8.4.18)$$

So far we have made no assumptions regarding the differentiability of $\phi(x)$. Let us formally differentiate equation (8.4.18) with respect to θ and then set (formally) $\theta = 0$. We obtain

$$Eg\left[\varphi(0)\sqrt{Y}\right]\left[\sqrt{Y}\frac{X}{Y}\varphi'(0) - \frac{1}{2}\sqrt{Y}\varphi(0)\right] = 0. \qquad (8.4.19)$$

If this formally constructed relationship were valid, (8.4.16) would follow easily from it. To see this, suppose that $\phi'(0) \leq 0$. Since $\phi(0) = \xi_0 \geq 0$, the left-hand member of (8.4.19) would be negative, which would contradict (8.4.19).

Thus we need to show that equation (8.4.19) is valid. Let us assume that for $x \geq 0$ the function $\phi(x)$ is measurable and has a finite one-sided derivative ξ'_0 at $x = 0$. Taking into consideration the probability densities for X and Y (see § 2) and denoting the left-hand member of (8.4.18) by $E(\theta)$, we obtain, for $\theta > 0$,

$$E(\theta) = \int_0^\infty dX \int_0^\infty dY G\left[\frac{\sqrt{Y}}{\sqrt{1+\theta}}\varphi\left(\theta\frac{X}{Y}\right)\right] \times p_1(X)\, p_2(Y) = 1 - \alpha, \tag{8.4.20}$$

where

$$p_1(X) = C_1 X^{\frac{n_1-3}{2}} \exp\left(-\frac{X}{2}\right),$$
$$p_2(Y) = C_2 Y^{\frac{n_2-3}{2}} \exp\left(-\frac{Y}{2}\right)$$

We note that $G(t) = O(1)$ for arbitrary real t. Thus $E(\theta)$ exists even at $\theta = 0$. We need to show that $E(\theta)$ is continuous, i.e. that $\lim_{\theta\to 0} E(\theta) = E(0)$, from which it will follow that $E(0) = 1 - \alpha$.

Let us make the substitutions $X\backslash Y = U$ and $Y = V$. Then $|\partial(X, Y)/\partial(U, V)| = V$ and for $\theta > 0$ we have, by virtue of (8.4.20),

$$E(\theta) = \int_0^\infty dU \int_0^\infty dV V G\left[\frac{\sqrt{V}}{\sqrt{1+\theta}}\varphi(\theta u)\right] p_1(UV)\, p_2(V) = 1 - \alpha. \tag{8.4.21}$$

Let us set $D(\theta) = E(\theta) - E(0)$ and represent this function as an integral of the corresponding difference; $D(\theta) = 1 - \alpha - E(0) = \text{const}$. Let us show that $D(\theta) \to 0$ as $\theta \to 0$, so that $D(\theta) = 0$ for $\theta \geq 0$. We break the integral with respect to U into two integrals by setting

$$D_0(\theta) = \int_0^{\theta^{-1+\epsilon}} dU \int_0^\infty dV(\quad), \quad D_1(\theta) = \int_{\theta^{-1+\epsilon}}^\infty dU \int_0^\infty dV(\quad).$$

Here ϵ is a small positive number.

As $\theta \to 0$ we have

$$D_1(\theta)$$
$$= O(1) \int_{\theta^{-1+\epsilon}}^\infty U^{\frac{n_1-3}{2}} dU \int_0^\infty V^{\frac{n_1+n_2-4}{2}} \exp\left[-\frac{1}{2}(1+U)V\right] dV =$$

$$= O(1) \int_{\theta^{-1+\epsilon}}^{\infty} dU \frac{U^{\frac{n_1-3}{2}}}{(1+U)^{\frac{n_1+n_2}{2}-1}} = O\left(\theta^{(1-\varepsilon)\frac{n_2-1}{2}}\right).$$

Let us require that $(n_2+1)/2 > 1$. Then $n_2 \geq 4$ and

$$D_1(\theta) = O\left(\theta^{\frac{\varepsilon}{2}-2\varepsilon}\right). \tag{8.4.22}$$

Now note that

$$D_0(\theta) = \int_0^{\theta^{-1+\epsilon}} dU \int_0^{\infty} dV\, V p_1(UV)\, p_2(V)$$
$$\times \left\{ G\left[\frac{\sqrt{V}}{\sqrt{1+\theta}} \varphi(\theta U)\right] - G\left[\sqrt{V}\varphi(0)\right]\right\}. \tag{8.4.23}$$

Here $0 \leq \theta U \leq \theta^{\epsilon}$. Since $\phi(x)$ is assumed to be differentiable from the right at $x = 0$, we have

$$\varphi(\theta U) = \varphi(0) + \theta U(\varphi'(0) + \eta),$$

where $\eta \to 0$ as $\theta \to 0$ (and hence as $\theta U \to 0$). Therefore

$$G\left[\frac{\sqrt{V}}{\sqrt{1+\theta}}\varphi(\theta U)\right] - G\left[\sqrt{V}\varphi(0)\right]$$
$$= \frac{1}{\sqrt{2\pi}} \int_{\sqrt{V}\varphi(0)}^{\frac{\sqrt{V}}{\sqrt{1+\theta}}\varphi(\theta U)} \exp\left[-\frac{\tau^2}{2}\right] d\tau$$
$$= \frac{1}{\sqrt{2\pi}} \int_{\sqrt{V}\varphi(0)}^{\frac{\sqrt{V}}{\sqrt{1+\theta}}(\varphi(0)+\theta U(\varphi'(0)+\eta))} \exp\left[-\frac{\tau^2}{2}\right] d\tau. \tag{8.4.24}$$

By virtue of the mean-value theorem, we have the following estimate for this integral when θ is sufficiently small:

$$O(1)\,\theta U \exp\left(-\frac{V\xi_0}{4}\right)\sqrt{V}, \text{where } \xi_0 = \varphi(0) > 0.$$

If we substitute this into (8.4.23), we obtain

$$D_0(\theta) = O(1)\,\theta \int_0^{\theta^{-1+\epsilon}} U^{\frac{n_1-1}{2}}\, dU \times$$

$$\times \int_0^\infty V_1^{\frac{n_1+n_2-3}{2}} \exp\left[-\frac{1}{2}\left(1+U+\frac{\xi_0}{2}\right)\right] V\,dV = O(\theta)$$

where $n_2 \geq 4$. Thus $\lim_{\theta\to 0} D_0(\theta) = 0$ and hence $D(\theta) = 0$ for all $\theta > 0$. Therefore by (8.4.22),

$$D_0(\theta) = O\left(\theta^{\frac{3}{2}-2\varepsilon}\right). \tag{8.4.25}$$

We can extend this reasoning to (8.4.24). For arbitrary $V > 0$,

$$\frac{\sqrt{V}}{\sqrt{1+\theta}}(\varphi(0)+\theta U(\varphi'(0)+\eta))$$
$$= \sqrt{V}\left(\varphi(0)-\frac{\theta}{2}\varphi'(0)+\theta U\varphi'(0)\right)+\eta\theta\left(U+\sqrt{V}\right). \tag{8.4.26}$$

where $\eta \to 0$ as $\theta \to 0$. Proceeding as we did in deriving the above estimate, we see that if we drop the term $\eta\theta(U+\sqrt{V})$ in (8.4.26), we obtain an error $O(\eta\theta)$ in (8.4.23). In other words, if we set

$$\varphi(0)-\frac{\theta}{2}\varphi(0)+\theta U\varphi'(0) = f(\theta, U),$$

then the expression

$$\frac{1}{\sqrt{2\pi}}\int_0^{\theta^{-1+\epsilon}} dU \int_0^\infty dV\, V p_1(UV)\, p_2(V)$$
$$\times \int_{\sqrt{V}\,\varphi(0)}^{\sqrt{V}\, f(\theta, U)} \exp\left[-\frac{\tau^2}{2}\right] d\tau \tag{8.4.27}$$

will differ from (8.4.23) by no more than $O(\eta\theta)$. If we extend the integral with respect to U in (8.4.27) from 0 to ∞, i.e. if we add to (8.4.27) the term

$$\frac{1}{\sqrt{2\pi}}\int_{\theta^{-1+\epsilon}}^\infty dU \int_0^\infty dV(\quad),$$

where the parentheses represent the same integrand, we obtain the error $O(\theta^{3/2-2\epsilon})$, just as in the derivation of (8.4.22). From this we finally obtain

$$D_0(\theta) = \frac{1}{\sqrt{2\pi}}\int_0^{\theta^{-1+\epsilon}} dU \int_0^\infty dV\, V p_1(UV)\, p_2(V).$$
$$\times \int_{\sqrt{V}\,\varphi(0)}^{\sqrt{V}\, f(\theta, U)} \exp\left[-\frac{\tau^2}{2}\right] d\tau + O(\eta\theta) \tag{8.4.28}$$

Furthermore, in accordance with (8.4.25),

$$E_0(\theta) - E_0(0) = O\left(\theta^{\frac{3}{2}-2\varepsilon}\right).$$

Let us divide both sides of this equation by θ and then let θ approach 0. This gives us the result that the quantity $(d/d\theta)\int_0^\infty dU \int_0^\infty dV\,(\;)$ vanishes at $\theta = 0$.

Here the quantity $\int_0^\infty dU \int_0^\infty dV\,(\;)$ is the integral in the right-hand member of (8.4.28) with $f(\theta, U)$ a linear function of θ and U. Therefore it is possible to differentiate under the integral sign in (8.4.28). When we do this we obtain

$$\frac{1}{\sqrt{2\pi}}\int_0^\infty dU \int_0^\infty dV\, V p_1(UV)\, p_2(V)$$
$$\times \exp\left[-\frac{(\sqrt{V}\varphi(0))^2}{2}\right]\sqrt{V}\left(-\frac{1}{2}\varphi(0)+U\varphi'(0)\right)=0. \quad (8.4.29)$$

This coincides with equation (8.4.19) and completes the proof of Lemma 8.4.3.

§5. COMPLETION OF THE PROOF OF THEOREM 8.3.1.

It is now easy for us to prove Theorem 8.3.1. Suppose that $n_2 \geq 4$. Then by Lemma 8.4.3 there will be tangency, under the hypothesis of Theorem 8.3.1, to the nondegenerate ellipse-critic $B(\xi, \eta) = D$, where $D > 1$. By (8.4.15) this ellipse will intersect the $(\xi = 0)$-axis at a point $(0, \eta'_0)$, where $1 < \eta'_0 \leq (3/2)(M+1)$. By the hypothesis of Theorem 8.3.1, $\phi(\eta)$ satisfies, for $1 < \eta < 2(M+1)$, a Lipschitz condition with exponent 1. By Lemma 8.4.1 the test $|\xi| \geq \phi(\eta)$ cannot be similar. This completes the proof of Theorem 8.3.1.

We note that this theorem can be strengthened by requiring merely that the test be similar in a countable set of pairs of values (σ_1, σ_2) such that the ratios σ_1/σ_2 are distinct and their values are bounded.

In fact, the principle of analytic continuation as applied to formula (8.4.1) for the points $\theta = n_2\sigma_1^2/n_1\sigma_2^2$ shows that this formula is valid for all $\theta > 0$.

CHAPTER IX

RANDOMIZED HOMOGENEOUS TESTS IN THE BEHRENS-FISHER PROBLEM. CHARACTERIZATION OF TESTS OF THE BARTLETT-SCHEFFÉ TYPE

§1. NONEXISTENCE OF "NULL-REGULAR" SIMILAR TESTS

We retain the previous notation for the different statistics and parameters used in the Behrens-Fisher problem. Homogeneous tests $\phi(\xi, \eta)$ will depend only on the two statistics $\xi = (\bar{x} - \bar{y})/s_2$ and $\eta = s_1/s_2$. It is natural for us to consider the statistic

$$t = \frac{\bar{x} - \bar{y}}{\sqrt{n_1 s_1^2 + n_2 s_2^2}}, \tag{9.1.1}$$

which, as is known (see for example [33]), is a fundamental instrument in the construction of similar tests for Student's problem (see Example 2, §1, Chapter III). For any homogeneous test $\phi(\xi, \eta)$ it is natural to require that it assume the null hypothesis with probability 1 if the fraction t is sufficiently small. To reject the null hypothesis H_0: $a_1 = a_2$ if the standardized difference of the means is sufficiently small is foolish. It is natural to consider tests $\phi(\xi, \eta)$ that lead to such solutions as being "irregular" in some sense or other. Let us make more precise what we mean by defining the property of null-regularity of a test $\phi(\xi, \eta)$ (see [44]).

The property of null-regularity of a test $\phi(\xi, \eta)$. Suppose that there exists a number $\eta_0 = 1 + \eta_1 > 1$ such that it is possible to draw a circle around every point on the segment $\xi = 0$, $0 \leq \eta \leq \eta_0$ with the property that, when the point (ξ, η) falls in that circle, the conditional probability of rejecting the null hypothesis H_0 on the basis of the test $\phi(\xi, \eta)$ is zero.

If the test is unrandomized, i.e. if the function $\phi(\xi, \eta)$ is always equal to 0 or 1 and if the boundary of the critical zone (the zone in which $\phi(\xi, \eta) = 1$)

consists of finitely many continuous curves, then the property of null-regularity of the test consists simply in the fact that it does not reject the null hypothesis if $\bar{x} - \bar{y} = 0$ and $s_1^2 + s_2^2 > 0$, which is an extremely simple and natural condition. This natural requirement on the test is, as it turns out, incompatible with the requirement of similarity of the test and *a fortiori* with the requirement of unbiasedness. If $\max(n_1, n_2) \geq 3$ we have

Theorem 9.1.1. *There are no randomized homogeneous similar (or a fortiori unbiased) tests for the Behrens-Fisher problem that possess the property of null-regularity.*

Let us make a few remarks concerning this theorem.

In a certain sense, it does not admit an improvement since the conclusion ceases to hold if we discard the requirement of null-regularity: similar randomized tests without this property do exist. In Chapter V we constructed ϵ-complete systems of such tests.

Let us give a simple example of the violation of the property of null-regularity. It might be noted that the distribution density of the statistics (ξ, η) that we shall write out in what follows is an even function of ξ for all η under the null hypothesis H_0. Let $\phi_1(\xi, \eta)$ denote a continuous function of ξ and η in the half-plane Ω: $-\infty < \xi < \infty$, $0 \leq \eta < \infty$, which for arbitrary values of η is an odd function of ξ and satisfies the condition

$$|\phi_1(\xi, \eta)| \leqslant \frac{1}{3}.$$

Let us set

$$\varphi(\xi, \eta) = \frac{1}{2} + \varphi_1(\xi, \eta).$$

Then by what was said above,

$$E\varphi_1(\xi, \eta) = 0, \quad E\varphi(\xi, \eta) = \frac{1}{2}$$

under the hypothesis H_0, so that $\phi(\xi, \eta)$ is a randomized similar test. However, the property of null-regularity is obviously violated.

We note also that from an analytical point of view this theorem can be strengthened. Instead of similarity of the test $\phi(\xi, \eta)$ with respect to all values of σ_1 and σ_2, we need only require its similarity with respect to an arbitrary countable set of distinct values contained in some finite segment.

Proof of Theorem 9.1.1. Let us apply the basic integral relationship of the type (8.2.3). For randomized tests this relationship is obviously written the same as for unrandomized ones, i.e. in the form

$$\iint_{\Omega} \varphi(\xi, \eta) \frac{\eta^{n_1-2}\, d\xi\, d\eta}{(\theta^2+\theta(1+\xi^2+\eta^2)+\eta^2)^N} = C'_{n_1 n_2} \alpha \theta^{-\frac{n_2}{2}} (1+\theta)^{-N+\frac{1}{2}}, \tag{9.1.2}$$

where $N = (n_1 + n_2 - 1)/2$, $\alpha\ (>0)$ is the level, $C'_{n_1 n_2} > 0$ and Ω is the half-plane $-\infty < \xi < \infty$, $0 < \eta < \infty$. Equation (9.1.2) holds for all positive values of θ.

Just as we did above, we shall carry out an analytic continuation with respect to $\theta = \sigma + i\tau$ onto the region Λ, which is the complex plane with a cut along the negative axis $\tau = 0$, $-\infty < \sigma < 0$. The left-hand member can also be extended to this region. The integral in (9.1.2) converges absolutely and uniformly in every closed subregion contained in Λ.

Thus equation (9.1.2) holds in the entire region Λ of values of the parameter θ.

We note that the same conclusion follows from the weaker assumption that the test $\phi(\xi, \eta)$ is similar for an arbitrary countable bounded set of values of σ_1/σ_2 and hence for values of θ. Elementary application of the principle of analytic continuation allows us in this case also to assert that equation (9.1.2) is valid for the entire region Λ of values of θ.

If we set

$$P(\theta, \xi, \eta) = (\theta + A(\xi, \eta))(\theta + B(\xi, \eta)),$$

we can always assume, as was shown in Chapter VIII, that

$$0 \leqslant A(\xi, \eta) \leqslant 1, \quad B(\xi, \eta) \geqslant 1.$$

We shall call the geometric loci defined by the equations

$$A(\xi, \eta) = D, \quad B(\xi, \eta) = D,$$

the *critical curves* (or simply *critics*) of the family of measures that is generated by our probability densities and the parameter θ. For $D < 1$ the equation $A(\xi, \eta) = D$ yields the family of hyperbolas

$$\frac{\eta^2}{D} - \frac{\xi^2}{1-D} = 1. \tag{9.1.3}$$

For $D = 1$ the hyperbola degenerates into the twice-covered ray $\xi = 0,\ 1 \leq \eta < \infty$. For $D = 0$ it degenerates into the $(\eta = 0)$-axis.

For $D > 1$ the equation $B(\xi, \eta) = D$ yields the family of confocal semi-ellipses

$$\frac{\xi^2}{D-1} + \frac{\eta^2}{D} = 1. \tag{9.1.4}$$

For $D = 1$ these semi-ellipses degenerate into the twice covered segment $\xi = 0$, $0 \leq \eta \leq 1$.

We can rewrite the basic equation (9.1.2) in the form

$$\iint_{\Omega} \varphi(\xi, \eta) \frac{\eta^{n_1-2}\, d\xi\, d\eta}{[(\theta + A(\xi, \eta))(\theta + B(\xi, \eta))]^N} = C'_{n_1 n_2} \alpha \theta^{-\frac{n_2}{2}} (1+\theta)^{-N+\frac{1}{2}}. \tag{9.1.5}$$

By the null-regularity of the test $\phi(\xi, \eta)$ there exists a number $\eta_0 = 1 + \eta_1 > 1$ such that it is possible to draw a circle around every point of the segment $\xi = 0$, $0 \leq \eta \leq \eta_0$ in which $\phi(\xi, \eta) = 0$ almost everywhere in the sense of Lebesgue measure. By the Heine-Borel theorem there exists a finite covering consisting of certain of these circles, so that we obtain an open region Γ containing our segment such that $\phi(\xi, \eta) = 0$ for almost all points $(\xi, \eta) \in \Gamma$.

Let us now look at the family of confocal semi-ellipses (9.1.4). Let us choose $D > 1$ so close to 1 that the corresponding semi-ellipse $B(\xi, \eta) = D$ lies entirely in the region Γ (the same will of course be true for smaller values of D). Now let us consider larger values of D. Let $D_0 (> 1)$ denote the least upper bound of the set of values of D for which $\phi(\xi, \eta) = 0$ almost everywhere inside the ellipse (9.1.4).

If $D_0 = \infty$ then $\phi(\xi, \eta) = 0$ almost everywhere on Ω. The test is then a trivial similar test. We disregard this case and assume that D_0 is a finite number.

The notation that we shall need is as follows. We denote by H a quantity (not always the same one) that remains bounded as the different parameters of the problem in question vary. We denote by $\zeta_0, \zeta_1, \cdots$ small positive constants each dependent on the preceding ones. We denote by $L_\theta^{(k)}$, where k is a non-negative integer, the operator for k differentiations with respect to θ at the point $\theta \in \Lambda$.

We have $D_0 > 1$. We set

$$D_0 - 1 = \zeta_0, \quad \zeta_1 = \min\left(\frac{\zeta_0}{2}, \frac{1}{2}\right).$$

Suppose that θ varies in a rectangle $Q \subset \Lambda$:

$$\theta = -D_0 + \delta + i\zeta, \quad 0 \leqslant \delta \leqslant \zeta_1, \quad 0 < \zeta < \zeta_1. \tag{9.1.6}$$

Our remaining considerations are based on the application of the operator $L_\theta^{(k)}$ to both sides of equation (9.1.5) with $\theta \in Q$. We shall see that, for suitable choice of $\theta \in Q$ and sufficiently large k, the order of growth with respect to k is not the same on the two sides of equation (9.1.5), which provides the desired contradiction.

From the definition of D_0, the function $\phi(\xi, \eta)$ is zero almost everywhere inside the semi-ellipse $B(\xi, \eta) < D_0$. We denote by Ω_0 the half-plane Ω from which we have removed the semi-ellipse referred to. We denote by $I_0(\theta)$ the corresponding integral, which coincides with the left-hand member of (9.1.5). We have

$$I_0(\theta) = \iint_{\Omega_0} \varphi(\xi, \eta) \frac{\eta^{n_1-2}\, d\xi\, d\eta}{[(\theta + A(\xi, \eta))(\theta + B(\xi, \eta))]^N}$$
$$= C'_{n_1 n_2} \alpha \theta^{-\frac{n_2}{2}} (1+\theta)^{-N+\frac{1}{2}}. \tag{9.1.7}$$

Note also that

$$0 \leqslant A(\xi, \eta) \leqslant 1, \tag{9.1.8}$$

so that for $\theta \in Q$,

$$|\theta + A(\xi, \eta)| \geqslant D_0 - \zeta_1 - 1 \geqslant \zeta_1 \tag{9.1.9}$$

by virtue of the definition of ζ_1.

We consider first the case in which the sizes n_1 and n_2 of the samples are of opposite parity. Then the number $N = (n_1 + n_2 - 1)/2$ is an integer, and the integrand in (9.1.7) is rational and has no branches.

From the integral $I_0(\theta)$, we take out the integral

$$I_1(\theta) = \iint_{B(\xi, \eta) \geqslant D_0 + 1} \tag{9.1.10}$$

and find an upper bound for the quantity

$$L_\theta^{(k)} I_1(\theta) \tag{9.1.11}$$

for $\theta \in Q$ and large values of k. By virtue of (9.1.9), we obtain

$$L_\theta^{(m)} \frac{1}{(\theta + A)^N} = H \frac{\Gamma(N+m)}{\zeta_1^m}. \tag{9.1.12}$$

(The meaning of the symbol H was explained above. In particular, $H \cdot H = H$ and $\zeta_1^{-N} = H$.)

Now let $B = B(\xi, \eta)$ vary in the region

$$D \leqslant B(\xi, \eta) < D + 1, \tag{9.1.13}$$

where $D \geq D_0 + 1$. Then

$$|\theta + B(\xi, \eta)| \geqslant D - D_0 \geqslant 1, \tag{9.1.14}$$

Consequently

$$L_\theta^{(m)} \frac{1}{(\theta + B)^N} = H \frac{\Gamma(N+m)}{(D - D_0)^{m+N}}. \tag{9.1.15}$$

Now let us find a bound for the quantity

$$L_\theta^{(k)} \frac{1}{(\theta + A)^N (\theta + B)^N} \tag{9.1.16}$$

where $\theta \in Q$ and $B = B(\xi, \eta)$ belongs to the region (9.1.14). To do this we use (9.1.12), (9.1.15) and the familiar formula for differentiating the product of two functions. We find that

$$L_\theta^{(k)} \frac{1}{(\theta + A)^N (\theta + B)^N} = H \sum_{m=0}^{k} C_k^m \frac{\Gamma(N+m)}{\zeta_1^m} \frac{\Gamma(N+k-m)}{(D - D_0)^{k-m+N}}. \tag{9.1.17}$$

Since $C_k^m = k!/m!\,(k-m)!$, we find, after some simple calculation, that

$$\begin{aligned} L_\theta^{(k)} \frac{1}{(\theta + A)^N (\theta + B)^N} &= H\Gamma(k+1)(D - D_0)^{-N} \zeta_1^{-k} k^{2N} \\ &= H\Gamma(k+1)(D - D_0)^{-N} \zeta_2^{-k}, \end{aligned} \tag{9.1.18}$$

where $\zeta_2 = \zeta_1/2$.

From $I_1(\theta)$ (see (9.1.10)) let us take out the integral over the region defined by (9.1.13); we denote this integral by I_{1D}. In this region (which is of the form of a horseshoe between two semi-ellipses), we have

$$\eta = H(D - D_0)^{\frac{1}{2}}.$$

The area of this horseshoe is $H(D-D_0)^{1/2}$. Remembering (9.1.18), we see that for $\theta \in Q$,

$$L_\theta^{(k)} I_{1D}(\theta) = H\Gamma(k+1)\zeta_2^{-k}(D-D_0)^{\frac{n_1}{2}-N-\frac{1}{2}}. \tag{9.1.19}$$

By the assumption that $\max(n_1, n_2) \geq 3$, we may assume that

$$n_1 \leqslant n_2, \quad \frac{n_1}{2} - N - \frac{1}{2} = \frac{n_2}{2} \geqslant \frac{3}{2}.$$

In this case we may sum the estimates (9.1.19) for the values $D = D_0 + 1$, $D_0 + 2, \cdots$. We obtain

$$L_\theta^{(k)} I_1(\theta) = H\Gamma(k+1)\zeta_2^{-k}. \tag{9.1.20}$$

Let us set

$$I_{00}(\theta) = I_0(\theta) - I_1(\theta) = \iint\limits_{D_0 \leqslant B(\xi,\eta) \leqslant D_0+1} . \tag{9.1.21}$$

Let $R > 100$ denote a large number (with value to be fixed later) such that $M = R\zeta_2^{-1}$ is an integer. Define

$$\Delta = \zeta_2 R^{-1},$$

$$F_{0,r}(\theta) = \iint\limits_{D_0 + r\Delta \leqslant B(\xi,\eta) < D_0 + (r+1)\Delta} , \tag{9.1.22}$$

so that

$$I_{00}(\theta) = F_{00}(\theta) + F_{01}(\theta) + \ldots + F_{0,M-1}(\theta). \tag{9.1.23}$$

We need to study the behavior of

$$L_\theta^{(k)} F_{0r}(\theta) \tag{9.1.24}$$

in the region

$$D_0 + r\Delta \leqslant B(\xi,\eta) \leqslant D_0 + (r+1)\Delta \tag{9.1.25}$$

with $\theta \in Q$. We have

$$\theta + B = r\Delta + \delta + i\zeta. \tag{9.1.26}$$

Let us define $\delta = \Delta/2$. Thus the value of the number $\theta = -D_0 + \delta + i\zeta$ depends on the number R and the number ζ, which we shall identify later. From (9.1.26) we obtain

$$|\theta + B| = \left(\left(r + \frac{1}{2}\right)^2 \Delta^2 + \zeta^2\right)^{\frac{1}{2}}.$$

Let us assume also that $0 < \zeta \leq \Delta$. Then in the region (9.1.25) and for the values of θ indicated we have

$$\left(r+\frac{1}{2}\right)\Delta \leqslant |\theta + B| \leqslant \Delta\left(\left(r+\frac{1}{2}\right)^2 + 1\right)^{\frac{1}{2}}. \tag{9.1.27}$$

Let us return now to (9.1.24). On the basis of (9.1.27) the expression (9.1.26) is replaced with

$$L_\theta^{(m)} \frac{1}{(\theta+B)^N} = H \frac{\Gamma(N+m)}{\left(\left(r+\frac{1}{2}\right)\Delta\right)^{m+N}}, \tag{9.1.28}$$

and the estimate (9.1.12) remains valid. Furthermore, for $k > 0$,

$$L_\theta^{(k)} \frac{1}{(\theta+A)^N(\theta+B)^N} = \frac{(-1)^k N(N+1)\dots(N+k-1)}{(\theta+B)^{N+k}(\theta+A)^N} + \gamma_k, \tag{9.1.29}$$

where

$$\gamma_k = H\sum_{m=1}^{k} C_k^m \frac{\Gamma(N+m)}{\zeta_1^{m+N}} \frac{\Gamma(N+k-m)}{|\theta+B|^{k-m+N}}$$

$$= H\frac{\Gamma(N+k)}{|\theta+B|^{k+N-1}} \frac{\left(r+\frac{1}{2}\right)\Delta}{\zeta_1}, \tag{9.1.30}$$

if $(r+1/2)\Delta < \zeta_1/2$. On the other hand, if $(r+1/2)\Delta \geq \zeta_1/2$ we conclude from (9.1.29) and the second of the three expressions (9.1.30) that

$$L_\theta^{(k)} \frac{1}{(\theta+B)^N(\theta+A)^N} = H\frac{\Gamma(N+k)}{\zeta_3^k}, \tag{9.1.31}$$

where $\zeta_3 = \zeta_2/2$.

If we now use (9.1.29) and (9.1.27), we obtain

$$\left| L_\theta^{(k)} \frac{1}{(\theta+A)^N(\theta+B)^N} \right| > c_0 \frac{\Gamma(N+k)}{(\theta+B)^{N+k}} \tag{9.1.32}$$

for

$$\left(r+\frac{1}{2}\right)\Delta < \zeta_4\zeta_1, \tag{9.1.33}$$

where c_0 is a positive constant.

Furthermore, if the number $i\zeta$ (see (9.1.6)) is sufficiently small, for example if $0 < \zeta < \Delta/10k^2$, then by (9.1.8) we have, for even $k + N$,

$$\operatorname{Re} L_\theta^{(k)} \frac{1}{(\theta + A)^N (\theta + B)^N} > \frac{1}{2} c_0 \frac{\Gamma(N+k)}{|\theta + B|^{N+k}}. \tag{9.1.34}$$

If $(r + 1/2)\Delta \geq \zeta_4\zeta_2$, we obtain the expression

$$L_\theta^{(k)} \frac{1}{(\theta + A)^N (\theta + B)^N} = H \frac{\Gamma(N+k)}{(\zeta_4\zeta_1)^k} \tag{9.1.35}$$

which is analogous to (9.1.31).

Let us now return to the quantities (9.1.22). By definition of the number D_0 we must have

$$\iint\limits_{D_0 \leqslant B(\xi, \eta) \leqslant D_0 + \Delta} \eta^{n_1 - 2} \varphi(\xi, \eta)\, d\xi\, d\eta > \beta_0(\Delta) > 0, \tag{9.1.36}$$

where $\beta_0(\Delta)$ is a positive number depending on Δ. Let us choose the number R sufficiently great and hence Δ sufficiently small that $\Delta/2 < \zeta_4\zeta_1$, so that condition (9.1.33) is satisfied when $r = 0$. Then from (9.1.32), (9.1.34), and (9.1.36) we obtain

$$\operatorname{Re} L_\theta^{(k)} F_{00}(\theta) > \frac{c_0 \beta_0(\Delta)}{2} \frac{\Gamma(N+k)}{\left(\Delta \sqrt{\frac{5}{4}}\right)^{N+k}}. \tag{9.1.37}$$

Let r_0 denote the largest integer satisfying condition (9.1.33). Then from (9.1.37), (9.1.34), and (9.1.33) we obtain

$$\operatorname{Re} L_\theta^{(k)} (F_{00}(\theta) + F_{01}(\theta) + \dots + F_{0, r_0}(\theta)) > \frac{c_0 \beta_0(\Delta)}{2} \times \frac{\Gamma(N+k)}{\left(\Delta \sqrt{\frac{5}{4}}\right)^{N+k}}. \tag{9.1.38}$$

By virtue of (9.1.35) we obtain

$$\operatorname{Re} L_\theta^{(k)} (F_{0, r_0+1}(\theta) + F_{0, r_0+2}(\theta) + \dots + F_{0, M-1}(\theta)) = H \frac{\Gamma(N+k)}{(\zeta_4\zeta_1)^k}. \tag{9.1.39}$$

If we keep (9.1.20) in mind and compare (9.1.29), (9.1.20), and (9.1.38), we conclude that, if Δ is chosen sufficiently small in comparison with $\zeta_4\zeta_1$ and ζ_2 (which leads to choice of a sufficiently large number R), for example if we give Δ a value $\le (1/1{,}000)\min(\zeta_4\zeta_1, \zeta_2)$ and then choose k sufficiently large and of the same parity as N, we obtain

$$\operatorname{Re} L_\theta^{(k)} I_0(\theta) > \frac{1}{4} c_0\beta_0(\Delta) \frac{\Gamma(N+k)}{\left(\Delta\sqrt{\frac{5}{4}}\right)^{N+k}} \tag{9.1.40}$$

for $\theta = -D_0 + \Delta/2 + i\Delta/10k^2$, but this inequality, as we shall see below, leads to a contradiction.

Let us now consider the case in which the numbers n_1 and n_2 are of like parity and let us show that an inequality analogous to (9.1.40) holds. In this case the number $N = (n_1 + n_2 - 1)/2$ is half of an odd number, and the expression

$$\frac{1}{[(\theta+A)(\theta+B)]^N},$$

in (9.1.7) has a branch point.

Keeping the previous notation, let us look at a sequence of large integers $\{k\}$, and for each k let us choose, just as before,

$$\theta = -D_0 + \frac{\Delta}{2} + i\frac{\Delta}{10k^2}. \tag{9.1.41}$$

For $\theta \in \Lambda$, if we assume that $\operatorname{arc}\theta = 0$ for $\theta > 0$ we find, by virtue of (9.1.8) and (9.1.26),

$$\frac{1}{[(\theta+A)(\theta+B)]^N} = i^N \frac{1}{[(-\theta-A)(\theta+B)]^N}; \tag{9.1.42}$$

Here, for values of θ defined by (9.1.42),

$$\operatorname{arc}(-\theta-A) = O\left(\frac{1}{(k^2)}\right), \tag{9.1.43}$$

and for values of $B = B(\xi, \eta)$ with $(\xi, \eta) \in \Omega_0$,

$$\operatorname{arc}(\theta+B) = O\left(\frac{1}{k^2}\right). \tag{9.1.44}$$

This means that for sufficiently large k,

$$\frac{1}{[(\theta+A)(\theta+B)]^N} = i^N \frac{1}{(-\theta-A)^N}\frac{1}{(\theta+B)^N}, \tag{9.1.45}$$

and we can apply our previous reasoning with minor modifications: instead of (9.1.12) we write the estimate

$$L_\theta^{(m)} \frac{1}{(-\theta - A)^N} = H \frac{\Gamma(N+m)}{\zeta_1^m}, \tag{9.1.46}$$

which follows from the familiar formula

$$\frac{\Gamma(N+m)}{\Gamma(N)} = N(N+1)\ldots(N+m-1).$$

Fromula (9.1.15) remains unchanged. We replace $\theta + A$ everywhere with $-\theta - A$ for the remainder of the discussion. Instead of formula (9.1.29), we obtain

$$\begin{aligned} i^{-N} L_\theta^{(k)} \frac{1}{(-\theta - A)^N (\theta + B)^N} &= \\ &= \frac{(-1)^k N(N+1)\ldots(N+k-1)}{(\theta + B)^{N+k}(-\theta - A)^N} + \gamma_k \end{aligned} \tag{9.1.47}$$

with the same estimate (9.1.30) for γ_k. Consequently,

$$(-1)^k i^{-N} \operatorname{Re} \frac{1}{(-\theta - A)^N (\theta + B)^N} > \frac{c_0}{2} \frac{\Gamma(N+k)}{|\theta + B|^{N+k}} \tag{9.1.48}$$

for $(r + 1/2)\Delta < \zeta_4 \zeta_1$.

Otherwise the reasoning proceeds as before except that, in inequalities (9.1.37), (9.1.38), and (9.1.40), we need to replace $\operatorname{Re} L_\theta^{(k)}(\ldots)$ with $\operatorname{Re}(-1)^k i^{-N} L_\theta^{(k)}(\ldots)$, As a result, we again arrive at a formula of the type (9.1.40):

$$(-1)^k i^{-N} \operatorname{Re} L_\theta^{(k)} I_0(\theta) > \frac{c_0 \beta_0(\Delta)}{4} \frac{\Gamma(N+k)}{\left(\Delta\sqrt{\frac{5}{4}}\right)^{N+k}} \tag{9.1.49}$$

for

$$\theta = -D_0 + \frac{\Delta}{2} + \frac{i\Delta}{10k^2}. \tag{9.1.50}$$

For values of θ satisfying equation (9.1.50), we obtain from (9.1.40) and (9.1.49) the inequality

$$|L_\theta^{(k)} I_0(\theta)| > \frac{c_0 \beta_0(\Delta)}{4} \frac{\Gamma(N+k)}{\left(\Delta\sqrt{\frac{5}{4}}\right)^{N+k}} \tag{9.1.51}$$

this time for arbitrary values of n_1 and n_2 except that $\max(n_1, n_2) \geq 3$.

Let us look at the application of the operator $L_\theta^{(k)}$ to the right-hand member of (9.1.7); in other words, let us consider the expression

$$L_\theta^{(k)}\alpha\theta^{-\frac{n_2}{2}}(1+\theta)^{-N+\frac{1}{2}}.$$

Using the considerations discussed at the beginning of this section, we see that

$$L_\theta^{(k)}\alpha\theta^{-\frac{n_2}{2}}(1+\theta)^{-N+\frac{1}{2}}=H\frac{\Gamma(N+k)}{\left(D_0-\frac{\Delta}{2}-1\right)^{N+k}}. \tag{9.1.52}$$

If Δ is chosen sufficiently small and then k is chosen sufficiently large, then (9.1.51) will satisfy (9.1.52), which completes the proof of our theorem.

We note that, when the sizes of the samples are of opposite parity and N is an integer, one can considerably simplify and shorten the proof of the theorem.

§2. BARTLETT-SCHEFFÉ TESTS

In Chapter V we saw how one can construct for the Behrens-Fisher problem a family of similar tests that is ϵ-complete in a certain sense. These tests were randomized and they depended only on sufficient statistics. By what was said in Chapter IV one can use these tests to construct unrandomized tests by leaving the σ-algebra of sufficient statistics. The family of such tests can also be made ϵ-complete.

However, certain tests of this type that are convenient and simple in application have long been known to statisticians. These are the Bartlett-Scheffé tests (see Neyman [58] and Scheffé [79]). Let us look at these tests.

Bartlett's test deals with the case when two samples $x_1, \cdots, x_{n_1} \in N(a_1, \sigma_1^2)$ and $y_1, \cdots, y_{n_2} \in N(a_2, \sigma_2^2)$ are of the same size $n = n_1 = n_2$. The critical zones for Bartlett's test are of the form

$$\frac{(n(n-1))^{1/2}\,|\overline{x}-\overline{y}|}{\left(\sum_{i=1}^{n}[(x_i-\overline{x})-(y_i-\overline{y})]^2\right)^{1/2}} > C \tag{9.2.1}$$

for arbitrary $C > 0$.

The test depends on the linear forms $X = \overline{x} - \overline{y}$ and $l_i = (x_i - \overline{x}) - (y_i - \overline{y})$, $i = 1, 2, \cdots, n$. These last forms are linearly dependent: $\Sigma_{i=1}^n l_i = 0$. The forms X and $l_1, \cdots, l_n$ are normal variables whose means are zero and whose variances are respectively $D(x) = (\sigma_1^2 + \sigma_2^2)/n$ and $D(l_i) = (1 - 1/n)\ (\sigma_1^2 + \sigma_2^2)$. If we divide both the numerator and the denominator in (9.2.1) by $(\sigma_1^2 + \sigma_2^2)^{1/2}$, we see that our test is similar. Below we shall show that the left-hand member of (9.2.1) is distributed like Student's fraction t_{n_1-1} with $n_1 - 1$ degrees of freedom.

Consider the case $n_1 \neq n_2$ (we assume without loss of generality that $n_1 < n_2$). Here we could use only n_1 observations $y_1, \cdots, y_{n_1}$ from the second sample and construct a similar test of the form (9.2.1). If we did this however, there would be a definite loss of information. Therefore Scheffé [79] proposed in 1943 a new variant of a test of the form (9.2.1) for this case.

We introduce the linear forms

$$l_i = x_i - \sum_{j=1}^{n_2} c_{ij} y_j \quad (i = 1, 2, \ldots, n_1). \tag{9.2.2}$$

These forms l_i constitute a normal vector, which we shall write as a column matrix

$$l = \begin{Vmatrix} l_1 \\ \cdot \\ \cdot \\ \cdot \\ l_{n_1} \end{Vmatrix} = x - Cy,$$

where

$$x = \begin{Vmatrix} x_1 \\ \cdot \\ \cdot \\ \cdot \\ x_{n_1} \end{Vmatrix}, \quad y = \begin{Vmatrix} y_1 \\ \cdot \\ \cdot \\ \cdot \\ y_{n_2} \end{Vmatrix}, \quad C = \| c_{ij} \|.$$

Suppose that $a_1 - a_2 = \delta$. For the equations

$$El = \begin{Vmatrix} \delta \\ \cdot \\ \cdot \\ \cdot \\ \delta \end{Vmatrix} = \tilde{\delta},\ E(l - \tilde{\delta})(l - \tilde{\delta})^T = \begin{Vmatrix} \sigma^2 & \ldots & 0 & \ldots & 0 \\ 0 & \ldots & \sigma^2 & \ldots & 0 \\ \cdot & \cdot & \cdot & \cdot & \cdot \\ 0 & \ldots & 0 & \ldots & \sigma^2 \end{Vmatrix} = \sigma^2 I_{n_1 n_1}$$

to hold for $\sigma^2 > 0$ (for calculation with random matrices, the reader is referred, for example, to [36]), it is necessary and sufficient that

$$a_1 - a_2 \sum_{j=1}^{n_2} c_{ij} = a_1 - a_2 = \delta, \tag{9.2.3}$$

so that

$$\sum_{j=1}^{n_2} c_{ij} = 1 \quad (i = 1, 2, \ldots, n_1), \tag{9.2.4}$$

$$E\left[(x - Cy - \tilde{\delta})(x - Cy - \tilde{\delta})^T\right] = \sigma^2 I_{n_1 n_1}. \tag{9.2.5}$$

Define

$$\xi = \begin{Vmatrix} x_1 - a_1 \\ \cdot \\ \cdot \\ \cdot \\ x_{n_1} - a_1 \end{Vmatrix}, \quad \eta = \begin{Vmatrix} y_1 - a_2 \\ \cdot \\ \cdot \\ \cdot \\ y_{n_2} - a_2 \end{Vmatrix}$$

Then if (9.2.3) is satisfied we may write (9.2.5) in the form

$$E\left[(\xi - C\eta)(\xi - C\eta)^T\right] = \\ = E\xi\xi^T - EC\eta\xi^T - E\xi\eta^T C^T + EC\eta\eta^T C^T \\ = \sigma_1^2 I_{n_1 n_1} + \sigma_2^2 CC^T \tag{9.2.6}$$

by virtue of the uncorrelatedness of $x_i - a_i$ and $y_j - a_j$. Thus, in addition to (9.2.4) we have the relation $CC^T = c^2 I_{n_1 n_1}$, where c^2 is a nonnegative number. Therefore

$$\sum_{k=1}^{n_2} c_{ik} c_{jk} = c^2 \delta_{ij},$$

where δ_{ij} is the familiar Kronecker delta. Furthermore,

$$D(l_i) = \sigma_1^2 + c^2 \sigma_2^2 = \sigma_c^2. \tag{9.2.7}$$

Under the null hypothesis H_0: $a_1 = a_2$ we have $D(l_i) = 0$. We define $L = \bar{l} = (l_1 + \cdots + l_{n_1})/n_1$ and $Q = \Sigma_{i=1}^{n_1}(l_1 - L)^2$. Then (see [33])

$$\frac{Q}{\sigma_2^2} = \chi^2_{n_1 - 1}, \quad \frac{\sqrt{n_1}\, L}{\sigma_c} \in N(0, 1),$$

and the quantity $\sqrt{n_1} L/(Q/(n_1 - 1))^{1/2} = t_{n_1 - 1}$ is Student's fraction with $n_1 - 1$ degrees of freedom. Therefore we can construct a similar test with critical zone

$$\frac{\sqrt{n_1}\, |L|}{(Q/(n_1 - 1))^{1/2}} > C. \tag{9.2.8}$$

If the hypothesis H_0 is not valid and $a_1 - a_2 = \delta \neq 0$, the random variable $(L - \delta)/(Q/n_1(n_1 - 1))^{1/2} = t_{n-1}$ is still Student's fraction. If for a given reliability γ we find a number t_γ such that $P(|t_{n_1 - 1}| < t_\gamma) = \gamma$ and if we construct the confidence interval

$$|\delta - L| \leqslant t_\gamma \left(\frac{Q}{n_1(n_1 - 1)}\right)^{1/2}, \tag{9.2.9}$$

we can see that the mean length of such a confidence interval is minimized when σ_c^2 is minimized. By virtue of the correspondence between the confidence intervals and the tests of the hypotheses that is described in [36], this property of the confidence interval will be a useful property of a similar test (9.2.8). We note on the basis of (9.2.7) that it is expedient to find matrices C that minimize

c^2. This minimization problem was studied in detail by Scheffé in [79]. He obtained the following solution for the matrix $C = \|c_{ij}\|$:

$$c_{ij}=\begin{cases}\delta_{ij}\left(\dfrac{n_1}{n_2}\right)^{1/2}-\dfrac{1}{(n_1n_2)^{1/2}}+\dfrac{1}{n_2} & (j\leqslant n_1),\\ \dfrac{1}{n_2} & (j>n_1).\end{cases}$$

It then follows that

$$l_i=x_i-\sum_{j=1}^{n_2}c_{ij}y_j=x_i-\left(\frac{n_1}{n_2}\right)^{1/2}y_i+(n_1n_2)^{-1/2}\sum_{j=1}^{n_2}y_j-\bar{y}.$$

$$L=\bar{x}-\bar{y}.$$

For $n_1 = n_2$ we obtain Bartlett's solution (the test (9.2.1)): $c_{ij} = \delta_{ij}$, $l_i = x_i - y_i$, $L = \bar{x} - \bar{y}$. Here $c' = 1$.

Thus, as stated above, the left-hand member of (9.2.1) is distributed like Student's fraction t_{n_1-1}.

§3. A HOMOGENEOUS RANDOMIZED TEST ASSOCIATED WITH BARTLETT'S TEST

If we "project" Bartlett's test with critical zone (9.2.1) onto the σ-algebra of sufficient statistics, we obtain a randomized test with the same power function. It turns out that this test has an extremely simple and characteristic form. Following the article [43], let us go through the corresponding calculations.

Let us assume that $n_1 = n_2 = n$. Just as above, we take $\xi = (\bar{x} - \bar{y})/s_2$ and $\eta = s_1/s_2$ and introduce the sample coefficient of correlation

$$r=\frac{\sum_{i=1}^{n}(x_i-\bar{x})(y_i-\bar{y})}{\left(\sum_{i=1}^{n}(x_i-\bar{x})^2\sum_{i=1}^{n}(y_i-\bar{y})^2\right)^{1/2}}.$$

This puts the test (9.2.1) in the form

$$\frac{\xi^2}{\eta^2-2r\eta+1}\geqslant C_1^2,\quad C_1^2=\frac{C^2}{n(n-1)}. \tag{9.3.1}$$

Since the normal samples $x_1, \cdots, x_n$ and $y_1, \cdots, y_n$ are independent, it follows (see Example 4, §5, Chapter IV) that the random variable r is stochastically independent of the random vector $(\bar{x}, \bar{y}, s_1, s_2)$ and *a fortiori* of the vector (ξ, η). Furthermore (see [33]), for the same reason, r has probability density

$$f_n(r) = \frac{\Gamma\left(\frac{n-1}{2}\right)}{\Gamma\left(\frac{n-2}{2}\right)\sqrt{\pi}}(1-r^2)^{\frac{n-4}{2}}, \quad -1 \leqslant r \leqslant 1. \tag{9.3.2}$$

It follows from (9.3.1) that

$$\frac{\xi^2}{C_1^2} \geqslant \eta^2 - 2r\eta - 1 \geqslant \eta^2 - 2\eta - 1 = (\eta - 1)^2. \tag{9.3.3}$$

Furthermore, from (9.3.1) we see that, when (9.3.3) is satisfied,

$$1 \geqslant r \geqslant \max\,[\tau(\xi, \eta), -1], \tag{9.3.4}$$

where

$$\tau(\xi, \eta) = \frac{1}{2}\left(\eta + \frac{1}{\eta} - \frac{\xi^2}{C_1^2\eta}\right). \tag{9.3.5}$$

Since r is stochastically independent of ξ and η, we obtain from what was said above a randomized test $\phi(\xi, \eta)$ defined (for given C_1) by

$$\left.\begin{aligned} \varphi(\xi, \eta) &= 0 && \text{for } |\xi| < C_1|\eta - 1|, \\ \varphi(\xi, \eta) &= \int_{\max[\tau(\xi, \eta), -1]}^{1} f_n(r)\,dr && \text{for } |\xi| \geqslant C_1|\eta - 1|, \end{aligned}\right\} \tag{9.3.6}$$

that is, under condition (9.3.3). We note that under this last condition $r(\xi, \eta) \leq 1$. The condition $|\xi| = C_1|\eta - 1|$ defines the boundary of the "randomized critical zone." Inside the quadrant $\xi \geq 0$, $\eta \geq 0$ this zone has the form of a right angle with vertex at the point (0,1) and bisector parallel to the ξ-axis. In the quadrant $\xi \leq 0$, $\eta \geq 0$ the boundary of the zone is the mirror image of this angle about the η-axis.

Returning to the sufficient statistics, we note that the boundary can be represented in the very simple form

$$\left|\frac{\bar{x} - \bar{y}}{s_1 - s_2}\right| = C_1. \tag{9.3.7}$$

If the sizes of the samples are all $n = 4$ then we see from (9.3.2) that $f_n(r)$ is constant and equal to $\Gamma(3/2)/\sqrt{\pi}\Gamma(1) = \frac{1}{2}$. Thus in this case the form of the test $\phi(\xi, \eta)$ is simplified.

The fact that the randomized critical zone extends to the point (0,1) is quite characteristic. It can be shown that this is necessary for a randomized homogeneous test. (In this connection see §3 of Chapter VIII, where a somewhat weaker assertion is proved in detail.)

§4. CHARACTERIZATION OF TESTS OF THE BARTLETT-SCHEFFÉ TYPE

In Chapter VIII, we used the method of analytic continuation with respect to a parameter to prove the possibility of various similar tests for the Behrens-Fisher problem. The same method can be applied to achieve purposes that in a way are just the opposite. One may seek to characterize the known similar tests by exhibiting certain simple properties of a qualitative nature that distinguish them from other tests. Apparently this can be done for many familiar tests in multivariant analysis and Fisher's analysis of variance. In this section we shall, following [47], do this for the classical Bartlett-Scheffé test described in the preceding section. In particular, for identical sizes of the samples $n_1 = n_2 = n$ we have Bartlett's unrandomized, mixed similar tests with critical zone Z_C:

$$\frac{(n(n-1))^{1/2}\,|\bar{x}-\bar{y}|}{\left(\sum_{i=1}^{n}[(x_i-\bar{x})-(y_i-\bar{y})]^2\right)^{1/2}} > C, \qquad (9.4.1)$$

where C is a constant.

The expression (9.4.1) can be represented in the form

$$\frac{X^2}{Q(l_1, \ldots, l)_\mu} > C_1, \qquad (9.4.2)$$

where $X = \bar{x} - \bar{y}$, $l_i = (x_i - \bar{x}) - (y_i - \bar{y})$ $(i = 1, 2, \cdots, \mu)$, $\mu = n - 1$, and $C_1 = C^2/n\,(n-1)$.

We note that the linear form X and the linear forms $l_1, \cdots, l_\mu$ of the observations x_i and y_i are linearly and stochastically independent. Furthermore, under the hypothesis H_0: $a_1 = a_2$ we have

$$E(X\,|\,H_0) = 0, \quad E(l_i\,|\,H_0) = 0, \qquad (9.4.3)$$

$$D(X) = \frac{\sigma^2}{n}, \quad D(l_i) = \frac{n-1}{n}\sigma_2, \quad i = 1, 2, \ldots, \mu, \qquad (9.4.4)$$

where $\sigma^2 = \sigma_1^2 + \sigma_2^2$

Thus the variances X and the linear forms l_i are also proportional for arbitrary values of σ_1 and σ_2. The test (9.4.1) is defined on the sample space W determined by the linear statistics $X, l_1, \cdots, l_\mu$. Using the results of §2, one can easily write the corresponding probability density.

Here we have a one-parameter exponential family with parameter $n/2\sigma^2 = n/2(\sigma_1^2 + \sigma_2^2)$ and with a single sufficient statistic

$$Q_1 = Q_1(X, l_1, \ldots, l_\mu), \tag{9.4.5}$$

which is a quadratic form in its variables. By virtue of Theorem 4.2.2 we see that the family of distributions generated by the sufficient statistic Q_1 is boundedly complete and that all similar zones are determined by the Neyman zones. Thus the crtiical zone Z_C is a Neyman zone, which one can also verify directly by noting that the left side of (9.4.1) is stochastically independent of Q_1.

We also mention a characteristic property of the test (9.4.1): there exists a positive constant ϵ (an arbitrary number less than C) such that the inequality

$$X^2 \leqslant \varepsilon_0 Q \tag{9.4.6}$$

implies that the test assumes the null hypothesis H_0. Thus the test that we are studying has the following two properties.

I. It is defined on the sample space defined by the forms $X, l_1, \cdots, l_\mu$.

II. It assumes the null hypothesis H_0 when inequality (9.4.6) is satisfied.

Here our test is a Neyman structure defined by $X, l_1, \cdots, l_\mu$. Therefore a complete description of tests of the Bartlett-Scheffé type reduces to construction of Neyman structures (see §2, Chapter IV).

The purpose of what follows is to prove that properties I and II already characterize the corresponding tests since they are Neyman structures for suitable exponential families.

Suppose that we have a randomized or unrandomized test $\phi = \phi(x_1, \cdots, x_{n_1}; y_1, \cdots, y_{n_2})$ for the Behrens-Fisher problem. Here, of course,

$$x_1, \ldots, x_{n_1} \in N(a_1, \sigma_1^2);\ y_1, \ldots, y_{n_2} \in N(a_2, \sigma_2^2).$$

Suppose that the test Φ is defined in the sample space generated by $\mu + 1$ linear forms $X, l_1, \cdots, l_\mu$, where $\mu \leq n_1 + n_2 - 1$, that are linearly and stochastically independent. Suppose that under the hypothesis H_0: $a_1 = a_2$ we have

$$E(X|H_0)=0, \quad E(l_i|H_0)=0, \quad i=1, 2, \ldots, \mu. \tag{9.4.7}$$

Here the variances of the statistics $X, l_1, \cdots, l_\mu$ can be various binary quadratic forms of the unknown parameters σ_1 and σ_2.

The first of these forms, X, will play a special role. Let us call this form the *leader* of the test. We now introduce the concept of *standardizer* of a test. By this we mean a continuous function $T(l_1, \cdots, l_\mu)$ that is homogeneous of degree $\nu > 0$ and that satisfies the conditions $T(l_1, \cdots, l_\mu) > 0$ for $(l_1, \cdots, l_\mu) \neq (0, \cdots, 0)$, with $T(l_1, \cdots, l_\mu) > c_0 > 0$, if at least one of the variables is equal to 1.

For arbitrary $s > 0$,

$$T(sl_1, \ldots, sl_\mu) = s^\nu T(l_1, \ldots, l_\mu). \tag{9.4.8}$$

In addition the standardizer T must be related to the leader X and the test ϕ as follows: there exists a positive number ϵ_0 such that the inequality

$$|X|^\nu < \varepsilon_0 T \tag{9.4.9}$$

implies the test ϕ assumes the null hypothesis H_0 with probability 1. Of course the test may assume this hypothesis even though inequality (9.4.9) is not satisfied.

Now we can formulate the fundamental theorem.

Theorem 9.4.1. *Suppose that a randomized similar test* $\phi = \phi(x_1, \cdots, x_{n_1}; y_1, \cdots, y_{n_2})$ *is defined on the sample space of linear forms* X, $l_1, \cdots, l_\mu$ *(satisfying the conditions listed above). Suppose that it has a leader* X *and a standardizer* T *such that it assumes the null hypothesis* H_0 *at least when condition* (9.4.9) *is satisfied. Then*

$$\frac{D(l_i)}{D(X)} = a_i \quad (i = 1, 2, \ldots, \mu), \tag{9.4.10}$$

where the a_i *are positive constants independent of* σ_1 *and* σ_2 *and our test is a Neyman structure for the exponential family generated by* $X, l_1, \cdots, l_\mu$.

We define $D(X) = \sigma^2 = f(\sigma_1, \sigma_2)$, where f is a binary quadratic form. Then $D(l_i) = a_i \sigma^2$ and the exponential family generated by $X, l_1, \cdots, l_\mu$ is a one-parameter family (the parameter being σ) and has the sufficient statistic

$$Q_0 = X^2 + \sum_{i=1}^{\mu} \frac{l_i^2}{a_i}.$$

Corollary. *Under the conditions of the theorem the test* ϕ *is a Neyman structure for the statistic* $Q_0 = X^2 + \Sigma_{i=1}^{\mu} l_i^2/a_i$. *If it is unrandomized, it is obtained by singling out the regions of given conditional probability on the surfaces of level* Q_0 *and "gluing" them together.*

In particular, for tests of the Bartlett-Scheffé type, the quantity $X = \bar{x} - \bar{y}$ serves as a leader and $T = \Sigma_{i=1}^{n} [(x_i - \bar{x}) - (y_i - \bar{y})]^2 = T(l_1, \cdots, l_{n-1})$, where $l_i = (x_i - \bar{x}) - (y_i - \bar{y})$ (for $i = 1, 2, \cdots, n-1$) serves as standardizer. Therefore the test is a Neyman structure of the special form (9.4.1).

We note also that if $T = T(l_1, \cdots, l_\mu)$ is a standardizer corresponding to a positive number ϵ_0, it can always be replaced with the standardizer $T_1 = l_1^2 + \cdots + l_\mu^2$ by replacing the number ϵ with $\epsilon_1 < \epsilon_0$. To see this, note that if a test ϕ assumes H_0 under the condition (9.4.9), it also assumes it under the condition

$$|X|^2 < \varepsilon_1 T_1$$

for sufficiently small $\epsilon_1 > 0$.

Let us now turn to the proof of our theorem. We have

$$D\{x_i\} = \sigma_1^2 \quad (i=1, \ldots, n_1); \quad D\{y_i\} = \sigma_2^2 \quad (i = 1, 2, \ldots, n_2).$$

Suppose that $X, l_1, \cdots, l_\mu$ (where $\mu \leq n_1 + n_2 - 2$) are linear forms in $x_1, \cdots, x_{n_1}; y_1, \cdots, y_{n_2}$ that satisfy the hypothesis of the theorem. Then obviously

$$D\{X\} = b_1\sigma_1^2 + b_2\sigma_2^2, \quad D\{l_i\} = c_{i1}\sigma_1^2 + c_{i2}\sigma_2^2. \tag{9.4.11}$$

(Here and in what follows the letters b, c, d and e with subscripts are positive constants.) We set

$$b_1\sigma_1^2 = \vartheta_1, \quad b_2\sigma_2^2 = \vartheta_2; \tag{9.4.12}$$

Thus

$$D\{X\} = \vartheta_1 + \vartheta_2, \quad D\{l_i\} = c_{i1}\vartheta_1 + c_{i2}\vartheta_2, \tag{9.4.13}$$

where the constants c_{ij} are reindexed.

Now let us consider a test $\phi = \phi(x_1, \cdots, x_{n_1}; y_1, \cdots, y_{n_2})$, that satisfies the hypotheses of the theorem. Then $\phi = \phi(X, l_1, \cdots, l_\mu)$, where X is the leader of the test. Our test must be similiar with respect to the parameters σ_1 and σ_2 when the hypothesis H_0: $a_1 = a_2$ is satisfied. Here $E\{X|H_0\} = E\{l_i|H_0\} = 0$ $(i = 1, 2, \cdots, \mu)$. For a given level of the test $\alpha > 0$, the assumption that the test is similar leads, by what was said above, to the integral relationship

$$\int_x \cdots \int \varphi_1(X, l_1, \ldots, l_\mu) \times \exp\left[-\frac{1}{2}\left(\frac{X^2}{\vartheta_1+\vartheta_2}+\sum_{i=1}^{\mu}\frac{l_i^2}{c_{i1}\vartheta_1+c_{i2}\vartheta_2}\right)\right] dX\, dl_1 \ldots dl_\mu = A_0(\vartheta_1+\vartheta_2)^{\frac{1}{2}}\prod_{i=1}^{\mu}(c_{i1}\vartheta_1+c_{i2}\vartheta_2)^{\frac{1}{2}}, \tag{9.4.14}$$

which is identically satisfied for $\vartheta_1, \vartheta_2 > 0$. Here $A_0 > 0$ is an absolute constant and X is the space of the variables $(X, l_1, \cdots, l_\mu)$. Let us set

$$\vartheta_1 = \theta_1\omega^{-1}, \quad \vartheta_2 = \theta_2\omega^{-1}, \tag{9.4.15}$$

where θ_1, θ_2, and ω are new positive parameters.

When we substitute these values into (9.4.14) we obtain the identity

$$\int_x \cdots \int \varphi_1(X, l_1, \ldots, l_\mu) \exp\left[-\frac{\omega}{2}\left(\frac{X^2}{\theta_1+\theta_2}+\sum_{i=1}^{\mu}\frac{l_i^2}{c_{i1}\theta_1+c_{i2}\theta_2}\right)\right] \omega^{\frac{\mu+1}{2}} dX\, dl_1 \ldots dl_\mu = A_0(\theta_1+\theta_2)^{\frac{1}{2}}\prod_{i=1}^{\mu}(c_{i1}\theta_1+c_{i2}\theta_2)^{\frac{1}{2}}, \tag{9.4.16}$$

which holds for arbitrary positive values of θ_1, θ_2 and ω. We multiply both sides of this identity by the expression $\omega^{\alpha}e^{a\omega/2}$, where $a > 0$ and $\alpha > -1$. We obtain

$$\int_x \cdots \int \varphi_1(X, l_1, \ldots, l_\mu) \exp\left[-\frac{\omega}{2}\left(\frac{X^2}{\theta_1+\theta_2}+\sum_{i=1}^{\mu}\frac{l_i^2}{c_{i1}\theta_1+c_{i2}\theta_2}+a\right)\right] \omega^{\alpha+\frac{\mu+1}{2}} dX\, dl_1 \ldots dl_\mu = A_0(\theta_1+\theta_2)^{\frac{1}{2}}\prod_{i=1}^{\mu}(c_{i1}\theta_1+c_{i2}\theta_2)^{\frac{1}{2}}\omega^{\alpha}e^{-\frac{a}{2}\omega} \tag{9.4.17}$$

For arbitrary $\eta > 0$ the integral of this function with respect to ω from η to ∞ converges absolutely. Furthermore

$$\int_0^\infty \omega^\alpha e^{-\frac{a}{2}\omega}\,d\omega = \left(\frac{a}{2}\right)^{-\alpha-1}\Gamma(1+\alpha) = a^{-\alpha-1}G(\alpha). \qquad (9.4.18)$$

where $G(\alpha)$ is a function that is regular in the half-plane $\operatorname{Re}\alpha > -1$. Let us suppose that θ_1 and θ_2 are positive numbers. If we integrate both sides of (9.4.17) with respect to ω from η to ∞ and then let $\eta \downarrow 0$ we obtain, by virtue of the boundedness and measurability of ϕ_1,

$$\int \cdots_{x} \int \varphi_1(X, l_1, \ldots, l_\mu)$$

$$\times\left(\frac{X^2}{\theta_1+\theta_2} + \sum_{i=1}^{\mu}\frac{l_i^2}{c_{i1}\theta_1 + c_{i2}\theta_2} + a\right)^{-\alpha-1-\frac{\mu+1}{2}} dX\,dl_1\ldots dl_\mu$$

$$= A_0 G(\alpha)\,a^{-\alpha-1}(\theta_1+\theta_2)^{1/2}\prod_{i=1}^{\mu}(c_{i1}\theta_1 + c_{i2}\theta_2)^{1/2}. \qquad (9.4.19)$$

In this equation we set $\theta_1 = \theta$, $\theta_2 = 1$ and $\alpha = -1 + \tau_0$, where τ_0 is a small positive number whose precise value we shall determine later. Now multiply both sides of (9.4.14) by $(1+\theta)^{-\tau_0-(\mu+1)/2}$. We get

$$\int \cdots_{x} \int \varphi_1(X, l_1, \ldots, l_\mu)$$

$$\times\left(X^2 + (1+\theta)\sum\left(\frac{l_i^2}{c_{i1}\theta + c_{i2}} + a\right)\right)^{-\tau_0-\frac{\mu+1}{2}} dX\,dl_1\ldots dl_\mu$$

$$= A_0 G_1(\tau_0)\,a^{-\tau_0}(1+\theta)^{-\tau_0-\frac{\mu}{2}}\prod_{i=1}^{\mu}(c_{i1}\theta + c_{i2})^{1/2}. \qquad (9.4.20)$$

Here a and θ are arbitrary positive numbers and $G_1(\tau_0)$ is a function that is regular for $\operatorname{Re}\tau_0 > 0$.

Now consider the fractions

$$r = \frac{1+\theta}{c_{i1}\theta + c_{i2}}, \qquad i = 1, 2, \ldots, \mu. \qquad (9.4.21)$$

These fractions may include some in which $c_{i1} = c_{i2}$, so that $r_i = 1/c_{i1} = 1/c_{i2}$ is independent of θ. This means that $D\{l_i\}/D\{X\}$ is independent of θ_1 and θ_2. If this property is satisfied for all values of $i = 1, 2, \cdots, \mu$, it will follow immediately that our theorem is valid. It is just this property that we wish to prove in what follows.

Let us suppose that this property is not satisfied for all indices $i = 1, 2, \cdots, \mu$. Suppose that it is satisfied for $i = 1, 2, \cdots, p$, where $0 \leq p < \mu$, and that it is not satisfied for $i = p + 1, \cdots, \mu$. (Of course this reindexing of the l_i does not restrict the generality.)

For $i = 1, 2, \cdots, p$, we write $b_i = 1/c_{i1} = 1/c_{i2}$. From (9.4.20), we obtain

$$\int_x \cdots \int \varphi_1(X, l_1, \ldots, l_\mu) \times \frac{dX\, dl_1 \ldots dl_\mu}{\left(X^2 + b_1 l_1^2 + \ldots + b_p l_p^2 + (1+\theta) \sum \left(\frac{l_i^2}{c_{i1}\theta + c_{i2}} + a\right)\right)^{\tau_0 + \frac{\mu+1}{2}}} = A_1 G_1(\tau_0)\, a^{-\tau_0} (1+\theta)^{-\tau_0 - \frac{\mu - p}{2}} \prod_{i=p+1}^{\mu} (c_{i1}\theta + c_{i2})^{1/2}. \tag{9.4.22}$$

Here $A_1 = A(b_1, \cdots, b_p)^{-1/2} > 0$. Also $\mu - p \geq 1$ since $p < \mu$.

The expression (9.4.22) is an analytic regular function of θ for $\operatorname{Re} \theta > 0$. Obviously it can be extended as a regular function to the complex θ-plane cut along the negative axis $-\infty < \operatorname{Re} \theta \leq 0$, $\operatorname{Im} \theta = 0$. We define its values as we approach the upper and lower edges of the cut in such a way as to preserve its continuity with the aid of rectifiable paths extending from points of the positive axis $\operatorname{Re} \theta > 0$, $\operatorname{Im} \theta = 0$. Then the two sides of (9.4.22) coincide. We define

$$Q_1 = X^2 + l_1^2 + \ldots + l_p^2,$$
$$Q_{2\theta} = \sum_{i=p+1}^{\mu} \frac{l_i^2}{c_{i1}\theta + c_{i2}} + a. \tag{9.4.23}$$

Let us set

$$\theta = -1 + i\zeta, \quad a = \zeta, \tag{9.4.24}$$

where ζ is a small positive number, which we shall make arbitrarily small in what follows.

We partition the region of integration $\mathfrak{X}$:

$$-\infty < X < \infty, \quad -\infty < l_i < +\infty \quad (i = 1, 2, \ldots, \mu)$$

into "layers:"

$$|X| \leqslant \zeta \tag{9.4.25}$$

and

$$|X| \in (2^{m-1}\zeta, 2^m\zeta] \quad (m = 1, 2, 3, \ldots). \tag{9.4.26}$$

Let us find an upper bound for the absolute value of the expression on the left-hand side of (9.4.22), where θ is the value indicated in (9.4.24). Consider the integral over the "layer" (9.4.25). On the basis of condition (9.4.9) regarding the leader and the standardizer of the test we have $\phi \equiv 0$ for $T \geq |X|^{\nu}/\zeta_0$. By the homogeneity of the standardizer T and its properties we then see that for $\phi \neq 0$

$$|l_i| \leqslant c_1 |X| \quad (i = 1, 2, \ldots, \mu), \tag{9.4.27}$$

where c_1 is a positive constant.

Now consider the integral (9.4.22) over the layer (9.4.25). By (9.4.27) we have in this case

$$|l_i| \leqslant c_1 \zeta. \tag{9.4.28}$$

Furthermore,

$$(1+\theta)\, a = i\zeta^2. \tag{9.4.29}$$

By (9.4.28) and (9.4.29), we have in the layer (9.4.25)

$$(1+\theta)\, Q_{2,\theta} = B\zeta X^2 + i\zeta^2, \tag{9.4.30}$$

where B is a bounded function. (In what follows the letter B will not always denote the same bounded function.) From this we see that for sufficiently small ζ,

$$|Q_1 + (1+\theta)\, Q_{2,\theta}| \geqslant |X^2 + B\zeta X^2 + i\zeta^2| \geqslant \left| \frac{1}{2} X^2 + i\zeta^2 \right| > \zeta^2. \tag{9.4.31}$$

everywhere in the layer (9.4.25). By virtue of condition (9.4.28) the integral over the layer (9.4.25) is bounded by

$$\frac{B\zeta^{\mu+1}}{(\zeta^2)^{\frac{\mu+1}{2}+\tau_0}} = B\zeta^{-2\tau_0}. \tag{9.4.32}$$

Now let us take the layer L_m defined by (9.4.26) for some value of $m = 1, 2, \cdots$. For the "layer" L_m we obtain on the basis of (9.4.27)

$$|l_i| \leqslant c_1 2^m \zeta,$$

$$|Q_1 + (1+\theta)\, Q_{2,\theta}| \geqslant |X^2 + B\zeta X^2 + i\zeta^2| \geqslant \frac{1}{2} X^2 \geqslant 2^{m-2}\zeta^2 \tag{9.4.33}$$

(assuming ζ sufficiently small). Hence the portion of the integral (9.4.22) over the layer L_m is bounded by

$$\frac{B\,(2^m\zeta)^{\mu+1}}{(2^{m-2}\zeta^2)^{\frac{\mu+1}{2}+\tau_0}} = B\zeta^{-2\tau_0}2^{m\tau_0}. \tag{9.4.34}$$

In this expression the number B is bounded by an absolute constant for all values of $m = 1, 2, \cdots$, and $\zeta > 0$. Now the number τ_0 can be chosen arbitrarily in the interval $0 < \tau_0 \leq 0.01$. Then if we sum the estimates (9.4.34) with respect to $m = 1, 2, \cdots$ and then the estimate (9.4.32), we obtain the following upper bound for the left-hand member of (9.4.22):

$$B\zeta^{-2\tau_0}. \tag{9.4.35}$$

Let us now look at the behavior of the right-hand member of (9.4.22). As was stated above, here we have $\mu - p \geq 1$. Furthermore, for $i = p+1, \cdots, \mu$ we have $c_{i1} \neq c_{i2}$, so that

$$\left| \prod_{i=p+1}^{\mu} (c_{i1}\theta + c_{i2})^{1/2} \right| > c_2 > 0$$

for $\theta = -1 + i\zeta$ and sufficiently small $\zeta > 0$. Thus the right-hand member of (9.4.22) has the following bound from below in absolute value:

$$c_3\zeta^{-2\tau_0}\zeta^{-\frac{\mu-p}{2}} \geqslant c_3\zeta^{-\frac{1}{2}-2\tau_0}. \tag{9.4.36}$$

Comparing this inequality with (9.4.35), we obtain a contradiction for $\zeta = 0$ sufficiently small.

Thus the inequality $\mu - p > 0$ is impossible and we have $p = \mu$. Hence the ratios

$$\frac{D\{l_i\}}{D\{X\}} = a_i$$

are independent of the values of θ_1 and θ_2. This establishes one of the assertions of the theorem.

In the space of the linear forms $(X, l_1, \cdots, l_\mu)$ the element of probability is therefore

$$c_0\sigma^{-(\mu+1)}\exp\left[-\frac{1}{2\sigma^2}\left(X^2 + \sum_{i=1}^{\mu}\frac{l_i^2}{a_i}\right)\right] dX\,dl_1\ldots dl_\mu,$$

where $c_0 > 0$ is a constant and $\sigma^2 = \vartheta_1 + \vartheta_2$.

Thus we have an exponential family with a single parameter σ^2 and a single sufficient statistic

$$Q_0 = X^2 + \sum_{i=1}^{\mu} \frac{l_i^2}{a_i},$$

so that in this case all similar tests are Neyman structures with probability

$$E\{\varphi \mid Q_0\} = \alpha$$

for all values of Q_0. This completes the proof of the theorem. Its corollary regarding the structure of unrandomized similar tests can be proven directly.

The Bartlett-Scheffé test is the special form of Neyman structure defined by the left-hand member of formula (9.2.8).

Special forms of linear forms l_i are chosen in such a way that property (9.4.10), which we have proven, is satisfied. Just which of the various Neyman structures are preferred in this sense depends on the requirements imposed on the power properties of the test.

We note that the method that we have just expounded is applicable to the study of possible construction of many tests in multivariant analysis.

CHAPTER X

AN UNRANDOMIZED HOMOGENEOUS SIMILAR TEST IN THE BEHRENS-FISHER PROBLEM

§1. CONSTRUCTION OF AN UNRANDOMIZED TEST

In the preceding chapter we discussed certain results dealing with unrandomized homogeneous similar tests in the Behrens-Fisher problem that have critical zone Z of the form $G(|\bar{x} - \bar{y}|/s_2,\ s_1/s_2) \geq 0$. These results are negative in nature and they point out the "poor" analytical properties of the boundary of the critical zone of such a test if such a test exists. However, if we require only that the boundary of a test be measurable, a test will in many cases still exist, as was shown in 1964 in the simultaneous articles [24] and [50]. Following the procedure used in [50], we shall now prove two theorems.

Theorem 10.1.1. *For an arbitrary level* $\alpha \in (0, 1)$ *and a pair of samples* n_1, n_2 *of opposite parity, there exists an unrandomized similar test for the Behrens-Fisher problem with critical zone defined by the values of* $|\bar{x} - \bar{y}|/s_2$ *and* s_1/s_2.

Theorem 10.1.1 can be strengthened somewhat.

Theorem 10.1.2. *Suppose that we have a finite number* K *of pairs of samples of sizes* n_{1i} *and* n_{2i} $(i = 1, \cdots, K)$ *of opposite parity. Then, there exists an unrandomized homogeneous test with critical zone*

$$G\left(\left|\frac{\bar{x}_i - \bar{y}_i}{s_{2i}}\right|,\ \frac{s_{1i}}{s_{2i}}\right) \geqslant 0$$

(*where* $i = 1, 2, \cdots, K$ *and the function* G *is the same for all values of* i) *that is similar for all pairs of samples and that has a prescribed level* $\alpha \in (0, 1)$.

Proof of Theorem 10.1.1. To construct a test $\Phi = \Phi(|\bar{x} - \bar{y}|/s_2,\ s_1/s_2)$, that is of level $\alpha \in (0, 1)$ and that assumes only the values 0 and 1, it will be sufficient to construct a cotest $\Psi = \Phi - \alpha$. The function Ψ must assume only the values $-\alpha$ and $1 - \alpha$. Define $\xi = (\bar{x} - \bar{y})/s_2$, $\eta = s_1/s_2$. Consider the critics $A(\xi, \eta)$ and $B(\xi, \eta)$. Let $\Phi(\xi, \eta)$ denote the characteristic function of the critical zone of the test sought and let Ω_1 denote the quadrant $\xi \geq 0$, $\eta \geq 0$. Then to find $\Phi(\xi, \eta)$ we have (see (8.4.2))

$$\iint_{\Omega_1} \frac{\eta^{n_1-2}\Phi(\xi, \eta)\, d\xi\, d\eta}{(\theta + A)^N (\theta + B)^N} = C_{n_1 n_2} \frac{\alpha}{2} \theta^{-\frac{n_2}{2}} (1+\theta)^{-N+\frac{1}{2}} \tag{10.1.1}$$

for arbitrary $\theta > 0$. Here $N = (n_1 + n_2 - 1)/2$. For the cotest $\Psi(\xi, \eta)$ we have the identity

$$\iint_{\Omega_1} \frac{\eta^{n_1-2}\Psi(\xi, \eta)\, d\xi\, d\eta}{(\theta + A)^N (\theta + B)^N} = 0 \tag{10.1.2}$$

for all $\theta > 0$. If the function $\Psi(\xi, \eta)$ assumes only the values $-\alpha$ and $1 - \alpha$ and if it satisfies (10.1.2), it will be the desired cotest.

By hypothesis the numbers n_1 and n_2 are of opposite parity. It follows that the number N is an integer. Let us assume that $n_1 \geq 2$ and $n_2 \geq 2$. Since these numbers are of opposite parity at least one of them is 3 or greater and $N \geq 2$.

In the quadrant $\Omega_1 : \xi \geq 0, \eta \geq 0$ the critics $A(\xi, \eta)$ and $B(\xi, \eta)$ constitute a coordinate system. Through each point of this quadrant passes one and only one critic of each of the two types. In examining the integral (10.1.2) we shall find it convenient to change to this coordinate system. To do this we need to find the Jacobian $\partial(\xi, \eta)/\partial(A, B)$. Since $(\theta + A)(\theta + B) = \theta^2 + \theta(1 + \xi^2 + \eta^2) + \eta^2$, we have $A + B = 1 + \xi^2 + \eta^2$ and $AB = \eta^2$. Thus $\xi = [(B-1)(1-A)]^{1/2}$ and $\eta = (AB)^{1/2}$. Therefore

$$\frac{\partial(\xi, \eta)}{\partial(A, B)} = \begin{vmatrix} -\frac{1}{2}(B-1)^{\frac{1}{2}}(1-A)^{-\frac{1}{2}} & \frac{1}{2}(B-1)^{-\frac{1}{2}}(1-A)^{\frac{1}{2}} \\ \frac{1}{2}A^{-\frac{1}{2}}B^{\frac{1}{2}} & \frac{1}{2}A^{\frac{1}{2}}B^{-\frac{1}{2}} \end{vmatrix}$$

$$= -\frac{1}{4}\frac{B-A}{(AB)^{\frac{1}{2}}(B-1)^{\frac{1}{2}}(1-A)^{\frac{1}{2}}}. \tag{10.1.3}$$

Inserting this expression into (10.1.2) and making the change of coordinates, we obtain an equation for finding the cotest $\Psi(\xi, \eta) = \Psi_1(A, B)$:

$$J(\theta) = \iint_{\Pi} \frac{\Psi_1(A, B)(AB)^{\frac{n_1-3}{2}}(B-A)\, dA\, dB}{\sqrt{1-A}\sqrt{B-1}\,(\theta + A)^N (\theta + B)^N} = 0. \tag{10.1.4}$$

Here Π is the half-strip $0 \le A \le 1$, $B \ge 1$ into which the quadrant Ω_1 is mapped.

Let us partition the half-strip Π into a sequence of disjoint half-strips:

$$\Pi_k : 1 - 2^{-k} \leqslant A < 1 - 2^{-k-1};\ 1 \leqslant B < \infty;\ k = 0, 1, 2, \ldots;$$

$$\Pi = \bigcup_{k=0}^{\infty} \Pi_k.$$

Let us construct the function $\Psi(A, B)$ in such a way that for every $k = 0, 1, 2, \cdots$ we have

$$J_k(\theta) = \iint_{\Pi_k} \frac{\Psi_1(A, B)(AB)^{\frac{n_1-3}{2}}(B - A)\,dA\,dB}{\sqrt{1-A}\sqrt{B-1}(\theta + A)^N(\theta + B)^N} = 0. \qquad (10.1.5)$$

Without loss of generality we can assume that $n_1 \ge 3$.

The function $(B - A)/\sqrt{1 - A}\sqrt{B - 1}$ has only an integral singularity at the point $(1, 1)$. Since $(n_1 - 3)/2 \ge 0$ it follows from equations (10.1.5) $(k = 0, 1, \cdots)$ that $J(\theta) = 0$. Since N is an integer and since $A \ne B$ in the half-strip Π_k, we have the decomposition into partial fractions

$$\frac{1}{(\theta + A)^N(\theta + B)^N} = \sum_{m=1}^{N} D_m \frac{(B - A)^{m-1}}{(B - A)^{2N-1}(\theta + A)^m} + \sum_{m=1}^{N} E_m \frac{(B - A)^{m-1}}{(B - A)^{2N-1}(\theta + B)^m}, \qquad (10.1.6)$$

where D_i and E_i $(i = 1, 2, \cdots, N)$ are constants. If we substitute (10.1.6) into the left-hand member of (10.1.5) we find that

$$J_k(\theta) = \sum_{m=1}^{N} D_m \int_{1-2^{-k}}^{1-2^{-k-1}} dA \times \int_1^{\infty} dB \frac{\Psi_1(A, B)(AB)^{\frac{n_1-3}{2}}(B - A)^m}{\sqrt{1-A}\sqrt{B-1}(B - A)^{2N-1}(\theta + A)^m}$$
$$+ \sum_{m=1}^{N} E_m \int_1^{\infty} dB \int_{1-2^{-k}}^{1-2^{-k-1}} dA \times \frac{\Psi_1(A, B)(AB)^{\frac{n_1-3}{2}}(B - A)^m}{\sqrt{1-A}\sqrt{B-1}(B - A)^{2N-1}(\theta + B)^m}. \qquad (10.1.7)$$

Now let us partition the strip Π_k into rectangles:

$$Q_{ks} : 1-2^{-k} \leqslant A < 1-2^{-k-1};\ s \leqslant B < s+1;\ s=1, 2, \ldots .$$

Define

$$P_m(A,\ B) = \frac{(AB)^{\frac{n_1-3}{2}}(B-A)^m}{\sqrt{1-A}\sqrt{B-1}(B-A)^{2N-1}},\quad m=1, 2, \ldots, N.$$

On each rectangle Q_{ks} we choose a constant $C_{ks} > 0$ such that the functions

$$p_{mks}(A,\ B) = \frac{1}{C_{ks}} P_m(A,\ B),\quad m=1,\ 2,\ \ldots,\ N,$$

will be probability densities with respect to Lebesgue measure on that rectangle. Obviously this is possible since $B > A$, so that $P_m(A, B) \geq 0$. Furthermore the $p_{mks}(A, B)$ are continuous functions on the closure $\overline{Q}_{ks}$ of the rectangle Q_{ks} for $k = 0, 1, 2, \cdots$.

At this point we apply the Romanovskiĭ-Sudakov theorem (which we shall prove in the next section) to the probability densities $p_{mks}(A, B)$ considered on $\overline{Q}_{ks}$ for each k, s, and N. For given k and s suppose that $\overline{Q}_{ks}$ is of the form $a \leq A \leq b$, $c \leq B \leq d$. By Theorem 10.2.1 we can construct a measurable dense set $\mathfrak{A} = \mathfrak{A}_{ks}$ such that, for $m = 1, 2, \cdots, N$,

$$\int_a^b I(A,\ B)\, p_{mks}(A,\ B)\, dA = \alpha \int_a^b p_{mks}(A,\ B)\, dA$$

for almost all $B \in [c,\ d]$,

$$\int_c^d I(A,\ B)\, p_{mks}(A,\ B)\, dB = \alpha \int_c^d p_{mks}(A,\ B)\, dB$$

for almost all $A \in [a,\ b]$,

where $I(A, B) = I_{ks}(A, B)$ is the characteristic function of the set $\mathfrak{A} = \mathfrak{A}_{ks}$ and $\alpha \in (0, 1)$ is the level of the test. The function $I(A, B)$ assumes only the values 0 and 1. Let us define $\Psi_1(A, B) = I(A, B) - \alpha$. Then $\Psi_1(A, B)$ assumes only the values $-\alpha$ and $1-\alpha$. Obviously, for $m = 1, 2, \cdots, N$,

$$\int_a^b \Psi_1(A,\ B)\, p_{mks}(A,\ B)\, dA = 0 \qquad \text{for almost all } B \in [c,\ d],$$

$$\int_c^d \Psi_1(A,\ B)\, p_{mks}(A,\ B)\, dB = 0 \qquad \text{for almost all } A \in [a,\ b].$$

Since the function p_{mks} differs from $P_m(A, B)$ only by a constant factor, if we extend the domain of definition of $\Psi_1(A, B)$ to the entire interior of the strip Π by setting it equal to the functions p_{mks} constructed on the individual Q_{ks} ($k = 1, 2, \cdots$; $s = 1, 2, \cdots$), we obtain

$$J_k(\theta) = 0 \quad (k = 0, 1, 2, \cdots) \text{ and } J(\theta) = 0 \text{ for all } \theta > 0.$$

This completes the proof of Theorem 10.1.1. To prove Theorem 10.1.2 we consider not the N functions $P_m(A, B)$ but the finitely many functions $P_{mi}(A, B)$, $i = 1, 2, \cdots, K$. For each i, $m = 1, 2, \cdots$, we have $N_i = (n_{1i} + n_{2i} - 1)/2$, where n_{1i} and n_{2i} are the sizes of the samples and

$$P_{mi}(A, B) = \frac{(AB)^{\frac{n_{1i}-3}{2}} (B - A)^m}{\sqrt{1 - A}\sqrt{B - 1}(B - A)^{2N_i - 1}}.$$

The corresponding integrals of the type $J(\theta)$ vanish simultaneously. The variables A and B are expressed in terms of $\bar{x}_i$, $\bar{y}_i$, s_{1i}, and s_{2i}.

§2. THE ROMANOVSKIĬ-SUDAKOV THEOREM [1)]

To give a firm foundation to the construction of an unrandomized similar test that we made in §1 we need to prove the following theorem, which we applied in that section and which was first published by Romanovskiĭ and Sudakov in [50] and [64].

Theorem 10.2.1. *Suppose that n probability densities $p_s(x, y)$, $s = 1, \cdots, n$, are defined on a rectangle M: $a \le x \le b$, $c \le y \le d$. Then, from given $\alpha \in [0, 1]$, one can construct a set $A = A_\alpha \subset M$ such that*

$$\left.\begin{aligned} &\int_a^b I_A(x, y)\, p_s(x, y)\, dx = \alpha \int_a^b p_s(x, y)\, dx \quad \text{for almost all } y \in [c, d] \\ &\text{and} \\ &\int_c^d I_A(x, y)\, p_s(x, y)\, dy = \alpha \int_c^d p_s(x, y)\, dy \quad \text{for almost all } x \in [a, b] \end{aligned}\right\} \qquad (10.2.1)$$

for $s = 1, \cdots, n$, where $I_A(x, y)$ is the characteristic function of the set A.

Without loss of generality we may assume that $a = c = 0$ and $b = d = 1$. The

---

1) This section was written by V. N. Sudakov.

general case reduces to this one by a linear transformation of coordinates.

Here we shall prove the general theorem on the values that the integrals in the left-hand members of equations (10.2.1) can both assume. Theorem 10.2.1 then follows easily.

Theorem 10.2.2. *Let* $p_s(x, y)$, $s = 1, \cdots, n$, *denote probability densities defined on the unit square* M *and let* A *denote a measurable subset of* M. *The collection, as* A *ranges over the class of all measurable subsets of* M, *of all sets each consisting of the* $2n$ *functions*

$$\left(\int_0^1 I_A(x, y)\, p_s(x, y)\, dx,\right.$$
$$\left.\int_0^1 I_A(x, y)\, p_s(x, y)\, dy, \quad s = 1, \ldots, n\right),$$

coincides with the collection, as $f(x, y)$ *ranges over the class of all functions that are measurable on the unit square and that satisfy the inequality* $0 \leq f(x, y) \leq 1$, *of the* $2n$ *sets*

$$\left(\int_0^1 f(x, y)\, p_s(x, y)\, dx,\right.$$
$$\left.\int_0^1 f(x, y)\, p_s(x, y)\, dy, \quad s = 1, \ldots, n\right).$$

To prove this we need to construct, for each function that is measurable on the unit square and that is bounded by zero and unity, the corresponding subset A of that square. For such a function $f(x, y)$ let D_f denote the set of measurable functions $g(x, y)$ (defined only up to the values they assume on a set of measure 0) that satisfy the conditions

$$\left.\begin{aligned} &\int_0^1 g(x, y)\, p_s(x, y)\, dx \\ &\qquad = \int_0^1 f(x, y)\, p_s(x, y)\, dx, \quad s = 1, \ldots, n, \\ &\int_0^1 g(x, y)\, p_s(x, y)\, dy \\ &\qquad = \int_0^1 f(x, y)\, p_s(x, y)\, dy, \quad s = 1, \ldots, n, \\ &\qquad\qquad 0 \leqslant g(x, y) \leqslant 1. \end{aligned}\right\} \tag{10.2.2}$$

Let us study the set D_f as a subset of the space $L_\infty(M)$ of bounded measurable functions defined on the square M. We know that the space $L_\infty(M)$ is dual to the space $L_1(M)$ of functions that are Lebesgue-summable on M. On the space $L_\infty(M)$ we introduce the weak topology [1)] of the dual space $\sigma(L_\infty, L)$. This means that $g_k(x, y) \longrightarrow g(x, y)$ if and only if for an arbitrary function $q(x, y) \in L_1(M)$

$$\iint_M q(x, y) g_k(x, y)\, dx\, dy \to \iint_M q(x, y) g(x, y)\, dx\, dy.$$

In the dual space every closed convex bounded subset is compact in the weak topology (see [8]).

Let us show that D_f is convex, closed and bounded and hence that it is compact in the topology of $\sigma(L_\infty, L_1)$. That D_f is bounded follows from the third of conditions (10.2.2). Each of conditions (10.2.2) defines a closed convex subset of L_∞. Proof of the convexity presents no difficulties. Also, the fact that the set defined by the third of conditions (10.2.2) is closed is also obvious. It remains to show that the first two of conditions (10.2.2) define closed subsets. The first of these conditions is equivalent to the condition that for an arbitrary bounded measurable function $\alpha(y)$, defined on $[0, 1]$.

$$\int_0^1 \alpha(y)\, dy \int_0^1 g(x, y) p_s(x, y)\, dx = \int_0^1 \alpha(y)\, dy \int_0^1 f(x, y) p_s(x, y)\, dx$$

or

$$\iint_M g(x, y) \alpha(y) p_s(x, y)\, dx\, dy = \text{const}.$$

---

1) A neighborhood of an element $\bar{g}(x, y)$ of $L_\infty(M)$ in the topology of $\sigma(L_\infty, L_1)$ is any set containing a set $V = V(\bar{g}; q_1, \cdots, q_k, \epsilon)$ consisting of all elements $g(x, y)$ of $L_\infty(M)$ such that

$$\left| \iint_M (\bar{g}(x, y) - g(x, y)) q_i(x, y)\, dx\, dy \right| < \varepsilon, \quad i = 1, \ldots, k$$

$$(q_i(x, y) \in L_1(M), \quad \varepsilon > 0).$$

We recall that a topological vector space is said to be *locally convex* if every neighborhood of every point in it contains a convex neighborhood of that point. Since the sets V are convex the space $L_\infty(M)$ is locally convex in the topology of $\sigma(L_\infty, L_1)$.

Since $\alpha(y)\,p_s(x, y) \in L_1(M)$, the set defined by the last equation is a closed hyperplane; hence the intersection of all such hyperplanes as $\alpha(y)$ ranges over the same set of all bounded measurable functions is also closed. The situation is analogous with the second of conditions (10.2.2). The set D_f, being the intersection of the sets examined above, is closed, convex and bounded; hence it is compact.

The proof of Theorem 10.2.2 will be complete if we can show that D_f contains functions $I(x, y)$ that assume only the values 1 and 0. Actually, not only does D_f contain such functions but it contains enough of them in the following sense.

A point a in a convex set D is called an *extremal point* of that set if the relations $b \in D$, $c \in D$ and $a = (b + c)/2$ imply that $a = b = c$. It is easy to see that each function $I(x, y) \in D_f$ that assumes only the values 0 and 1 is an extremal point of D_f. On the other hand, from the well-known Kreĭn-Milman theorem [8], every convex compact set in a locally convex Hausdorff space is the closed convex hull of the set of its extremal points. In particular this applies to the compact set D_f in the space $L_\infty(M)$ equipped with the weak topology of the dual space. Since the set D_f is nonempty, the set of its extremal points is also nonempty and even "sufficiently rich". It remains to show that the role of extremal points of D_f can be played only by functions that assume only the two values 0 and 1. To do this we construct, for every function $g(x, y) \in D_f$ for which the measure of the set $[(x, y) \in M;\ 0 < g(x, y) < 1]$ is positive, a nonzero function $h(x, y)$ such that $g + h \in D_f$ and $g - h \in D_f$. This will show that g is not an extremal point of D_f.

We begin our construction of such a function $h(x, y)$ by choosing a set $B \subset M$ of positive measure on which the function $g(x, y)$ is separated from both zero and unity by a positive amount:

$$B = \{(x,\ y) : 0 < \beta \leqslant g(x,\ y) \leqslant 1 - \beta\}.$$

Let us choose natural numbers k and l such that $k_l > n(k + l)$.

We know (see [66], p. 129) that almost all points of a plane measurable set B are points of density of that set. From this it follows that there exists a set of distinct numbers $\epsilon_0 = 0, \epsilon_1, \cdots, \epsilon_{k-1}$; $\delta_0 = 0, \delta_1, \cdots, \delta_{l-1}$ such that

$$\operatorname{mes} \bigcap_{i,\ j=0}^{\substack{i=k-1\\ j=l-1}} (B - \{(\varepsilon_i,\ \delta_j)\}) > 0.$$

(Here $B - \{(\xi, \eta)\} = \{(x - \xi, y - \eta);\ (x, y) \in B\}$.) In the set

$$\bigcap_{i,\, j=0}^{i=k-1,\ j=l-1} (B - \{(\varepsilon_i, \delta_j)\}),$$

let us choose a subset C of positive measure with diameter not exceeding the smallest of the differences $|\epsilon_p - \epsilon_q|$, $|\delta_p - \delta_q|$ for $p \neq q$. Obviously $(C + \{(\epsilon_{i_1}, \delta_{j_1})\}) \cap (C + \{(\epsilon_{i_2}, \delta_{j_2})\}) = \Lambda$ for $(i_1, j_1) \neq (i_2, j_2)$ and $C + \{(\epsilon_i, \delta_j)\} \subset B$ for arbitrary i, j.

For an arbitrary fixed point $(x, y) \in C$ let us consider the homogeneous linear system of equations in the unknowns h_{ij}:

$$\left.\begin{aligned} &\sum_{i=0}^{k-1} p_s(t_{ij}) h_{ij} = 0, \ s = 1, \dots, n, \ j = 0, \dots, l-1, \\ &\sum_{j=0}^{l-1} p_s(t_{ij}) h_{ij} = 0, \ s = 1, \dots, n, \ i = 0, \dots, k-1, \end{aligned}\right\} \tag{10.2.3}$$

where $t_{ij} = t_{ij}(x, y) = (x, y) + (\epsilon_i, \delta_j)$. Obviously the rank of the matrix of coefficients of this system does not exceed $k(k + l - 1)$.

Let us now reindex the unknowns h_{ij} in a single sequence by assigning to each a single index number: $h_1, h_2, \cdots, h_{kl}$ (where kl represents a single integer). To every point $(x, y) \in C$ let us assign the number $N(x, y)$ such that the rank of the submatrix of coefficients of the system (10.2.3) consisting of the coefficients of $h_1, \cdots, h_N$ is equal to N and the rank of the submatrix of coefficients of the system (10.2.3) consisting of the coefficients of $h_1, \cdots, h_{N+1}$ is also equal to N. If all the coefficients of h_1 are equal to 0 we take $N = 0$.

Now consider a subset C_1 of C that is of positive measure and on which the function $N(x, y)$ assumes a constant value. By the choice of N the system (10.2.3) will be satisfied if we assign to the unknown h_{N+1} an arbitrary value $h_{N+1} = \epsilon$ by setting $h_{N+2} = \cdots = h_{kl} = 0$ and then define unambiguously $h_1, \cdots, h_N$ in the system (10.2.3). Here the functions $h_1 = h_1(x, y, \epsilon), \cdots, h_N = h_N(x, y, \epsilon)$ are obviously measurable with respect to all the arguments and they are linearly independent with 1. Let us suppose that a single ϵ is chosen for all $(x, y) \in C_1$.

Now on the set

$$C_2 = \bigcup_{i,\, j} (C_1 + \{(\varepsilon_i, \delta_j)\})$$

we consider the measurable function $h(x, y)$ obtained by setting $h(x+\epsilon_i, y+\delta_j) = h_p(x, y, \epsilon)$, where $h_p(x, y, \epsilon)$ is the value of the unknown h_{ij} constructed as described above for the system (10.2.3) after its coefficients are evaluated at the point (x, y) and where the unknown h_{N+1} is assigned the value ϵ. Since the function $\tilde{h}$ is almost everywhere finite, it is bounded everywhere on C_2 except on a subset of arbitrarily small measure. Suppose that $|h(x, y)| \leq Q$ everywhere except on a set C_3 such that $\operatorname{mes} C_3 < \operatorname{mes} C_1$. Then

$$\operatorname{mes} C_1 \setminus \bigcup_{i,\, j=0}^{\substack{i=k-1\\ j=l-1}} (C_3 - \{(\varepsilon_i, \delta_j)\}) > 0.$$

Finally, we define the sets

$$C_4 = C_1 = \bigcup_{i,\, j=0}^{\substack{i=k-1\\ j=l-1}} (C_3 - \{(\varepsilon_i, \delta_j)\}),$$

$$C_5 = \bigcup_{i,\, j=0}^{\substack{i=k-1\\ j=l-1}} (C_4 + \{(\varepsilon_i, \delta_j)\}),$$

and we define on the square M the function

$$\tilde{\tilde{h}}(x, y) = \begin{cases} h(x, y), & (x, y) \in C_5, \\ 0, & (x, y) \in M \setminus C_5. \end{cases}$$

This function $\tilde{\tilde{h}}(x, y)$ is now defined on the entire square, and as one can easily see $|\tilde{\tilde{h}}(x, y)| \leq Q$. Let us show that the function

$$h(x, y) = \frac{B}{Q} \tilde{\tilde{h}}(x, y)$$

is the function we have been seeking. Let l denote the straight line $x = x_0$. If l intersects the set $C_4 + \epsilon_0$, then

$$\int_Q^1 \tilde{\tilde{h}}(x_0, y)\, p_s(x_0, y)\, dy$$

$$= \sum_{j=0}^{l-1} \int_{(C_4 + \{(\varepsilon_{i_0}, 0)\}) \cap l} \tilde{\tilde{h}}(x_0, y+\delta_j)\, p_s(x_0, y+\delta_j)\, dy.$$

This last expression is equal to

$$\int_{(C_4+\{(\varepsilon_{i_0}, 0)\})\cap l} p_s(t_{i_0 j})\, h_{i_0 j}\, d\mathfrak{y} = 0,$$

since the integrand is identically equal to 0 by virtue of the second group of equations (10.2.3). We treat analogously the integrals over straight lines of the form $y = y_0$ that intersect the set C_5. In the case of lines that do not intersect the set C_5, the function $\approx{h}$ is identically equal to 0, and on these, of course,

$$\int_0^1 \tilde{\tilde{h}}(x_0, y)\, p_s(x_0, y)\, dy = \int_0^1 \tilde{\tilde{h}}(x, y_0)\, p_s(x, y_0)\, dx = 0.$$

Similar equations hold for the function $h(x, y)$, which differs from $\tilde{\tilde{h}}(x, y)$ only by a constant factor. Consequently if the function $g(x, y) \in D_f$ is different from both zero and unity on a set of positive measure, then $g(x, y) + h(x, y) \in D_f$ and $g(x, y) - h(x, y) \in D_f$, so that $g(x, y)$ is not an extremal point of the convex compact set D_f. This completes the proof of the theorem.

CHAPTER XI

THE PROBLEM OF MANY SMALL SAMPLES

§1. STATEMENT OF THE PROBLEM

By the problem of many small samples we mean the problem of verifying a hypothesis H_0 regarding a class of distributions defined on a measurable space $(\mathfrak{X}, \mathfrak{A})$ with measure $\mu(\alpha)$ in the form of a family of densities

$$f_\theta(x) = f(x, \theta_1, \cdots, \theta_s),$$

where $\theta_1, \cdots, \theta_s$ are parameters that vary in E_s. They are not assumed to be interdependent. The hypothesis H_0 is the hypothesis that the samples belong to such a class under suitable choice of the parameters $\theta_1, \cdots, \theta_s$. This hypothesis is verified on the basis of q independent samples $O_1, \cdots, O_q$ of sizes $n_1, \cdots$ $\cdots, n_q$. The parameters $\theta_1, \cdots, \theta_s$ are assumed to remain constant for each of the samples but they vary from sample to sample. The sizes of the samples $n_1, \cdots, n_s$ are in general assumed to be small, so that we cannot use them to estimate the parameters $\theta_1, \cdots, \theta_s$ with the necessary accuracy.

Such a problem is encountered naturally in problems of continuous control of industrial production when it is desirable to use many small samples to verify the normality of a random deviation of some aspect of the production from the technological standard, in order to be able afterwards to establish some plan of continuous control on the basis of a normal law.

To verify the hypothesis H_0 it is natural to use the above-described methods of eliminating parameters and constructing similar statistics. However an important role is played here by the question as to what alternatives the hypothesis H_0 has. The choice of such statistics depends on this.

Let us consider the case in which the observations have an exponential-type distribution of the form

$$p_\theta(x) = C(\theta) \exp [-(\theta_1 T_1(x) + \cdots + \theta_s T_s(x))] \rho(x), \qquad (11.1.1)$$

on a measurable space $(\mathfrak{X}, \mathfrak{A})$, where $\rho(x)$ is the density with respect to the measure $\mu(x)$ on $(\mathfrak{X}, \mathfrak{A})$. Here the parameters are not assumed to be interdependent.

If no other choice of alternative is prescribed by the statement of the problem itself, we can take as an alternative a family with probability density of the form

$$p_\theta(x, \varepsilon, V) = C(\theta, \varepsilon) \times \exp[-(\theta_1 T_1(x) + \ldots + \theta_s T_s(x) + \varepsilon V(x))] \rho(x), \tag{11.1.2}$$

where $V(x)$ is a statistic and ϵ is a small number. Here we assume that $p_\theta(x, \epsilon, V)$ is a probability density for small values of ϵ. For $\epsilon = 0$ we obtain the hypothesis H_0. We denote the alternative for a given value of ϵ by H_ϵ. For an arbitrary admissible value of ϵ the function $T = (T_1(x), \cdots, T_s(x))$ is a sufficient statistic for $\theta = (\theta_1, \cdots, \theta_s)$. Since the parameters are assumed to be indepenent for each admissible value of ϵ, the family of sufficient statistics T is complete (see §2, Chapter IV).

If our independent samples $O_1, \cdots, O_q$ are of the same size n, we can consider the conditional distributions of the samples for fixed sufficient statistics in each sample. Such conditional distributions depend only on ϵ, V, and the values of the fixed sufficient statistics, and we can construct a criterion for distinguishing the hypotheses H_0 and H_ϵ of a given level that has minimal conditional probabilities of errors of the second kind.

For series of repeated samples $O_1, \cdots, O_q$ from one-dimensional distributions containing only parameters of displacement and dilation (these distributions are not necessarily assumed to be exponential), the construction of similar statistics can be achieved by constructing quite simple affine invariants of the sample space. Some interesting studies have been made by Petrov [60] in this direction. We shall expound these results in the following section.

§2. A. A. PETROV'S INVESTIGATIONS

Suppose that a repeated sample $x_1, \cdots, x_n$ is taken from a distribution with probability density $f(x)$ with respect to Lebesgue measure. We define

$$S = \sum_{i=1}^{n} (x_i - \bar{x})^2, \quad y_i = \frac{x_i - \bar{x}}{S}.$$

Then

$$\sum_{i=1}^{n} y_i = 0, \quad \sum_{i=1}^{n} y_i^2 = 1. \tag{11.2.1}$$

Thus the distribution of the random vector $\eta = (y_1, \cdots, y_n)$ is concentrated on the $(n-2)$-dimensional sphere $\mathfrak{S}$ defined by (11.2.1). Let us derive the distribution of η on the sphere $\mathfrak{S}$. We introduce the new coordinate system $\bar{x}, S, \phi_1, \cdots, \phi_{n-2}$, where $\phi_1, \cdots, \phi_{n-2}$ are the usual spherical coordinates for $\mathfrak{S}$. In the new coordinate system the element of probability is of the form

$$f(x_1), \ldots, f(x_n)\,dx_1\ldots dx_n$$
$$= \prod_{i=1}^{n} f(\bar{x}+Sy_i)\left|\frac{\partial(x_1, \ldots, x_n)}{\partial(\bar{x}, S, \varphi_1, \ldots, \varphi_{n-2})}\right| d\bar{x}\,dS\,d\varphi_1\ldots d\varphi_{n-2}.$$

The probability density η for a point $\phi_1, \cdots, \phi_{n-2}$ of the sphere $\mathfrak{S}$ is

$$p(\eta) = \int_0^\infty dS \int_{-\infty}^{\infty} \prod_{i=1}^{n} f(\bar{x}+Sy_i)\frac{\partial(x_1, \ldots, x_n)}{\partial(\bar{x}, S, \varphi_1, \ldots, \varphi_{n-2})}. \tag{11.2.2}$$

From this we easily obtain

$$p(\eta) = \sqrt{n}\int_0^\infty dS \int_{-\infty}^{\infty} S^{n-2} \prod_{i=1}^{n} f(\bar{x}+Sy_i)\,d\bar{x}. \tag{11.2.3}$$

In particular, for a normal sample

$$f(x;\ a,\ \sigma) = \frac{1}{\sqrt{2\pi}\sigma}\exp\left[-\frac{(x-a)^2}{2\sigma^2}\right],$$

the corresponding density $p_0(\eta)$ is of the form

$$p_0(\eta) = \frac{\sqrt{n}}{(\sqrt{2\pi})^n}\int_0^\infty dS \int_{-\infty}^{\infty} S^{n-2}\exp\left(-\frac{n\bar{x}^2+S^2}{2}\right)d\bar{x} = \frac{\Gamma\left(\frac{n-1}{2}\right)}{2\pi^{\frac{n-1}{2}}}.$$

Thus for a normal sample, the density η is constant.

The question naturally arises as to whether this property determines the normality of the original sample. For the case of a continuous original density $f(x)$ and sample size $n \geq 6$, a surprising answer has been obtained by Zinger [21]. A strengthened form of this assertion can be found in [22].

Since the vector $(y_1, \cdots, y_n)$ is an affine invariant, instead of $f(x;\ a,\ \sigma)$ we may consider $f_0(x) = f(x;\ 0,\ 1)$. As an alternative, let us take the type of densities that is represented by a density of the form

$$f_\varepsilon(x) = f_0(x)(1+\varepsilon H(x)+R(x,\ \varepsilon)), \tag{11.2.4}$$

where ϵ is a small number, $H(x)$ is a polynomial, and the function $R(x,\ \epsilon)$ satisfies, for all values of α, the inequality

$$|R(x,\ \varepsilon)| \leqslant \varepsilon^2 \bar{R}(x). \tag{11.2.5}$$

where $\bar{R}(x)$ is a polynomial that assumes nonnegative values for all real x. Let us show that for the corresponding probability density $p_\epsilon(\eta)$ of the vector η on the sphere $\mathfrak{S}$,

$$p_\varepsilon(\eta) = p_0(\eta) + \varepsilon\sqrt{n}\int_0^\infty dS \int_{-\infty}^\infty S^{n-2}\sum_{i=1}^n(\bar{x}+Sy_i)$$

$$\times\prod_{i=1}^n f_0(\bar{x}+Sy_i)\,dx + O(\varepsilon^2) \tag{11.2.6}$$

uniformly with respect to η.

The expression (11.2.3) for $p_\epsilon(\eta)$ contains the product

$$\prod_{i=1}^n f_\varepsilon(\bar{x}+Sy_i) = \Pi_1\Pi_2,$$

where

$$\Pi_1 = \prod_{i=1}^n f_0(\bar{x}+Sy_i) = \frac{1}{(2\pi)^{\frac{n}{2}}}\exp\left(-\frac{n\bar{x}^2+S^2}{2}\right).$$

$$\Pi_2 = \prod_{i=1}^n(1+\varepsilon H(\bar{x}+Sy_i) + R(\bar{x}+Sy_i,\ \varepsilon)).$$

For brevity we write

$$H_i = H(\bar{x}+Sy_i),\quad R_n = R(\bar{x}+Sy_i,\ \varepsilon)\quad (i=1,\ 2,\ \ldots,\ n).$$

Then

$$\Pi_2 = \prod_{i=1}^n(1+\varepsilon H_i + R_i) = 1+\varepsilon\sum_{i=1}^n H_i \neq \tilde{R}.$$

Here the quantity $\tilde{R}$ is the sum of finitely many terms of the form $\epsilon^p H_{i_1}\cdots H_{i_p} R_{j_1}\cdots$ $\cdots R_{j_q}$, where $p = 0, 1, \cdots, n$; $q = 0, 1, \cdots, n$; $i_1, \cdots, i_p$, $j_1, \cdots, j_q$ are distinct, $p + q \leq n$ and $p + 2q \geq 2$. Returning to (11.2.3) we now find that

$$p_\varepsilon(\eta) = \sqrt{n}\int_0^\infty dS\int_{-\infty}^\infty S^{n-2}\left(1+\varepsilon\sum_{i=1}^n H_i + \tilde{R}\right)\Pi_1\,d\bar{x}$$

$$= p_0(\eta) + \varepsilon\sqrt{n}\int_0^\infty dS\int_{-\infty}^\infty S^{n-2}\sum_{i=1}^n H(\bar{x}+Sy_i)$$

$$\times\prod_{i=1}^n f_0(\bar{x}+Sy_i)\,d\bar{x} + \sqrt{n}\int_0^\infty dS\int_{-\infty}^\infty S^{n-2}\tilde{R}\Pi_1\,d\bar{x}.$$

To prove (11.2.6) it only remains to show that

$$\tilde{\tilde{R}}=\sqrt{n}\int_0^\infty dS\int_{-\infty}^\infty S^{n-2}\tilde{R}\Pi_1 d\bar{x}$$

$$=\frac{\sqrt{n}}{(2\pi)^{\frac{n}{2}}}\int_0^\infty dS\cdot\int_{-\infty}^\infty S^{n-2}\tilde{R}\exp\left(-\frac{n\bar{x}^2+S^2}{2}\right)d\bar{x}+O(\varepsilon^2) \quad (11.2.7)$$

uniformly with respect to η (for fixed sample size n). It follows from what we said above that $\tilde{\tilde{R}}$ is the sum of finitely many terms of the form

$$I=C\int_0^\infty dS\int_{-\infty}^\infty S^{n-2}\alpha^r H_{i_1}\ldots H_{i_p}R_{j_1}\ldots$$

$$\ldots R_{j_q}\exp\left(-\frac{n\bar{x}^2+S^2}{2}\right)d\bar{x},$$

where $C>0$ depends only on n. Let us show that each of these integrals is $O(\epsilon^2)$.

By virtue of (11.2.5),

$$|I|=O(1)|\varepsilon|^{p+2q}\int_0^\infty dS\int_{-\infty}^\infty S^{n-2}\prod_{k=1}^p|H(\bar{x}+Sy_{i_k})|$$

$$\times\prod_{k=1}^q\bar{R}(\bar{x}+Sy_{j_k})\exp\left(-\frac{n\bar{x}^2+S^2}{2}\right)d\bar{x}.$$

Let us denote by $\bar{I}$ the double integral in this expression. Since $p+2q\geq 2$, we need only show that

$$\bar{I}=\int_0^\infty dS\int_{-\infty}^\infty S^{n-2}\prod_{k=1}^p|H(\bar{x}+Sy_{i_k})|$$

$$\times\prod_{k=1}^q\bar{R}(\bar{x}+Sy_{j_k})\exp\left(-\frac{n\bar{x}^2+S^2}{2}\right)d\bar{x}\leqslant\bar{K},$$

where $\bar{K}$ is independent of η. To do this we choose a polynomial $P(x)$ such that for all x

$$P(x)\geqslant 1,\quad P(x)\geqslant\bar{R}(x),\quad P(x)\geqslant|H(x)|.$$

For such a polynomial we may choose, for example, the polynomial

$$P(x)=\bar{R}(x)+(H(x))^2+1.$$

Then

$$\bar{I}\leqslant\int_0^\infty dS\int_{-\infty}^\infty S^{n-2}\prod_{i=1}^n\left[P(\bar{x}+Sy_i)\exp\left(-\frac{(\bar{x}+Sy_i)^2}{2}\right)\right]d\bar{x}.$$

Let us choose the constant K_1 in such a way that

$$\left|P(x)\exp\left[-\frac{x^2}{2}\right]\right| \leqslant K_1 \exp\left[-\frac{x^2}{4}\right].$$

Then

$$\overline{I} \leqslant K_1^n \int_0^\infty dS \int_{-\infty}^{\infty} S^{n-2}\exp\left[-\frac{S^2}{4}\right]\exp\left[-\frac{n\overline{x}^2}{4}\right]d\overline{x}$$

$$= K_1^n 2^{n-2}\sqrt{\pi n}\,\Gamma\left(\frac{n-1}{2}\right) = \overline{K},$$

which completes the proof of (11.2.7).

Let $\eta_1, \cdots, \eta_q$ denote q independent observations on the vector η. Suppose that we are verifying the hypothesis H_0: $p(\eta) = p_0(\eta)$ as opposed to the alternative H_ϵ: $p(\eta) = p_\epsilon(\eta)$ for given $\epsilon \neq 0$. According to the familiar Neyman-Pearson criterion, the test that is optimal in a certain sense is based on the statistic

$$\gamma(\eta_1, \ldots, \eta_q) = \ln \frac{p_\epsilon(\eta_1), \ldots, p_\epsilon(\eta_q)}{p_0(\eta_1), \ldots, p_0(\eta_q)}. \tag{11.2.8}$$

Here it should be noted that, by what was shown above, $p_\epsilon(\eta_i) \neq 0$ for $\eta_i \in \mathfrak{G}$ and sufficiently small ϵ ($i = 1, 2, \cdots, q$). The test will have a critical zone $\gamma(\eta) > c$ with level

$$\alpha = P(\gamma > c|H_0)$$

and with probability of error of the second kind

$$\beta = P(\gamma \leq c|H_\epsilon).$$

We consider the behavior of (11.2.8) as $q \to \infty$ with fixed n. We define $\gamma_k = \ln(p_\epsilon(\eta_k)/p_0(\eta_k))$, so that

$$\gamma(\eta_1, \ldots, \eta_q) = \sum_{k=1}^{q} \gamma_k.$$

The quantities γ_k ($k = 1, 2, \cdots, q$) are independent and identically distributed. We shall show that they have a finite third moment both under the hypothesis H_0 and under the hypothesis H_ϵ. Then the central limit theorem (see [33]) can be applied to the random variable γ. We have

$$P\left(\frac{\gamma - qa_0}{\sigma_0\sqrt{q}} < t\,|\,H_0\right) = G(t) + \frac{\theta_0\rho_0}{\sqrt{q}}, \tag{11.2.9}$$

$$P\left(\frac{\gamma - qa_\epsilon}{\sigma_\epsilon\sqrt{q}} < t\,|\,H_\epsilon\right) = G(t) + \frac{\theta_1\rho_\epsilon}{\sqrt{q}}, \tag{11.2.10}$$

where

$$G(t)=\int_{-\infty}^{t} f_0(x)\,dx, \quad a_0=E(\gamma_k\,|\,H_0); \quad a_\varepsilon=E(\gamma_k\,|\,H_\varepsilon),$$
$$\sigma_0^2=D(\gamma_k\,|\,H_0), \quad \sigma_\varepsilon^2=D(\gamma_k\,|\,H_\varepsilon),$$
$$\rho_0=\frac{1}{\sigma_0^3}E(|\gamma_k-a_0|^3|H_0), \quad \rho_\varepsilon=\frac{1}{\sigma_\varepsilon^3}E(|\gamma_k-a_\varepsilon|^3|H_\varepsilon),$$
$$|\theta_0|\leqslant 2, \quad |\theta_1|\leqslant 2.$$

From this we obtain the expression for the level of the test α and the probability β of error of the second kind.

$$\alpha=1-G\left(\frac{c-qa_0}{\sigma_0\sqrt{s}}\right)-\frac{\theta_0\rho_0}{\sqrt{q}},$$
$$\beta=G\left(\frac{s-a_0}{\sigma_\varepsilon\sqrt{q}}\right)+\frac{\theta_\varepsilon\rho_\varepsilon}{\sqrt{q}}.$$

We denote by t_z the number satisfying the equation

$$1-G(t_z)=z.$$

Suppose that α, β and ϵ are given. Using the usual symbol for asymptotic equality, we obtain, as $q\to\infty$,

$$t_\alpha\sim\frac{c-qa_0}{\sigma_0\sqrt{q}}, \quad t_{1-\alpha}=-t_\alpha\sim\frac{c-qa_0}{\sigma_\varepsilon\sqrt{q}},$$
$$c\sim qa_0+t_\alpha\sigma_0\sqrt{q}, \quad c\sim qa_\varepsilon-t_\beta\sigma_\varepsilon\sqrt{q}.$$

Therefore $t_\alpha\sigma_0+t_\beta\sigma_\epsilon\sim\sqrt{q}(a_\epsilon-a_0)$ and

$$q\sim\frac{(t_\alpha\sigma_0+t_\beta\sigma_\varepsilon)^2}{(a_\varepsilon-a_0)^2}. \tag{11.2.11}$$

If α and ϵ are fixed, the probability of error of the second kind $\beta(q, \alpha, \epsilon)=P(\gamma\le c|H_\epsilon)\to 0$ as $q\to\infty$. For given β let $\bar{q}=\bar{q}(\alpha, \beta, \epsilon)$ denote the smallest integer such that $\beta(q, \alpha, \epsilon)\le\beta$. Then q is the number of observations necessary to distinguish the hypotheses H_0 and H_ϵ, H_0 and H_α on the basis of a sample of constant size q with probabilities α and β of errors of the first and second kinds respectively. We define

$$q^*=q^*(\alpha, \beta, \varepsilon)=\frac{(t_\alpha\sigma_0+t_\beta\sigma_\varepsilon)^2}{(a_\varepsilon-a_0)^2}.$$

To compare q^* and $\bar{q}$ we use the following fact. Suppose that $q^*(\alpha, \beta, \epsilon)\to\infty$ as $\epsilon\to 0$ for fixed α and β and that the quantities ρ_0 and ρ_ϵ remain bounded. Then

$$\frac{\bar{q}(\alpha, \beta, \varepsilon)}{q^*(\alpha, \beta, \varepsilon)}\to 1. \tag{11.2.12}$$

To prove this we note that by virtue of the expressions for α and β we have $c = qa_0 + t_\alpha\sigma_0(1+\lambda)$, where $\lambda = \lambda(\epsilon, q) \to 0$ as $\epsilon \to 0$ and $q \to \infty$; also

$$\beta(q, \alpha, \varepsilon) = G\left(\sqrt{q}\,\frac{a_0 - a_\varepsilon}{\sigma_\varepsilon} + t_\alpha \frac{\sigma_0}{\sigma_\varepsilon}(1+\lambda)\right) + \frac{\theta_\alpha \rho_\varepsilon}{\sqrt{q}}.$$

Let β denote a given admissible probability of error of the second kind. Let us fix $\delta > 0$ arbitrarily small and set

$$q' = \frac{(t_\alpha\sigma_0(1-\delta) - t_{\beta+\delta}\sigma_\varepsilon)^2}{(a_\varepsilon - a_0)^2}, \quad q_1 = [q'],$$

$$q'' = \frac{(t_\alpha\sigma_0(1+\delta) + t_{\beta-\delta}\sigma_\varepsilon)^2}{(a_\varepsilon - a_0)^2}, \quad q_2 = [q''] + 1$$

where the brackets denote the integral part of the number contained in them.

The quantities q_1 and q_2 approach ∞ as $\epsilon \to 0$; the quantities $\lambda(\epsilon, q_1)$ and $\lambda(\epsilon, q_2)$ depend only on ϵ and they approach 0 as ϵ approaches 0. When q_1 observations are made we have

$$\beta(q_1, \alpha, \varepsilon) \leqslant G\left(\sqrt{q''}\,\frac{a_0 - a_\varepsilon}{\sigma_\varepsilon} + t_\alpha \frac{\sigma_0}{\sigma_\varepsilon}(1+\delta)\right) + \frac{\theta_\varepsilon \rho_\varepsilon}{\sqrt{q'}}$$
$$= \beta - \delta + \frac{\theta_\varepsilon \rho_\varepsilon}{\sqrt{q'}} < \beta$$

for sufficiently small ϵ. This means that $\overline{q}(\alpha, \beta, \epsilon) \leq q_1$ for sufficiently small ϵ.

Analogously, one can show that under the same condition

$$\overline{q}(\alpha, \beta, \epsilon) \geq q_2.$$

Hence

$$\min\left\{(1-\delta)^2, \frac{t^2_{\beta+\delta}}{t^2_\beta}\right\} - \frac{1}{q^*} \leqslant \frac{q_2}{q^*} \leqslant \frac{\overline{q}(\alpha, \beta, \varepsilon)}{q^*(\alpha, \beta, \varepsilon)}$$
$$\leqslant \frac{q_1}{q^*} \leqslant \max\left\{(1+\delta)^2, \frac{t^2_{\beta-\delta}}{t^2_\beta}\right\} + \frac{1}{q^*}$$

and

$$\min\left\{(1-\delta)^2, \frac{t^2_{\beta+\delta}}{t^2_\beta}\right\} \leqslant \underline{\lim}_{\varepsilon \to 0} \frac{\overline{q}}{q^*} \leqslant \overline{\lim}_{\varepsilon \to 0} \frac{\overline{q}}{q^*}$$
$$\leqslant \max\left\{(1+\delta)^2, \frac{t^2_{\beta-\delta}}{t^2_\beta}\right\}.$$

Since δ is arbitrarily small, the relation (11.2.12) follows.

The above reasoning is valid for rather arbitrary random objects η. Let us apply it to a particular given vector $\eta = (y_1, \cdots, y_n)$. Here by (11.2.7) we have

$$p_\varepsilon(\eta) = p_0(\eta)(1 + u(\eta)\varepsilon + O(\varepsilon^2)),$$

where

$$u(\eta)=b\int_0^\infty dS\int_{-\infty}^{\infty} S^{n-2}\exp\left(-\frac{n\bar{x}^2+S^2}{2}\right)\sum_{i=1}^{n} H(\bar{x}+Sy_i)\,d\bar{x};$$

$$b=\frac{\sqrt{n}}{2^{\frac{n-2}{2}}\sqrt{\pi}\,\Gamma\left(\frac{n-1}{2}\right)}.$$

The function $u(\eta)$ is bounded by the quantity G. Thus for sufficiently small ϵ

$$\gamma=\ln\frac{p_\varepsilon(\eta)}{p_0(\eta)}=\ln[1+\varepsilon u+O(\varepsilon^2)]=\varepsilon u+O(\varepsilon^2).$$

If we now set $\mu^2 = E(u^2|H_0)$, we obtain

$$a_\varepsilon-a_0=\varepsilon^2\mu^2+O(\varepsilon^3).$$

It is easy to see that $E(u|H_0)=0$:

$$E(1\,|\,H_\varepsilon)=E(1\,|\,H_0)+\varepsilon E(u\,|\,H_0)+O(\varepsilon^2),$$

or

$$1=1+\varepsilon E(u\,|\,H_0)+O(\varepsilon^2),\quad E(u\,|\,H_0)=O(\varepsilon),$$

that is, $E(u|H_0)=0$. Therefore $a_0=E(\epsilon u+O(\epsilon^2)|H_0)=O(\epsilon^2)$; and $a_0^2=O(\epsilon^4)$; analogously $a_\epsilon^2=O(\epsilon^4)$. It is then easy to show that $\sigma_0^2=\epsilon^2\mu^2+O(\epsilon^3)$ and $\sigma_\epsilon^2=\epsilon^2\mu^2+O(\epsilon^3)$, so that, from what was said above,

$$\bar{q}\sim\frac{(t_\alpha+t_\beta)^2}{\varepsilon^2\mu^2}. \tag{11.2.13}$$

If we set $c_n=n/\mu^2$ we have

$$\bar{q}\sim c_n\frac{(t_\alpha+t_\beta)^2}{\varepsilon^2 n}. \tag{11.2.14}$$

Thus to get an idea how many independent samples of size n should be made to verify H_0 as opposed to H_ϵ, we need to calculate c_n in the two cases. Various representative examples of such calculations are presented by Petrov [60]. Other interesting calculations are made by Volodin [12].

APPENDIX

UNSOLVED QUESTIONS

In the analytical formulation of certain problems presented in this book dealing with the elimination of nuisance parameters in statistical problems, certain questions and unsolved problems arise rather naturally. Here we arrange these according to the chapters dealing with the topic in question.

Chapter II. 1. Continuation of the work of Koopman, Dynkin and Brown on the classification of distributions admitting sufficient statistics of finite rank for the case in which the distribution densities can vanish. Weakening of the condition for existence of exponential families in the case when the densities do not vanish on the sample space.

2. Investigation of exponential families with "sliding carrier", i.e. families whose carrier depends on the parameters.

Chapter III. 1. Consider a repeated normal sample $x_1, \cdots, x_n$, where $x_i \in N(a, \sigma^2)$. Let us partition all almost-everywhere-continuous functions $\gamma(a, \sigma)$ into two classes:

I. "Verifiable functions" for which the hypothesis H_0: $\gamma(a, \sigma) = \gamma_0$ admits an invariant verification.

II. The remaining functions.

How can one describe the class I?

2. The same question for an infinite sample $x_1, x_2, \cdots$, and the application of sequential analysis, where the mean number of steps is bounded up to the final solution.

3. Generalization of question 1. Suppose that we have an exponential family of the form (5.2.3). Let us partition the collection of functions $\gamma(\theta_1, \cdots, \theta_s)$ into "verifiable" and "unverifiable" classes, just as in question 1. How can we describe the class of the form I?

Chapter IV. 1. How can we describe all the similar zones of distribution families of the form (4.3.5)?

2. Let $x_1, \cdots, x_n$ ($x_i \in N(0, 1)$) denote a repeated normal sample. Let $P_1(x_1, \cdots, x_n)$ and $P_2(x_1, \cdots, x_n)$ denote two independent polynomial statistics. Is it always possible to "uncouple" them by means of an orthogonal transformation, i.e. to reduce them to two statistics depending on the completeness of the different variables? (For a given sample size n and given degrees m_1 and m_2 of the polynomials the question can be solved in a finite number of steps for all such polynomials (see [41]).)

Chapter V. 1. Is it possible to weaken the condition for complexification of the relations (condition (Y_I) in §8)?

2. Is it possible to weaken the condition on the Jacobians (Y_{II})?

3. How can one describe in terms of generalized Laplace transformations (in the sense of the theory of generalized functions) the construction of nonsmooth cotests?

These questions are related to the following question in the theory of analytic functions.

4. Let Z denote a simply-connected complex polycylinder. Let O' denote a ring of functions that are holomorphic and that have a holomorphic continuation from the polycylinder to the Cartesian product of the half-planes containing Z. How can we describe the structure of the ideals of the ring O'? Under what conditions will the basis of the ideals be finite?

5. How can we construct cotests for the case in which the function h vanishes in a given region?

Formula (5.8.6) leads to the following analytic problem: Let $A_j(T_1, \cdots, T_s)$, $j = 1, 2, \cdots, r$; $r < s$, denote sufficiently smooth functions defined in the Euclidean space E_s of the arguments $T_1, \cdots, T_s$. Let U denote a simply-connected region contained in E_s and bounded by smooth surfaces. How can we describe the system of all functions $H_1, \cdots, H_r$ for which the sum of the convolutions

$$A_1 * H_1 + \cdots + A_r * H_r$$

exists and vanishes on U?

6. Is it possible to describe smooth cotests for exponential families with "sliding carrier" (cf. question 2 to Chapter II)?

7. The development of computational methods, in particular linear programming

methods, for finding optimal similar tests.

Chapter VI. 1. Generalization of Wijsman's results on the construction of all similar tests of the hypothesis H_0: $a/\sigma = \gamma_0$ for a repeated normal sample $x_1, \cdots, x_n$ where $x_j \in N(a, \sigma^2)$. In what problems associated with tests of hypotheses on normal samples do parabolic differential equations appear? What can we do in other cases?

Chapter VII. 1. Questions of unbiased estimates of zero (UEZ's) analogous to the questions in Chapter V on cotests, i.e. questions 1–5 of Chapter V with the word "cotest" replaced by the expression "unbiased estimate of zero".

2. Generalization of the results of §2 dealing with inadmissible unbiased estimates. Ways of obtaining more general forms of inadmissible unbiased estimates for incomplete exponential families on a compact set of values of the parameters. Cases of inadmissibility for noncompact sets of values of the parameters.

3. Replacement of the variance as a function of the loss of an unbiased estimate with other sufficiently smooth functions. Conditions for inadmissibility of unbiased estimates.

4. Classification of the best unbiased estimates for incomplete exponential families from a standpoint of variance. From an analytic point of view this problem is connected with the following question from functional algebra. Let O denote the ring of all holomorphic functions defined on a compact simply-connected polycylinder and let I denote an ideal contained in O. Find the ring K of all linear differential operators L with constant coefficients such that $LI \subset I$.

5. Problems analogous to the preceding ones for the case in which we are using not the variance but other sufficiently smooth functions of the loss.

6. Construction of an analogue to Theorem 7.3.1 dealing with inadmissibility of a sample mean for scale parameters. Extension of Theorem 7.3.1 to observations connected with a homogeneous Markov chain.

Chapter VIII. 1. Investigation of the question of nonexistence, in general, for incomplete exponential families of similar zones that depend on sufficient statistics with sufficiently smooth boundaries.

2. Weakening of the conditions of Theorem 8.3.1. Does there exist a similar Fisher-Welch-Wald test if we require only continuity of the test boundary or only satisfaction of a Lipschitz condition for it?

Chapter IX. 1. Can we construct a test analogous to the simple randomized test in §3 for the case of nonidentical sample sizes?

2. Generalization of Theorem 9.4.1 on the characterization of the Bartlett-Scheffé test. Weakening of the conditions on the basic space of linear forms. Construction of an analogue of this theorem to characterize the simplest tests of the method of least squares with unknown weights.

Chapter X. 1. How can one construct a homogeneous unrandomized similar test for the Behrens-Fisher problem in the case of like parity of the sizes of the samples?

Chapter XI. 1. Investigations analogous to those made by Petrov for the scheme of many small samples from complete exponential families referred to in §1.

SUPPLEMENT

NEW RESULTS IN THE THEORY OF ESTIMATION AND TESTING HYPOTHESES FOR PROBLEMS WITH NUISANCE PARAMETERS

A. M. KAGAN AND V. P. PALAMODOV[1)]

The results expounded in this Supplement are mostly answers to questions put at the end of the book (see queries: 1 to Chapter III, 3 and 5 to Chapter V, 4 and 6 to Chapter VII). But §5 is an exception; there the problem of estimating a location parameter is considered when the "nuisance parameter" is the form of the function $F(x-\theta)$ itself, for which only several first distribution moments for $\theta = 0$ are known.

The results of §§1 and 2 are due to V. P. Palamodov [6, 7],[2)] of §3 to A. M. Kagan and V. P. Palamodov [2, 3], of §4 to A. M. Kagan [4, 12], of §5 to A. M. Kagan and A. L. Ruhin [5].

§1. INVARIANT VERIFICATION OF FUNCTIONS WHICH ARE POLYNOMIALS IN a AND $1/\sigma^2$ FOR NORMAL SAMPLES

This section gives the description of all polynomials $P(a, 1/\sigma^2)$ admitting invariant verification in the sense of §2, Chapter III (strictly speaking, that sense will be modified a little), on the evidence of a repeated sample $(x_1, \cdots, x_n)$ from a normal population $N(a, \sigma^2)$. The method of this section is purely analytical, in contrast, for instance, to that of E. Lehmann [13] establishing the non-verifiability of certain functions as a consequence of the indistinguishability of the corresponding families of distributions.

1) Editor's note. The translation of the Supplement was provided by the authors.

2) Authors' note. These refer to the Bibliography at the end of this Supplement, not to the main Bibliography.

Let $x_1, \cdots, x_n \in N(a, \sigma^2)$ be independent normal variables. In studying the question of the invariant verification we can restrict ourselves, without loss of generality, to those tests $\phi(\bar{x}, s)$ depending only upon the sufficient statistics

$$\bar{x} = \frac{1}{n} \sum_{i=1}^{n} x_i; \quad s^2 = \frac{1}{n} \sum_{i=1}^{n} (x_i - \bar{x})^2.$$

We have

$$E_{a,\sigma} \phi(\bar{x}, s) = \bar{\phi}(a, \sigma^2) = \check{\phi}(a, \xi)$$

$$= C_0 \int_{-\infty}^{\infty} \int_0^{\infty} \xi^{n/2} \exp[-\xi(s + (\bar{x} - a)^2)] s^{(n-3)/2} \phi(\bar{x}, s)\, d\bar{x}\, ds. \qquad \text{(S.1.1)}$$

Here we denoted $\xi = 1/\sigma^2$; C_0 is a constant.

Note that the integral (S.1.1) can be continued as an analytic function of two complex variables to the product of the complex plane C of the values of a and the complex halfplane C^+ of the values of ξ with $\mathrm{Re}\,\xi > 0$. This enables us to introduce the following definition.

Definition. A function $f(a, \xi)$ defined in $\mathrm{C} \times \mathrm{C}^+$ is called C-verifiable if there exists a test ϕ such that

$$\check{\phi}(a, \xi) = \psi(f(a, \xi)); \quad a \in \mathrm{C}, \quad \xi \in \mathrm{C}^+$$

for some $\psi \not\equiv \text{const}$.

Any real C-verifiable function is obviously verifiable in the sense of §2, Chapter III. The converse is not true; in [6] we give sufficient conditions for the C-verifiability of functions which are verifiable in the sense of §2, Chapter III.

Lemma S.1.1. *For any test* $\phi(\bar{x}, s)$

$$|\check{\phi}(a, \xi)| \le C_0 \cdot \left|\frac{\xi}{\mathrm{Re}\,\xi}\right|^{n/2} \exp\left[\frac{|\xi|^2}{\mathrm{Re}\,\xi} |\mathrm{Im}\, a|^2\right] \qquad \text{(S.1.2)}$$

for $a \in \mathrm{C}$, $\xi \in \mathrm{C}^+$ $(C_0, C_1, \cdots$ *in what follows are positive constants*).

Proof.

$$|\check{\phi}(a, \xi)| = C_0 |\xi|^{n/2} \times$$

$$\times \exp(-\operatorname{Re}(\xi a^2)) \int_{-\infty}^{\infty} \int_0^{\infty} \exp[-\xi(s+\bar{x}^2) + 2\xi a\bar{x}] s^{(n-3)/2} \phi(\bar{x}, s)\, ds\, d\bar{x}$$

$$\leq C_0 |\xi|^{n/2} \exp(-\operatorname{Re}(\xi a^2)) \int_{-\infty}^{\infty} \int_0^{\infty} \exp[-\operatorname{Re}\xi(s+\bar{x}^2) + 2\operatorname{Re}(\xi a)\bar{x}] s^{(n-3)/2}\, ds\, d\bar{x} \tag{S.1.3}$$

because $0 \leq \phi(\bar{x}, s) \leq 1$. The quantity

$$C_0 |\operatorname{Re}\xi|^{n/2}$$

$$\cdot \exp\left[-\frac{\operatorname{Re}(\xi a)^2}{\operatorname{Re}\xi}\right] \int_{-\infty}^{\infty} \int_0^{\infty} \exp[-\operatorname{Re}\xi(s+\bar{x}^2) + 2\operatorname{Re}(\xi a)\bar{x}] s^{(n-3)/2}\, ds\, d\bar{x}$$

is the power function of the trivial test $\phi \equiv 1$ at the point $(\operatorname{Re}\xi, \operatorname{Re}(\xi a)/\operatorname{Re}\xi)$. Taking this into account, we can write the right side of (S.1.3) in the form

$$C_0 \left|\frac{\xi}{\operatorname{Re}\xi}\right|^{n/2} \exp\left[\frac{\operatorname{Re}(\xi a)^2}{\operatorname{Re}\xi} - \operatorname{Re}(\xi a^2)\right].$$

Putting $a = \alpha + i\beta$, $\xi = \zeta + i\eta$ we get

$$\frac{\operatorname{Re}(\xi a)}{\operatorname{Re}\xi} - \operatorname{Re}\xi a^2$$

$$= \frac{1}{\zeta}[(\alpha\zeta - \beta\eta)^2 - \zeta(\zeta(\alpha^2\beta^2) - 2\eta\alpha\beta)] = \frac{1}{\zeta}(\zeta^2 + \eta^2)\beta^2 = \frac{|\xi|^2}{\operatorname{Re}\xi} |\operatorname{Im} a|^2,$$

which leads to (S.1.2).

We shall say that $r(a, \xi) \in R$ if

$$r(a, \xi) = \frac{p(a, \xi)}{q(a, \xi)} = \frac{p_m(\xi) a^m + \cdots + p_0(\xi)}{q_k(\xi) a^k + \cdots + q_0(\xi)} \tag{S.1.4}$$

where p_i, q_j $(i = 0, 1, \cdots, m;\ j = 0, \cdots, k)$ belong to the ring A of the functions holomorphic in C^+ and $p_m q_k \neq 0$. Without loss of generality we suppose that either $m > k$ or $m = k$ but $p_m(\xi)/q_k(\xi) \neq \text{const}$. Moreover, we can suppose that the number m is the least possible and the functions p_i and q_i do not vanish simultaneously.

Theorem S.1.1. 1°. *If a function* $r \in R$ *with* $m > 0$ *is* C-*verifiable, then* $k = 0$ *in* (S.1.4), *and the function* $q_0(\xi) \neq 0$ *on* C^+. *Moreover, the power function of the test* $\psi(r(a, \xi))$ *is an entire function of order not exceeding* $2/m$.

2°. *If, moreover, the functions* $q_0, p_0, \dots, p_m$ *and* p_0/q_0 *are analytic in the vicinity of zero, the function is of the form*

$$r(a, \xi) = r_2(\xi)a^2 + r_1(\xi)a + r_0(\xi); \; r_i = \frac{p_i}{q_0} \in A$$

times a constant; also, $r_2(0) = r_1(0) = 0$; $r_2'(0) = 1$ *and the image of the mapping* $r_2\colon \mathbf{C}^+ \to \mathbf{C}$ *belongs to* $\mathbf{C}^+$. *The function* ψ *is bounded in any closed angle belonging to* $\mathbf{C}^+$.

In what follows we shall denote by $\mathbf{C}^*$ the complex plane compactified in the usual way.

Lemma S.1.2. *Let* $\rho\colon \omega \to \mathbf{C}^*$ *be a function meromorphic in a domain* $\omega \subset \mathbf{C}$ *and not a constant; let* Ω *be the range of its values. If* $\psi(\rho(\zeta))$, $\zeta \in \omega$ *is analytic in* ω, *then* ψ *is analytic in* Ω.

Proof. Let $z_0 \in \Omega$ be an arbitrary point distinct from ∞, and let $\zeta_0 \in \omega$ be a point at which $\rho(\zeta_0) = z$. If $\rho'(\zeta_0) \neq 0$, then in the neighborhood of the point z_0 we have $z = \rho(w(z))$ where $w(z)$ is holomorphic in z_0. Hence $\psi(z) = \psi(\rho(w(z)))$ which implies that ψ is holomorphic in z_0.

Since $\rho \not\equiv \text{const}$, the zeros of its derivative are isolated. Let ζ_0 be one of the zeros of ρ' and $U \subset \omega$ a closed bounded neighborhood of the point ζ_0, containing no other zeros of ρ' and no poles of ρ. Its image $V = \rho(U)$ is a bounded neighborhood of the point $z_0 = \rho(\zeta_0)$.

In the domain $V \setminus \{z_0\}$ the function ψ is bounded and was proved to be analytic. Hence it is analytic at the point z_0 also.

Now let ζ_0 be a pole of the function ρ. Choose a bounded closed neighborhood $U \subset \omega$ of the point ζ_0. Its image V is a neighborhood of ∞, and in the domain $V \setminus \{z_0\}$ the function ψ is bounded and analytic. Hence it is analytic at the point z_0 also. The lemma is proved.

We pass now to the proof of Theorem S.1.1. We suppose that $\psi \not\equiv \text{const}$.

It is easy to see that the range of values of the function $r\colon \mathbf{C} \times \mathbf{C}^+ \to \mathbf{C}^*$ always contains $\mathbf{C}$; hence by Lemma S.1.2 the function ψ is entire.

If $k > 0$, then for a suitable ξ_0 the image of the mapping $r(a, \xi_0)\colon \mathbf{C} \to \mathbf{C}^*$ contains the point ∞. Hence by Lemma S.1.2 the function ψ is analytic at that point. Since it is entire it must be a constant, which contradicts our assumptions. Hence $k = 0$. In a similar way we prove that q_0 has no zeros in $\mathbf{C}^+$. Hence we get

$$r(a, \xi) = r_m(\xi)a^m + \cdots + r_0(\xi); \quad r_i = \frac{p_i}{q_0} \in A.$$

We fix the point ξ so that $r_m(\xi) \neq 0$. Then $r(a, \xi) \sim r_m(\xi)a^m$ for $a \to \infty$. The inequality (S. 1.2) implies for certain values of the constants A and B that

$$|\psi(r(a, \xi))| \leq C_0 \exp(A|\operatorname{Im} a|^2) \leq C_0 \exp(B\,|r(a, \xi)|^{2/m});$$

hence we deduce that the order of the entire function ψ does not exceed $2/m$. The first assertion of Theorem S. 1.1 is proved.

We pass to the proof of the second assertion. The functions $r_0, p_1, \cdots, p_m, q_0$ are analytic in the vicinity of zero. Hence the functions $r_i = p_i/q_0$ are analytic in the vicinity of zero, except perhaps at zero itself, where the only possible singularity is a pole. We shall show that in fact the coefficients r_i $(i = 1, \cdots, m)$ are analytic in the vicinity of zero and $r_i(0) = 0$. For each $i = 1, \cdots, m$ we have in the vicinity of zero

$$r_i(\xi) = \rho_i \xi^{\alpha_i} + \rho_i' \xi^{\alpha_i + 1} + \cdots,$$

with certain $\rho_i \neq 0$ and α_i. If all $\alpha_i > 0$ our assertion is proved. Suppose that $\alpha_i < 0$ for a certain i; consider the quantity

$$\alpha = \max_{i \geq 1} \left\{ -\frac{\alpha_i}{i} \right\}.$$

Let k be the largest number for which $-\alpha_k/k = \alpha$. For $\lambda \in \mathbf{C}$ put

$$a_\lambda(\xi) = \left\{ \frac{\lambda}{r_m(\xi)} \right\}^{1/k}, \quad \xi > 0,$$

where the branch of the root can be chosen arbitrarily. Since $\alpha_k < 0$, the quantity $a_\lambda(\xi)$ is bounded for $\xi \to 0$ for any $\lambda \in \mathbf{C}$. Since $\xi > 0$, we get from (S. 1.2):

$$|\psi(r(a_\lambda(\xi), \xi))| \leq C \exp(B(\lambda)\,\xi). \tag{S. 1.5}$$

On the other hand,

$$r(a_\lambda(\xi), \xi) = \sum_{j>k} r_j(\xi) a_\lambda^j(\xi) + \lambda + \sum_{j<k} r_j(\xi) a_\lambda^j(\xi). \tag{S. 1.6}$$

In the first sum on the right-hand side of (S. 1.6)

$$|r_j(\xi)a_\lambda^j(\xi)| = \left[\frac{|r_j(\xi)|^{1/j}}{|r_k(\xi)|^{1/k}}\right]|\lambda|^{j/k}$$

$$\sim C\,\xi^{j(\alpha_j/j - \alpha_k/k)} \to 0$$

for $\xi \to 0$, in view of the choice of k. In the second sum

$$|r_j(\xi)a_\lambda^j(\xi)| \le C\xi^{j(\alpha_j/j - \alpha_k/k)}|\lambda|^{j/k} \le C(|\lambda|^{(k-1)/k} + 1)$$

for $0 < \xi < 1$, since $\alpha_j/j \ge \alpha_k/k$ for $j < k$ and the function $r_0(\xi)$ is bounded for $\xi \to 0$. We choose ξ so small that the first sum in (S. 1.6) is smaller than 1. Then

$$|r(a_\lambda(\xi), \xi) - \lambda| \le C(|\lambda|^{(k-1)/k} + 1). \tag{S.1.7}$$

We take now a sufficiently large R. If the point λ runs over the circumference with the radius R, (S. 1.7) implies that the point $r(a_\lambda(\xi), \xi)$, which depends continuously on λ, runs over a curve homeomorphic to that circumference and containing the circle

$$|z| \le R' = R - C(R^{(k-1)/k} + 1).$$

On the other hand, from (S. 1.5) it follows that on this curve the function ψ is bounded by the constant $C\exp(B(\lambda)\xi)$. Since ξ can be chosen as small as we please, ψ is bounded by the constant C, which does not depend upon λ on the curve described above and hence in the circle of radius R'. As $R \to \infty$ so does R'; hence ψ is bounded on the whole plane; hence $\psi \equiv \text{const}$. Since we assumed that $\psi \not\equiv \text{const}$ we see that all $\alpha_i > 0$.

Lemma S. 1.3. *For each $\xi \in \mathbf{C}^+$ there exists an $\epsilon > 0$ such that the function ψ is bounded inside the angle*

$$|\operatorname{arc} z - \operatorname{arc} r_m(\xi)| \le \epsilon.$$

and, for odd m, also inside the angle

$$|\operatorname{arc} z - \operatorname{arc}(-r_m(\xi))| \le \epsilon. \tag{S.1.8}$$

Proof. Take an arbitrary $\xi \in \mathbf{C}^+$. Since $r_m(0) = 0$, it follows that $r_m(\xi) \not\equiv$ const. Therefore in the vicinity of the point ξ we can find points ξ_+ and ξ_- such that

$$\operatorname{arc} r_m(\xi_+) > \operatorname{arc} r_m(\xi) > \operatorname{arc} r_m(\xi_-),$$

$$\operatorname{arc} r_m(\xi_+) - \operatorname{arc} r_m(\xi_-) = \delta < \pi/2. \tag{S.1.9}$$

We shall show now that the function ψ is bounded inside the angle

$$\operatorname{arc} r_m(\xi_+) - \delta/3 \geq \operatorname{arc} z \geq \operatorname{arc} r_m(\xi_-) + \delta/3. \tag{S.1.10}$$

Consider the curves

$$\theta_\pm(a) = r(a, \xi_\pm), \quad \xi \in [0, \infty). \tag{S.1.11}$$

We have

$$\theta_\pm(a) - r_m(\xi_\pm)a^m = O(a^{m-1}); \quad \xi \to \infty.$$

This implies that the domain between the curves $\xi_\pm(a)$, taken together with a sufficiently large circle, contains the angle (S.1.10). By the inequality (S.1.2) the function $\psi(r(a, \xi_\pm))$ is bounded for real values of a. Hence the function ψ is bounded on the curves (S.1.11). But by (S.1.9) these curves are contained inside an angle less than $\pi/2$, while ψ was shown above to be an entire function of order not exceeding $2/m \leq 2$. Hence by the Phragmén-Lindelöf principle, the function ψ is bounded in the domain lying between the curves (S.1.11), as was to be proved. The case of odd m is dealt with by analogy with the preceding one.* To complete the proof of Theorem S.1.1, consider the set

$$\{\operatorname{arc} r_m(\xi), \ \xi \in C^+\}. \tag{S.1.12}$$

Since

$$r_m(\xi) = \rho_m \xi^{\alpha_m} + \rho_m' \xi^{\alpha_m + 1} + \cdots; \quad \alpha_m > 0$$

for sufficiently small ξ's, we have

$$\operatorname{arc} r_m(\xi) \sim \alpha_m \operatorname{arc} \xi + \operatorname{arc} \rho_m \quad \text{for } \xi \to 0. \tag{S.1.13}$$

Therefore if $\alpha_m > 1$, the set (S.1.12) contains an interval of length 2π. By Lemma S.1.3 it then follows that the function ψ is bounded inside the angle $2\pi - \epsilon$ for any $\epsilon > 0$. As ψ is an entire function of a finite order, we have by the Phragmén-Lindelöf principle: $\psi = \text{const}$. Hence $\alpha_m = 1$. Multiplying $r(a, \xi)$ into $1/p_m$, we get $r_m'(0) = 1$.

---

* With ξ replaced by $-\xi$.

In this case it follows from (S. 1.13) that the set (S. 1.2) contains the interval $(-\pi/2, \pi/2)$. Then by Lemma S. 1.3 the function ψ is bounded inside the angle $|\text{arc } z| \leq \pi/2 - \epsilon$ for any $\epsilon > 0$. Hence ψ is a function of order at least 1 and of finite type. On the other hand, we have proved that its order does not exceed $2/m$. Hence, $m \leq 2$. The case $m = 1$ is excluded because for $m = 1$ Lemma S. 1.3 would imply the boundedness of the function ψ also in the angle $|\text{arc } z - \pi| \leq \pi/2 - \epsilon$ for any $\epsilon > 0$, which is impossible for nonconstant functions of finite order.

It remains only to verify that the image of the mapping $r_2\colon \mathbf{C}^+ \to \mathbf{C}$ belongs to $\mathbf{C}^+$. Suppose it is not so, and that for a certain $\xi \in \mathbf{C}^+$, $|\text{arc } r_2(\xi)| > \pi/2$. By Lemma S. 1.3 the function ψ is bounded inside the angle $|\text{arc } z - \text{arc } r_2(\xi)| \leq \epsilon$. Since this function is of the first order and bounded inside any angle of the form $|\text{arc } z| \leq \pi/2 - \epsilon$, again the Phragmén-Lindelöf principle implies that it vanishes identically. Theorem S. 1.1 is proved.

Theorem S. 1.2. *Let a polynomial* $p(a, \xi)$ *which is not a function of* ξ *only be* C*-verifiable. To be so, it is necessary and sufficient for* $p(a, \xi)$ *to be representable as a linear form of*

$$\xi(a^2 + Aa + B) \tag{S. 1.14}$$

where A *and* B *are real and* $A^2 - 4B \leq 0$.

Proof. Necessity. Let $p(a, \xi)$ be a C-verifiable polynomial, and let the test ϕ be such that $\check{\phi}(a, \xi) = \psi(p(a, \xi))$ where $\psi \not\equiv \text{const}$. By Theorem S. 1.1 the function ψ is an entire one and $p(a, \xi)$, after multiplication by a suitable constant, is of the form

$$p(a, \xi) = p_2(\xi)a^2 + p_1(\xi)a + p_0(\xi),$$

where

$$p_2(0) = p_1(0) = 0; \quad p_2'(0) = 1. \tag{S. 1.15}$$

By the conditions of the theorem $p_2(\xi)$, $p_1(\xi)$, $p_0(\xi)$ are polynomials. We shall show that their order does not exceed 1. Suppose this is not true; then for a certain $a = a_0$ the polynomial $p(a_0, \xi)$ is of order $k > 1$ as a polynomial in ξ. Hence

$$p(a_0, \xi) = C\xi^k + O(|\xi|^{k-1}); \quad \xi \to \infty. \tag{S. 1.16}$$

Let the point ξ move inside the angle $|\text{arc}\,\xi| \leq \pi/2 - \epsilon$. Then from (S. 1.2) it follows that $\psi(p(a_0, \xi))$ is bounded. On the other hand, the first term in (S. 1.16) runs over all the values of the angle $|\text{arc}\,\xi| \leq \pi - 2\epsilon$. Hence the function $p(a_0, \xi)$ runs over all the values of the complex plane, except perhaps in a certain circle and in the angle $|\text{arc}\, z - \pi| \leq 3\epsilon$. Hence the entire function ψ is bounded on the whole plane except perhaps in an angle $|\text{arc}\, z - \pi| \leq 3\epsilon$, which is as small as we please. Since this function is of finite order, the Phragmén Lindelöf principle implies that it is bounded on the whole plane and therefore is a constant. This contradiction proves that $p_2(\xi)$, $p_1(\xi)$ and $p_0(\xi)$ are linear functions of ξ.

From (S. 1.15) we get that $p_2(\xi) = \xi$; $p_1(\xi) = A\xi$, where A is a constant. Subtracting a suitable constant from $p(a, \xi)$ we can also annul the constant term of $p_0(\xi)$. Then $p(a, \xi)$ takes the form (S. 1.14).

Consider now the function $\text{arc}\,(a^2 + Aa + B)$ for real values of a. Suppose it to be nonconstant and let ϕ_1 and ϕ_2 be two distinct values of it. In that case, if ξ runs over the angle $|\text{arc}\,\xi| \leq \pi/2 - \epsilon$, and the point a runs over the real axis, the point $\xi(a^2 + Aa + B)$ will take on all the values inside the angles

$$|\text{arc}\, z - \phi_1| \leq \pi/2 - \epsilon \text{ and } |\text{arc}\, z - \phi_2| \leq \pi/2 - \epsilon.$$

From the inequality (S. 1.2) it follows that ψ is bounded inside these angles for any $\epsilon > 0$. Since $\phi_1 \neq \phi_2$ and ψ is an entire function of the first order, this is impossible in view of the Phragmén-Lindelöf principle. Hence $\text{arc}\,(a^2 + Aa + B) = \text{const}$. On the other hand, $\text{arc}\,(a^2 + Aa + B) = 0$ for $a \to \infty$. Hence $\text{arc}(a^2 + Aa + B) = 0$, A and B are real numbers and $a^2 + Aa + B$ is nonnegative for all real a, so that $A^2 - 4B \leq 0$.

Sufficiency. Consider the test

$$\phi(\bar{x}, s) = \begin{cases} (1 - t(\bar{x}^2/s))^{(n-3)/2}; & s \geq t\bar{x}^2 \\ 0 & ; s < t\bar{x}^2 \end{cases}$$

for a given $t > 0$. It can be shown that for a certain pair of numbers (y, s_0) with $s_0 > 0$ we have

$$E_{a,\xi}\, \phi(\bar{x} - y, s - s_0) = C_0 \exp(-\tau\xi^2(a^2 + Aa + B))$$

where $\tau \in (0, 1)$, so that $\xi(a^2 + Aa + B)$ is verifiable. We omit the details; they can be found in [6].

§2. THE DESCRIPTION OF ALL COTESTS FOR A CLASS OF EXPONENTIAL FAMILIES WITH POLYNOMIAL RELATIONS

This is closely related to §8, Chapter V, and we use here the notation and terminology of that chapter.

Consider the exponential family of densities with respect to Lebesgue measure

$$p(T, \theta) = C(\theta) h(T) \exp (\theta_1 T_1 + \cdots + \theta_s T_s) \tag{S. 2.1}$$

with $\theta, T \in R^s$. Before stating the conditions imposed upon $h(T)$ and the parametric set, we introduce the set $\omega \subset R^s$ determined in the following way. Construct the set of $\theta \in R^s$ for which for a certain $C = C(\theta)$ the condition

$$\{(\theta, T) < C\} \supset \operatorname{Supp} h = \mathcal{T}$$

holds.

This set is a cone whose interior we denote by ω. It is important to remark that if $\theta \in \omega$ we can choose a constant B such that for a suitable $\epsilon > 0$

$$(\theta, T) \leq -\epsilon|T| + B; \quad T \in \mathcal{T}. \tag{S. 2.2}$$

We shall suppose that the family (S. 2.1) satisfies the following requirements.

1.1. $\mathcal{T} = \operatorname{Supp} h$ is a convex set.

1.2. ω is nonvoid.

1.3. For any $\epsilon > 0$

$$\| h(T) \exp(-\epsilon|T|)\|_{L_2} < \infty. \tag{S. 2.3}$$

(Note that from (S. 2.2) and (S. 2.3) it follows that for any $\theta \in \omega$, $p(T, \theta)$ can be considered as a probability density.)

2.1. The parameter θ takes on values in an everywhere dense subvariety of a real algebraic variety $\Pi \cap \omega$.

2.2. The polynomial ideal I formed by the real polynomials of θ vanishing on $\Pi \cap \omega$ is a principal one, i. e. it consists of all polynomials of the form

$$G(\theta) P(\theta)$$

for a suitable $P(\theta)$.

Denote by N the variety of complex roots of $P(\theta)$ and by $\omega \times R^s_y$ the set $(\theta \in C^s;\ \operatorname{Re}\theta \in \omega)$.

3.1. Each connected component of the intersection $N \cap (\omega \times R^s_y)$ has at least one real point where $\operatorname{grad} P(\theta) \neq 0$.

Theorem S.2.1. *Under Conditions* 1.1–3.1 *each cotest can be represented uniquely by the formula*

$$\phi(t) = \frac{1}{h(T)} P(-\mathfrak{D})\Psi(T); \quad \mathfrak{D} = \left(\frac{\partial}{\partial T_1}, \ldots, \frac{\partial}{\partial T_s}\right) \tag{S.2.4}$$

where $\Psi(T)$ *is an (ordinary) function such that*

$$\left.\begin{array}{c} \operatorname{Supp}\Psi \subset \operatorname{Supp} h \\ \|\Psi \exp(-\epsilon|T|)\|_{L_2} < \infty \text{ for any } \epsilon > 0 \end{array}\right\}. \tag{S.2.5}$$

The function Ψ *is uniquely determined.*

Conversely, each function of the type (S.2.4) *with* $\Psi(T)$ *satisfying the requirements* (S.2.5) *and lying between* α *and* $1-\alpha$ *for a value of* $\alpha \in (0, 1)$, *is a cotest.*

Proof. 1. Take a point $a \in \omega$. Then for certain constants $C > 0$ and B the condition

$$(a, T) \leq -C|T| + B \tag{S.2.6}$$

is fulfilled on the set $\operatorname{Supp} h$.

2. Set

$$\|\Phi\|_\epsilon = \|\Phi(T)\exp\epsilon|T|\,\|_{L_2}.$$

By a theorem of Hörmander [11], for any differential operator with constant coefficients there is a fundamental solution (in general, a generalized function) $E(T)$ under the condition:

$$\|E * \Phi\|_{-\epsilon} \leq C\|\Phi\|_\epsilon \tag{S.2.7}$$

for any $\Phi(T)$ for which the right-hand side of the formula is finite. Let $E(T)$ be such a solution for $P(-\mathfrak{D} + a)$.

Let $\phi(T)$ be a given cotest, put $\check{\phi} = \phi h$ and

$$\Psi(\tau) = E(T)\exp[-(a, T)] * \check{\phi}(T)$$
$$= \int E(T)\exp[-(a, T)]\check{\phi}(\tau - T)\,dT$$
$$= \exp[-(a, \tau)]\int E(T)\exp(a, \tau - T)\check{\phi}(\tau - T)\,dT.$$

In view of the inequality (S. 2.6), the function $\exp(a, T)\check{\phi}(T)$ belongs to the space L_2 with the weight $\exp(\epsilon|T|)$ for a certain $\epsilon > 0$. Therefore the property (S. 2.7) of the function E implies the inequality

$$\|\exp(a, \tau)\Psi(\tau)\|_{-\epsilon}$$
$$= \|E * \exp(a, T)\check{\phi}(T)\|_{-\epsilon} \leq C\|\exp(a, T)\check{\phi}(T)\|_{\epsilon}$$
$$\leq C'\|\check{\phi}\|_{-\epsilon} \quad \text{for all small } \epsilon > 0. \tag{S. 2.8}$$

Let us check the equality (S. 2.1). To this end we shall establish first that the function $E(T)\exp[-(a, T)]$ is the fundamental solution for the operator $p(-\mathfrak{D})$. By the Leibnitz differentiation formula we have

$$P(-\mathfrak{D})\{E(T)\exp(-(a, T))\}$$
$$= \exp(-(a, T))\sum_i \frac{a^i}{i!}P^{(i)}(-\mathfrak{D})E$$
$$= \exp(-(a, T))P(-\mathfrak{D} + a)E = \exp -(a, T)\cdot\delta = \delta,$$

from which we easily obtain

$$P(-\mathfrak{D})\Psi = P(-\mathfrak{D})(E(T)\exp(-(a, T)) * \check{\phi}) = \delta * \check{\phi} = \check{\phi}.$$

3. We must prove now that $\operatorname{Supp}\Psi \subset \operatorname{Supp} h$. Let $\tau \overline{\in} \operatorname{Supp} h$; then from the convexity of $\operatorname{Supp} h$ it follows that for a certain λ we have $(\lambda, T) > (\lambda, \tau)$ for $T \in \operatorname{Supp} h$. The construction of $E(T)$ leads to the relation

$$P(\mathfrak{D}_T + a)E(\tau - T) = 0 \text{ in the half-space } (\lambda, \tau - T) < 0.$$

By a theorem of V. P. Palamadov [8, 9] we have the representation

$$E(\tau - T) = \int_{\substack{P(\theta + a) = 0 \\ |\operatorname{Re}\theta| < 2\epsilon}} \exp(\theta, T)\mu(d\theta), \tag{S. 2.9}$$

where the condition $|\operatorname{Re}\theta| < 2\epsilon$ is secured by the special choice of the fundamental solution $E(T)$ for which $\|E * \phi\|_{-\epsilon} < C\|\phi\|_{\epsilon}$ (the property (S. 2.7)). The measure μ in (S. 2. 9) is such that the integral (S. 2. 9) converges absolutely as a generalized function in the half-space $(\lambda, \tau - T) < 0$.

We now have

$$\Psi(\tau) = \int E(\tau - T) \exp(-(a, \tau - T)) \check{\phi}(T)\, dT$$

$$= \int_{\operatorname{Supp} h = \mathfrak{T}} E(\tau - T) \exp(-(a, \tau - T)) \check{\phi}(T)\, dT$$

$$= \exp(-(a, \tau)) \int_{\substack{P(\theta + a) = 0 \\ |\operatorname{Re}\theta| < 2\epsilon}} (\exp(\theta + a, T), \check{\phi}(T)) \mu(d\theta). \qquad \text{(S.2.10)}$$

Here we have set

$$(\exp(\theta + a, T), \check{\phi}(T)) = \int_{\mathfrak{T}} \exp(\theta + a, T) \check{\phi}(T)\, dT.$$

The interchanging of the order of integration is permissible because for a sufficiently small $\epsilon' > 0$, in view of $|\operatorname{Re}\theta| < 2\epsilon$, we have:

$$\exp(\theta + a, T) = O(\exp(-\epsilon' |T|)) \text{ for } T \in \mathfrak{T}.$$

We shall now show that

$$(\exp(\theta_0 + a, T), \check{\phi}(T)) = 0 \text{ for } p(\theta_0 + a) = 0, \ |\operatorname{Re}\theta_0| < 2\epsilon.$$

For sufficiently small a we have $\theta_0 + a \in \omega \times R^s_y$. Let N' be a connected component of the intersection $N \cap (\omega \times R^s_y)$, containing the point $\theta_0 + a$. By the conditions of the theorem N' has at least one real point ζ at which grad $P(\zeta) \neq 0$. Since $P(\theta)$ is a polynomial with real coefficients, the condition grad $P(\zeta) \neq 0$ implies that the set N' contains an open $(s - 1)$-dimensional part ν of the set of the real zeros of $P(\theta)$. Now the cotest condition

$$E_\theta \phi = 0, \quad \theta \in \Pi \cap \omega$$

implies that

$$(\exp(\theta, T), \check{\phi}(T)) = 0 \text{ for } \theta \in \nu. \qquad \text{(S.2.11)}$$

Since $(\exp(\theta, T), \check{\phi}(T))$ is analytic in $\omega \times R^s_y$, the relation (S.2.11) holds by the principle of analytic continuation for the whole N', i.e.

$$(\exp(\theta_0 + a, T), \check{\phi}(T)) = 0$$

for $p(\theta + a) = 0$. Hence

$$\operatorname{Supp} \Psi \subset \mathfrak{T}.$$

4. We shall show that for any $\epsilon > 0$

$$\|\Psi\|_{-\epsilon} < \infty.$$

(S. 2.8) implies that

$$\|\exp(a, T)\Psi\|_{-\epsilon} < \infty.$$

First we shall prove that Ψ does not depend upon the choice of the point $a \in \omega$. Take

$$\Psi_a = \Psi \text{ and } \Psi_{a'} = \Psi'.$$

We have $\|\Psi \exp(a, T)\|_{-\epsilon} < \infty$ and $\|\Psi' \exp(a', T)\|_{-\epsilon} < \infty$. From this and from the convexity of the cone ω we deduce that the integrals

$$\widetilde{\Psi} = \int \Psi \exp(\theta, T)\, dT \text{ and } \widetilde{\Psi}' = \int \Psi' \exp(\theta, T)\, dT$$

converge absolutely for $\theta = b + c + a'$; $b \in \omega$. But

$$P(-\mathfrak{D})\Psi = P(-\mathfrak{D})\Psi'$$

implies that

$$P(\theta)(\widetilde{\Psi} - \widetilde{\Psi}') = 0 \text{ for } \theta \in \omega + a + a'.$$

Hence $\widetilde{\Psi} = \widetilde{\Psi}'$ and $\Psi = \Psi'$, and therefore

$$\|\Psi \exp(a, T)\|_{-\epsilon} < \infty \text{ for any } a \in \omega.$$

Since $|a|$ in ω can be made as small as we need, we have

$$\|\Psi\|_{-\epsilon} < \infty.$$

The argument of this section also shows that the function Ψ in (S. 2.1) is uniquely determined by the conditions (S. 2.5).

5. *Application to the Behrens-Fisher problem.* For the Behrens-Fisher problem we have

$$p(T, \theta) = C(\theta)\, h(T) \exp(\theta_1 T_1 + \cdots + \theta_4 T_4);$$

$$\omega = (\theta_2 < 0,\ \theta_4 < 0);$$

$$h(T) = (T_2 - T_1^2)^{(n-3)/2}(T_4 - T_3^2)^{(m-3)/2}.$$

The corresponding ideal is a principal one and is generated by the polynomial

$$P(\theta) = \theta_1\theta_4 - \theta_2\theta_3.$$

Thus the Conditions 1.1–2.2 obviously hold. Let us check the requirement 3.1. We take an arbitrary point $\theta \in N \cap (\omega \times R^s_y)$ and let it move into the real subspace while remaining during the whole motion in one and the same connected component.

1) Suppose that $\theta_3 \neq 0$. Consider the path formed by the points:

$$\theta_t = (\theta_1, \theta_2, t\theta_3, t\theta_4); \quad t \in \left[1, \frac{\theta_1}{\theta_3}\right].$$

Clearly all points θ_t belong to the same component of N. Consider now the path

$$\theta'_t = (\mathrm{Re}\,\theta_1 + it\,\mathrm{Im}\,\theta_1, \mathrm{Re}\,\theta_2 + it\,\mathrm{Im}\,\theta_2, \mathrm{Re}\,\theta_1 + it\,\mathrm{Im}\,\theta_1, \mathrm{Re}\,\theta_2 + it\,\mathrm{Im}\,\theta_2); \quad t \in (0, 1).$$

This path takes θ_t into the point $(\mathrm{Re}\,\theta_1, \mathrm{Re}\,\theta_2, \mathrm{Re}\,\theta_3, \mathrm{Re}\,\theta_4)$ belonging to the real subspace.

2) If $\theta_3 = 0$ but $\theta_4 \neq 0$ the argument must be changed in an obvious way.

3) If $\theta_3 = 0$, $\theta_4 = 0$, the path

$$\theta_t = (\mathrm{Re}\,\theta_1 + it\,\mathrm{Im}\,\theta_1, \mathrm{Re}\,\theta_2 + it\,\mathrm{Im}\,\theta_2, 0, 0); \quad t \in (1, 0)$$

takes the point $(\theta_1, \theta_2, 0, 0)$ into $(\mathrm{Re}\,\theta_1, \mathrm{Re}\,\theta_2, 0, 0)$. It remains only to remark that for $P(\theta) = \theta_1\theta_4 - \theta_2\theta_3$, we have $\mathrm{grad}\,P(\theta) = 0$ only at the origin, which does not belong to ω.

Hence for the Behrens-Fisher problem all cotests are of the form

$$\phi = \frac{1}{h}\left[\frac{\partial^2}{\partial T_1 \partial T_4} - \frac{\partial^2}{\partial T_1 \partial T_3}\right]\Psi.$$

6. We can describe all the cotests for exponential families with an arbitrary number of polynomial relations. In that case the ideal I is not necessarily a principal one. Each cotest can be written in the form

$$\phi = \frac{1}{h}\sum_j P_j(-\mathfrak{D})\Psi_j,$$

where $P_j(\theta)$ are the generators of the ideal I and $\Psi_j(T)$ are in general, generalized functions with supports in $\mathcal{T}$.

§3. CONDITIONS OF OPTIMAL UNBIASED ESTIMATION FOR INCOMPLETE EXPONENTIAL FAMILIES WITH POLYNOMIAL RELATIONS

We consider the problem of unbiased estimation of parametric functions of n independent observations of a random variable with the density with respect to Lebesgue measure:

$$f(x;\alpha) = \exp\{t_0(x) + c_1(\alpha)t_1(x) + \cdots + c_s(\alpha)t_s(x) + c_0(\alpha)\}; \tag{S. 3.1}$$

here $s \le n$; the abstract parameter $\alpha \in A$. Introduce the natural parameters

$$\theta_i = c_i(\alpha)$$

and suppose that for $\alpha \in A$ the point $\theta = (\theta_1, \cdots, \theta_s)$ runs over an everywhere dense subset of an algebraic subvariety of the domain $\Omega \subset R^s$. This subvariety we shall write in the form $\Omega \cap \Pi$, where Π is an algebraic variety in R^s given by the polynomial relations

$$\left.\begin{array}{l} \Pi_1(\theta_1, \cdots, \theta_s) = 0 \\ \cdots\cdots\cdots\cdots\cdots\cdots \\ \Pi_r(\theta_1, \cdots, \theta_s) = 0 \end{array}\right\} \tag{S. 3.2}$$

with $r < s$.

The distribution of the repeated sample $(x_1, \cdots, x_n) = \vec{x}$ from the set (S. 3.1) is given in R^n by the density with respect to Lebesgue measure

$$f^{(n)}(\vec{x};\theta) = (C(\theta))^n \exp\left\{\sum_1^n t_0(x_i) + \theta_1 \sum_1^n t_1(x_i) + \cdots + \theta_s \sum_1^n t_s(x_i)\right\} \tag{S. 3.3}$$

where $C(\theta)$ is determined by the norming condition. The sufficient statistics for the family (S. 3.3) are

$$T_1 = \sum_1^n t_1(x_i); \cdots ; T_s = \sum_1^n t_s(x_i).$$

We shall suppose that $T_1, \cdots, T_s$ are functionally independent; then the distribution of the vector $T = (T_1, \cdots, T_s)$ is given in R^s by the density with respect to Lebesgue measure

$$p(T;\theta) = C(\theta)h(T)\exp(\theta_1 T_1 + \cdots + \theta_s T_s), \tag{S. 3.4}$$

where $h(T) \ge 0$; $\theta \in \Omega \cap \Pi$. We denote by $\operatorname{Supp} h$ the support of the function

$h(T)$, i.e. $\operatorname{Supp} h = \{T: h(T) > 0\}$; let $\mathfrak{T} = \operatorname{int} \operatorname{Supp} h$. Our condition for $h(T)$ consists in the following requirements:

$$\left.\begin{array}{c} h(T) \text{ is infinitely differentiable on } \mathfrak{T} \\ \operatorname{mes} \operatorname{Supp} h = \operatorname{mes} \mathfrak{T}. \end{array}\right\} \tag{S.3.5}$$

Let $g(T)$ be an unbiased estimate for a certain function $\gamma(\theta)$ depending only upon the vector of sufficient statistics T and having a finite variance for all values of $\theta \in \Omega \cap \Pi$. We shall investigate the conditions which are implied for the variety Π and the estimate itself by the optimality property of the estimate for all $\theta \in \Omega \cap \Pi$ in the class of unbiased estimates of the function $\gamma(\theta)$ with finite variances. Throughout this section we take the variance for the quality measure of the estimate. It is clear that the behavior of the function $g(T)$ outside $\mathfrak{T}$ is of no importance for its properties as an estimate of $\gamma(\theta)$; therefore we shall suppose $g(T)$ to be defined only on $\mathfrak{T}$. Denote by N the least (complex) algebraic variety in $\mathbf{C}^s$ containing $\Pi \cap \Omega$. Since Π is itself an algebraic variety, $\Pi \cap \Omega = N \cap \Omega$.

Theorem S.3.1. *In order for the function $g(T)$, $T \in \mathfrak{T}$ for which $E_\theta g^2 < \infty$; $\theta \in \Omega \cap \Pi$ to be the best unbiased estimate of the function $E_\theta g = \gamma(\theta)$ for the exponential family* (S.3.5), $\theta \in \Omega \cap \Pi$, *under condition* (S.3.5) *it is necessary and sufficient that:*

1) *In the space $\mathbf{C}^s$ of variables $\theta_1, \dots, \theta_s$ there exists a linear system of coordinates $\theta'_1, \dots, \theta'_s$ in which N is a cylinder of type $L \times \nu$, where L is the coordinate space $\theta'_1 = \dots = \theta'_m = 0$ and ν is a certain set in the space: $\theta'_{m+1} = \dots = \theta'_s = 0$; $0 \leq m \leq s$.*

2) *In the corresponding system of coordinates $T'_1, \dots, T'_s$ in the space R^s of values $(T_1, \dots, T_s)$ the function $g(T)$ depends only on $T'_{m+1}, \dots, T'_s$.*

Proof. Let $\chi(T)$ be an arbitrary unbiased estimate of zero with a finite variance, i.e.

$$E_\theta(\chi) = 0; \ E_\theta \chi^2 < \infty; \ \theta \in \Omega \cap \Pi.$$

Lemma S.3.1. *For $g(T)$ to be the optimal unbiased estimate of $\gamma(\theta)$, $\theta \in \Omega \cap \Pi$, it is necessary and sufficient that for each unbiased estimate of zero with a finite variance*

$$E_\theta(g\chi) = 0; \ \theta \in \Omega \cap \Pi.$$

In fact, let $g(T)$ be the best unbiased estimate of $\gamma(\theta)$ and $\chi(T)$ an arbitrary unbiased estimate of zero. Then for an arbitrary constant c

$$\mathfrak{D}_\theta(g+c\chi) = \mathfrak{D}_\theta(g) + 2c\,E_\theta(g\chi) + c^2\mathfrak{D}_\theta(\chi). \qquad \text{(S. 3.6)}$$

As for all values of c

$$\mathfrak{D}_\theta(g+c\chi) \geq \mathfrak{D}_\theta(g).$$

then it is easy to deduce from (S. 3.6) that $E_\theta(g\chi) = 0,\ \theta \in \Omega \cap \Pi$. Conversely, let $E_\theta(g\chi) = 0,\ \theta \in \Omega \cap \Pi$ for any unbiased estimate of zero $\chi(T)$. If the estimate $g_1(T)$ is such that

$$E_\theta g_1 = \gamma(\theta),\ E_\theta g_1^2 < \infty,\ \theta \in \Omega \cap \Pi,$$

then $\chi = g_1 - g$ will be an unbiased estimate of zero. We have:

$$\mathfrak{D}_\theta(g_1) = E_\theta(g - \gamma(\theta) + \chi)^2 = \mathfrak{D}_\theta(g) + \mathfrak{D}_\theta(\chi) \geq \mathfrak{D}_\theta(g),$$

which proves Lemma S. 3.1.

We can now prove the sufficiency of the conditions of the theorem. Without loss of generality, we can assume that in the initial system itself the subspace L is given by the equations $\theta_1 = \cdots = \theta_m = 0$, and ν is a subset of the subspace $\theta_{m+1} = \cdots = \theta_s = 0$, and that the function $g(T)$ depends only upon $T_{m+1}, \cdots, T_s$. Let $\chi(T)$ be an arbitrary unbiased estimate of zero with $E_\theta(\chi^2) < \infty,\ \theta \in N \cap \Omega$. Take any point $\theta_0 = (\theta_{10}, \cdots, \theta_{s0}) \in N \cap \Omega$, denote by ω the section of Ω by the surface $\theta_1 = \theta_{10}, \cdots, \theta_m = \theta_{m0}$ and put

$$\Psi(T) = \Psi(T_{m+1}, \cdots, T_s) = \int \chi(T) h(T) \exp(\theta_{10}T_1 + \cdots + \theta_{m0}T_m)\,dT_1 \cdots dT_m.$$

The relation $N = L \times \nu$ implies that, together with the point θ_0, the set N contains the points $(\theta_{10}, \cdots, \theta_{m0}, \theta_{m+1}, \cdots, \theta_s)$ for all values of $\theta_{m+1}, \cdots, \theta_s$. Hence and from the condition $E_\theta(\chi) = 0;\ \theta \in \Omega \cap N$ we deduce that

$$\int \Psi(T) \exp(\theta_{m+1}T_{m+1} + \cdots + \theta_s T_s)\,dT_{m+1} \cdots dT_s = 0$$

for all $(\theta_{m+1}, \cdots, \theta_s) \in \omega$. By the uniqueness theorem for Laplace transforms, $\Psi(T) = 0$. We have further for $\theta_0 \in \Omega \cap N$:

$$E_{\theta_0}(g\chi) = C(\theta_0)\int g(T_{m+1}, \dots, T_s)\,dT_{m+1}\cdots dT_s$$

$$\times \int \chi(T)h(T)\cdot \exp(\theta_{10}T_1 + \cdots + \theta_{m0}T_m)\,dT_1\cdots dT_m$$

$$= C(\theta_0)\int g(T_{m+1},\dots,T_s)\exp(\theta_{m+1,0}T_{m+1} + \cdots + \theta_{s,0}T_s)\,dT_{m+1}\cdots dT_s$$

$$\times \int \chi(T)h(T)\exp(\theta_{10}T_1 + \cdots + \theta_{m0}T_m)\,dT_1\cdots dT_m$$

$$= C(\theta_0)\int g(T)\Psi(T)\exp(\theta_{m+1,0}T_{m+1} + \cdots + \theta_{s0}T_s)\,dT_{m+1}\cdots dT_s = 0.$$

By Lemma S. 3.1, $g(T)$ is the best unbiased estimate of the function $\gamma(\theta)$.

The proof of the necessity of the conditions of Theorem S. 3.1 is more complicated and the following lemmas are required. Denote by $\mathfrak{D} = \mathfrak{D}(\mathfrak{T})$ the space of infinitely differentiable functions of T with compact supports lying in $\mathfrak{T}$. For any function $\phi(T) \in \mathfrak{D}$ the Laplace transform

$$\tilde{\phi}(\theta) = \int \exp(\theta, T)\,\phi(T)\,dT$$

is an entire function in $\mathbf{C}^s$.

Directly from Lemma S. 3.1 it follows that if $g(T)$ is the best unbiased estimate of the function $\gamma(\theta)$, $\theta \in \Omega \cap \Pi$ and $\chi(T) \in \mathfrak{D}$ is such that

$$\tilde{\chi}(\theta) = 0, \quad \theta \in \Omega \cap \Pi,$$

then

$$g\tilde{\chi}(\theta) = 0, \quad \theta \in \Omega \cap \Pi. \tag{S. 3.7}$$

Lemma S. 3.2. *If an entire function $\phi(\theta)$ is equal to zero on $\Omega \cap \Pi$, then it is equal to zero on the whole set N.*

Proof. Let M be an irreducible algebraic variety in $\mathbf{C}^s$. By a theorem of Whitney [17] there exists a proper subvariety M_* with the following property: the set $(M\setminus M_*) \cap R^s$ in the vicinity of each of its points θ is a real analytic subvariety of dimension d, and the vectors grad $f(\theta)$ (where $f(\theta)$ is an arbitrary real polynomial vanishing on M) form an $(s-d)$-dimensional space.

Now represent the variety N as the sum of irreducible subvarieties

$$N = N^1 \cup \cdots \cup N^l.$$

Applying to each N^λ the theorem of Whitney, we separate in it the exclusive subvariety M_*^λ. We remark that each of the sets $N^\lambda \setminus N_*^\lambda$ has a nonvoid intersection with Ω, since otherwise $\Omega \cap \Pi$ would be a part of $\bigcup_{\mu \neq \lambda} N^{\mu} \cup N^{\lambda}$, and N would not be the least algebraic variety containing $\Omega \cap \Pi$. We now fix the λ and choose a point $\theta \in (N^{\lambda} \backslash N^{\lambda}_{*}) \cap \Omega$. By the theorem of Whitney, in the vicinity of the point θ the set $N^{\lambda} \cap \Omega$ is a real analytic variety of dimension d; moreover there exist real polynomials $f_1, \cdots, f_{s-d}$, vanishing on $N^{\lambda} \cap \Omega$, whose gradients are linearly independent at the point θ. The polynomials $f_1, \cdots \cdots, f_{s-d}$ vanish on N^{λ}, since otherwise N would not be the least variety containing $\Omega \cap \Pi$. Hence there exists a complex neighborhood U of the point θ such that $N^{\lambda} \cap U$ is a complex analytic variety of dimension d. Since $\phi(\theta) = 0$ for $\theta \in \Omega \cap N^{\lambda}$, it follows that $\phi(\theta) = 0$ for $\theta \in U \cap N^{\lambda}$. But any complex irreducible variety is a connected analytic variety with the exception of a nowhere dense set. Hence $\phi(\theta) \equiv 0$, $\theta \in N^{\lambda}$. Since λ is any of the numbers $1, \cdots, l$, $\phi \equiv 0$ on N. Lemma S.3.2 is proved.

From (S.3.7) and Lemma S.3.2 it follows that for any $\chi \in \mathfrak{D}$, if $\widetilde{\chi}(\theta) = 0$, $\theta \in \Omega \cap \Pi$ then $g\widetilde{\chi}(\theta) = 0$, $\theta \in N$.

Let A be the ring of all polynomials of $\theta \in \mathbf{C}^s$ with complex coefficients. If $\mathfrak{J}$ is an ideal in A, and $\zeta \in \mathbf{C}^s$, we shall denote by $\mathfrak{J}_{\zeta}$ the ideal formed by the polynomials $p(\theta + \zeta)$, where $p(\theta) \in \mathfrak{J}$. Consider the set $\mathfrak{A} \subset A$ of polynomials $p(\theta)$ for which $g(T)$ is the best unbiased estimate of the function $\gamma(\theta)$ and is also a generalized solution in $\mathfrak{T}$ of the equation

$$p(\mathfrak{D})g = 0, \quad \mathfrak{D} = \left(\frac{\partial}{\partial T_1}, \cdots, \frac{\partial}{\partial T_s}\right).$$

It is easy to see that $\mathfrak{A}$ is an ideal. Denote by I the ideal in A formed by all polynomials vanishing on N.

Lemma S.3.3. *If* $\zeta \in N$, *then* $I_{\zeta} \subset \mathfrak{A}$.

Proof. Let $p(\theta) \in I_{\zeta}$ i.e. $p(\theta - \zeta) \in I$. Considering $g(T)$ as a generalized function in $\mathfrak{T}$, we have for an arbitrary function $\phi \in \mathfrak{D}$:

$$(p(\mathfrak{D})g, \phi \exp(\zeta, T)) = (g, p(-\mathfrak{D})\phi \exp(\zeta, T))$$
$$= (g, \exp(\zeta, T)p(-\mathfrak{D}-\zeta)\phi)(T) = \int g(T)\exp(\zeta, T)p(-\mathfrak{D}-\zeta)\phi(T)\,dT.$$

Since $p(\theta - \zeta) \in I$, the function $\widetilde{p(-\mathfrak{D}-\zeta)\phi} = p(\theta - \zeta)\widetilde{\phi}(\theta)$ vanishes on $\Omega \cap \Pi$. But then we have

$$\int g(T)\exp(\zeta, T)p(-\mathfrak{D}-\zeta)\phi(T)\,dT = 0; \quad \theta \in \Omega \cap \Pi$$

because the expression to the left is equal to $g\widetilde{\chi}(\theta)$, where $\chi = p(-\mathfrak{D} - \zeta)\phi$ satisfies the condition $\widetilde{\chi}(\theta) = 0$, $\theta \in \Omega \cap \Pi$. Now from (S. 3.8) we deduce that the generalized function $p(\mathfrak{D})g$ vanishes on the functions of the form

$$\phi \exp(\zeta, T).$$

As any function from $\mathfrak{D}$ can be represented in this form, $p(\mathfrak{D})g = 0$ in $\mathfrak{T}$ i.e. $p(\theta) \in \mathfrak{A}$. Lemma S. 3.3 is thus proved.

Consider the ideal $\mathfrak{I} = \{I_\zeta\}_{\zeta \in N}$. By Lemma S. 3.3, each summand I_ζ belongs to $\mathfrak{A}$; hence $\mathfrak{I} \subset \mathfrak{A}$. Moreover, each polynomial from $\mathfrak{I}$ vanishes on the set $L = \bigcap_{\zeta \in N}(N - \zeta)$. But if $\theta \overline{\in} L$, then for a certain $\zeta \in N$, $\theta \overline{\in} (N - \zeta)$. Hence we can find a polynomial $f \in I_\zeta \subset \mathfrak{I}$ such that $f(\theta) \neq 0$. Thus L is the set of common roots of polynomials from $\mathfrak{I}$ and therefore is an algebraic variety.

Lemma S. 3.4. *The set* $L = \bigcap_{\zeta \in N}(N - \zeta)$ *is a linear subspace, and for each point* $\zeta \in L$

$$\mathfrak{I}_\zeta = \mathfrak{I}. \tag{S. 3.9}$$

Proof. Let $\zeta \in L$; if $\theta \in N$, then $\theta + \zeta \in N$ because $\zeta \in N - \theta$. Hence

$$\mathfrak{I}_\zeta = \{I_{\theta + \zeta}\}_{\theta \in N} \subset \{I_\theta\}_{\theta \in N} = \mathfrak{I}, \tag{S. 3.10}$$

i.e. $\mathfrak{I}_\zeta \subset \mathfrak{I}$. But the set of all roots of the polynomials from $\mathfrak{I}_\zeta$ is $L - \zeta$ and so from (S. 3.10) it follows that $L - \zeta \supset L$ i.e. $L \supset L + \zeta$ for all $\zeta \in L$. Hence L is a semigroup with respect to the operation of vector addition in $\mathbf{C}^s$. We shall now show that L is a linear subspace in $\mathbf{C}^s$. Let Λ be the largest linear subspace contained in L (we remark that L contains at least the origin of coordinates). If $\Lambda \neq L$, there exists a point $\zeta \in L \setminus \Lambda$. Now take an arbitrary point $\theta \in \Lambda$. Since L is a semigroup, it must contain the points $\theta + k\zeta$; $k = 1, 2, \cdots$.

Let $p(\theta)$ be an arbitrary polynomial vanishing on L; since $p(\theta + k\zeta) = 0$, $k = 1, 2, \cdots$, $p(\theta + k\zeta) = 0$ for all $\lambda \in \mathbf{C}^1$. The set of all straight lines $\theta + \lambda\zeta$, $\lambda \in \mathbf{C}^1$ forms a linear subspace Λ' spread on Λ and ζ. Since L is an algebraic variety, $\Lambda' \subset L$; i.e. Λ is not the largest linear subspace contained in L. This contradiction proves that $L = \Lambda$.

Since L is a linear subspace, it follows from $\zeta \in L$ that $-\zeta \in L$. Hence for any points $\theta \in N$ and $\zeta \in L$, we have $\theta - \zeta \in N$. Hence in (S. 3.10) we have an inverted inclusion and (S. 3.9) is proved.

Lemma S. 3.5. *The set of vectors*

$$\operatorname{grad} p(0), \quad p \in \mathcal{I} \tag{S. 3.11}$$

coincides with the space $L^{\perp}$ *orthogonal to* L.

Proof. It is obvious that all the vectors (S. 3.11) belong to $L^{\perp}$ and that they form a linear space. We shall show that this linear space contains all the vectors from $L^{\perp}$. Suppose this is not so; then there exists a vector $a \in L^{\perp}$ which is orthogonal to all the vectors (S. 3.11). Let q be an arbitrary polynomial from I. For each point $\zeta \in N$ the polynomial $q(\theta + \zeta)$ belongs to $\mathcal{I}$ and so the vector $\operatorname{grad} q(\zeta)$ is orthogonal to a. Hence $q'_a(\zeta) = 0$ on N i.e. $q'_a \in I$. Applying this argument to the polynomial q'_a we see that $q''_a = 0$ on N, as was to be proved. Hence at each point $\zeta \in N$ all derivatives of the polynomial q in the direction of the vector a are equal to zero. This means that q vanishes on the whole line $\zeta + \lambda a$, $\lambda \in C^1$, On the other hand, N is the set of the roots of all polynomials from I. Hence the set N, together with each of its points ζ, contains the whole straight line $\{\zeta + \lambda a\}$. Hence the set $L \in \bigcap_{\zeta \in N} (N - \zeta)$ contains the straight line $\{\lambda a\}$, in contradiction to the orthogonality of a to L. This contradiction proves the lemma.

Lemma S. 3.6. *Each polynomial* $p(\theta)$ *vanishing on* L *belongs to* $\mathcal{I}$.

Proof. Select the polynomials $p_1, \cdots, p_m \in \mathcal{I}$ in such a way that the vectors $\operatorname{grad} p_i(0)$; $i = 1, \cdots, m$ form a basis in the space $L^{\perp}$. Construct a regular system of coordinates $(w_1, \cdots, w_m)$ in the vicinity of the origin so that $w_1 = p_1, \cdots, w_m = p_m$. Since the polynomial p vanishes on the coordinate subspace $w_1 = \cdots = w_m = 0$ (i.e. on L), it can be written in the form

$$p = \sum_1^m w_i f_i = \sum_1^m p_i f_i$$

where f_i are certain functions holomorphic at the origin. By (S. 3.9) we can also obtain the representation

$$p = \sum p'_i f_i; \quad p'_i \in \mathcal{I}, \tag{S. 3.12}$$

where f_i are certain functions holomorphic at an arbitrary point ζ of L. Such a representation can obviously be obtained for each point $\zeta \in \overline{L}$, since we can always find a polynomial $p' \in \mathcal{I}$ that does not vanish at ζ.

We fix now an arbitrary point $\zeta \in C^s$. Denote by R_ζ the ring of all rational functions in C^s whose denominators do not vanish at the point ζ; by $\mathcal{H}_\zeta$ we denote the ring of all functions holomorphic in ζ. As is well known, the pair $(R_\zeta, \mathcal{H}_\zeta)$ is flat (see [15]). This means that $\mathcal{H}_\zeta$ is a flat R_ζ-module [15] and that for each R_ζ-module E the natural operation $E \to E \otimes_{R_\zeta} \mathcal{H}_\zeta$ is injective. We now take for E the factor-ring $R_\zeta | \mathcal{I}R_\zeta$ where $\mathcal{I}R_\zeta$ is the ideal in R_ζ formed by all the functions of the form

$$\Sigma q_j r_j; \; q_j \in \mathcal{I}; \; r_j \in R_\zeta .$$

Since $\mathcal{H}_\zeta$ is a flat R_ζ-module, we have

$$E \times \mathcal{H}_\zeta \cong \mathcal{H}_\zeta | \mathcal{I}\mathcal{H}_\zeta$$

where $\mathcal{I}\mathcal{H}_\zeta$ is an analogous ideal in $\mathcal{H}_\zeta$. Hence we can assert that the natural mapping

$$R_\zeta | \mathcal{I}\mathcal{H}_\zeta \to \mathcal{H}_\zeta | \mathcal{I}\mathcal{H}_\zeta$$

is injective. By (S. 3.12) the polynomial p belongs to $\mathcal{I}\mathcal{H}_\zeta$ and hence belongs to $\mathcal{I}\mathcal{R}_\zeta$, in view of the injectivity of this mapping. Hence

$$q_\zeta p \in \mathcal{I} \tag{S. 3.13}$$

where q_ζ is a certain polynomial distinct from zero in the point ζ. Consider an ideal in A generated by the polynomials q_ζ. By the Hilbert Nullstellensatz (see for instance [1]) the unity of the ring A belongs to that ideal i. e. $\Sigma_1^k h_i q_{\zeta_i} = 1$ with certain $h_i \in A$. Putting $\zeta = \zeta_i$ in (S. 3.13), multiplying by h_i and adding, we finally obtain $p \in \mathcal{I}$, which proves the lemma.

We can now complete the proof of the theorem. In the space C^s choose a rectangular system of coordinates $\theta'_1, \cdots, \theta'_s$ in such a way as to make L coincide with the coordinate subspace in which $\theta'_1 = \cdots = \theta'_m$. In that case $\theta'_1, \cdots, \theta'_m$ vanish on L and by Lemma S. 3.6 belong to the ideal $\mathcal{I}$. Since $\mathcal{I} \subset \mathfrak{A}$, in the corresponding system of coordinates in R^s we have in the domain $\mathcal{T}$ the equations

$$\frac{\partial g}{\partial T'_1} = \cdots = \frac{\partial g}{\partial T'_m} = 0.$$

These equations in the space of generalized functions mean that the function g, up to its values on a set of measure zero, is constant with respect to the variables $T'_1, \cdots, T'_m$ in the domain $\mathfrak{T}$. The representation $N = L \times \nu$ is obvious, because it follows from $L \subset N - \zeta$ that together with the point ζ the set N contains the whole variety $L + \zeta$.

Theorem S. 3.1 is proved.

It is interesting to compare the situation with respect to optimal unbiased estimation of parametric functions for complete and incomplete exponential families. For the former, by the Rao-Blackwell-Kolmogorov theorem, every statistic $g(T)$ depending only upon sufficient statistics is an optimal estimate of its mathematical expectation $E_\theta g$. For the incomplete exponential families, in view of Theorem S. 3.1, the optimality of $g(T)$ as an estimate of $E_\theta g$ means, roughly speaking, a kind of quasi-completeness with respect to some of parameters (the representation $N = L \times \nu$ is analogous to the completeness). As regards the optimal estimate itself, it depends on sufficient statistics, having the same indices as the parameters with the "quasi-completeness" property.

§4. THE SAMPLE MEAN AS THE ESTIMATE OF SCALE PARAMETERS

In this section we study the families of distribution functions of the form $F(x/\sigma)$ on the half-line $(0, \infty)$ depending upon the scale parameter $\sigma \in (0, \infty)$. Our purpose is to study the unbiased estimation of σ on the evidence of the sample $(x_1, \cdots, x_n)$ from the population characterized by $F(x/\sigma)$. As usual, we take the quadratic loss function.

Suppose the condition

$$\int_0^\infty x^2 dF(x) = \alpha_2 < \infty \tag{S. 4.1}$$

holds. If we set

$$\alpha_1 = \int_0^\infty x\, dF(x),$$

the statistics $\alpha_1^{-1}\bar{x}$ will be an unbiased estimate of the parameter with a finite variance by (S. 4.1).

We remark first of all that in the class of all (not only unbiased) estimates of σ, the $\alpha_1^{-1}\bar{x}$ are always inadmissible except in the case of a degenerate distribution $F(x)$. In fact, we have

$$E_\sigma(c_1\bar{x} - \sigma)^2 = \sigma^2 E_1(c_1\bar{x} - 1)^2$$

and $\min_{c_1} E_1(c_1\bar{x} - 1)^2$ is attained, as is easily verified, for

$$c_1 = \frac{\alpha_1}{\alpha_1^2 + (\alpha_2 - \alpha_1^2)/n} .$$

Since $\alpha_2 \geq \alpha_1^2$ and, moreover, the equality sign holds for only the degenerate ones in the class of all estimates (except in the degenerate case), $\alpha_1^{-1}\bar{x}$ is always inadmissible.

When is $\alpha_1^{-1}\bar{x}$ admissible in the class of unbiased estimates of the scale parameter σ? To answer this question we shall first prove Lemma S. 4.1, where we use the notation $y = (x_2/x_1, \cdots, x_n/x_1)$.

Lemma S. 4.1. *Let*

$$s_n = c_n\bar{x}\frac{E_1(\bar{x}\,|\,y)}{E_1(\bar{x}^2|y)} \tag{S. 4.2}$$

where the constant c_n *is determined from the condition*

$$c_n E_1\left\{\frac{E_1(\bar{x}\,|\,y)^2}{E_1(\bar{x}^2|y)}\right\} = 1. \tag{S. 4.3}$$

Then

$$E_\sigma s_n = \sigma;\ E_\sigma(s_n - \sigma)^2 \leq E_\sigma(\alpha_1^{-1}\bar{x} - \sigma)^2$$

and equality holds in (S. 4.4) *if and only if, with probability* 1,

$$\frac{E_1(\bar{x}\,|\,y)}{E_1(\bar{x}^2|y)} = b_n;\ b_n = \alpha_1^{-1}c_n^{-1}.$$

Proof. Put $\gamma = \gamma(y) = E_1(\bar{x}|y)/E_1(\bar{x}^2|y)$. We have

$$E_\sigma(s_n) = c_n E_\sigma(\bar{x}\gamma) = c_n\sigma E_1(\gamma E_1(\bar{x}\,|\,y)) = \sigma c_n E_1\left\{\frac{E_1(\bar{x}\,|\,y)^2}{E_1(\bar{x}^2\,|\,y)}\right\} = \sigma$$

by (S. 4.3);

$$E_\sigma(\alpha_1^{-1}\bar{x} - \sigma)^2 = E_\sigma(\alpha_1^{-1}\bar{x} - s_n)^2 + E_\sigma(s_n - \sigma)^2 + 2E_\sigma\{(\alpha_1^{-1}\bar{x} - s_n)(s_n - \sigma)\}.$$

Further,

$$\begin{aligned}
&E_\sigma((\alpha_1^{-1}\bar{x} - s_n)(s_n - \sigma)) \\
&= \sigma^2 E_1((\alpha_1^{-1} - c_n\gamma)(c_n\gamma E_1(\bar{x}^2|y) - E_1(\bar{x}\,|y))) \\
&= \sigma^2(c_n - 1)\left\{\alpha_1^{-1}E_1(E_1(\bar{x}\,|y)) - c_n E_1\left\{\frac{E_1(\bar{x}\,|y)^2}{E_1(\bar{x}^2|y)}\right\}\right\} \\
&= \sigma^2(c_n - 1)\left\{1 - c_n E_1\left[\frac{E_1(\bar{x}|y)^2}{E_1(\bar{x}^2|y)}\right]\right\} = 0,
\end{aligned}$$

by (S. 4.3). Thus

$$E_\sigma(\alpha_1^{-1}\bar{x} - \sigma)^2 = E_\sigma(\alpha_1^{-1}\bar{x} - s_n)^2 + E_\sigma(s_n - \sigma)^2 \geq E(s_n - \sigma)^2 \quad \text{(S.4.5)}$$

and the equality sign in (S. 4.5) holds for all $\sigma \in (0, \infty)$ simultaneously if and only if with probability 1

$$\alpha_1^{-1}\bar{x} = c_n\bar{x}\,\frac{E_1(\bar{x}\,|y)}{E_1(\bar{x}^2|y)};$$

i. e. if

$$E_1(\bar{x}\,|y) = b_n E_1(\bar{x}^2\,|y), \qquad b_n = \alpha_1^{-1}c_n^{-1}.$$

This proves Lemma S. 4.1.

From now on we shall suppose that all the moments of $F(x)$

$$\int_0^\infty x^k\,dF(x), \qquad k = 1, 2, \cdots, \quad \text{(S.4.6)}$$

are finite.

Theorem S. 4.1. *Let the function $F(x)$ satisfy the condition* (S. 4.6). *In order for the statistic $\alpha_1^{-1}\bar{x}$ to be an admissible estimate of σ in the class of the unbiased estimates in the sample sizes $n = n_1$, $n = n_2$ (n_1, n_2 are any numbers with $n_2 > n_1 \geq 3$) from the population given by $F(x/\sigma)$, it is necessary and sufficient for $F(x)$ to be either degenerate:*

$$F(x) = \begin{cases} 0, & x < x_0 \\ 1, & x_0 \leq x < \infty \end{cases} \qquad \textit{for some } x_0 > 0,$$

or a gamma-distribution:

$$F(x) = \begin{cases} 0, & x \leq 0 \\ (\gamma^m/\Gamma(m)) \int_0^x x^{m-1} e^{-\gamma x}\, dx; & 0 < x < \infty \end{cases}$$

for some $m > 0$, γ.

Theorem S.4.2. *If* $F(x)$ *satisfies the condition* (S.4.6) *and* $\alpha_1^{-1}\bar{x}$ *is optimal in the class of the unbiased estimates of the parameter* σ *for a sample size* $n \geq 3$, *then* $F(x)$ *is either degenerate or a gamma-distribution.*

We omit the proofs, which proceed by means of functional equations. They are given in detail in [5].

§5. NONPARAMETRIC APPROACH TO THE ESTIMATION OF LOCATION PARAMETERS

Let $x_1, \cdots, x_n$ be a repeated sample from the population with the distribution function $F(x - \theta)$ satisfying the conditions

$$\int x\, dF = 0, \qquad \int x^2 dF < \infty. \tag{S.5.1}$$

In this case the parameter $\theta \in R^1$ to be estimated on the evidence of the sample $(x_1, \cdots, x_n)$ means the mathematical expectation. For the well-known type of distribution function $F(x)$ E. Pitman [14] introduced, as early as 1938, the following estimate for θ:

$$t_n = \bar{x} - E_0(\bar{x} \mid x_2 - x_1, \cdots, x_n - x_1). \tag{S.5.2}$$

For an absolutely continuous $F(x)$, $F(x) = \int_{-\infty}^x f(u)\, du$ the Pitman estimate, as mentioned in §3, Chapter VII, can be written in the form

$$t_n = \frac{\int_{-\infty}^{\infty} \xi \prod_1^n f(x_i - \xi)\, d\xi}{\int_{-\infty}^{\infty} \prod_1^n f(x_i - \xi)\, d\xi}. \tag{S.5.3}$$

C. Stein proved [16] that under the condition

$$\int |x|^3\, dF(x) < \infty$$

the Pitman estimate is absolutely admissible.

The loss function is assumed to be quadratic throughout this section. We shall now consider in detail the situation when the form of the function $F(x)$ is unknown. If we suppose that $F(x)$ is allowed to be *arbitrary*, satisfying only the condition (S. 5.1), it is easy to see that there is no better estimate for θ than $\bar{x}$. However, in the case when for some integer $k > 1$ the first $2k$ moments of the distribution function $F(x)$ are known:

$$\mu_l = \int x^l \, dF(x), \qquad l = 1, 2, \cdots, 2k \tag{S. 5.5}$$

and

$$\mu_{2k} = \int x^{2k} \, dF < \infty \tag{S. 5/5}$$

the information about $F(x)$ contained in the moments (S. 5.4) can be used more efficiently than $\bar{x}$ for construction of the estimates of the parameter θ.

Under condition (S. 5.5) the set of all polynomials $Q(x_1, \cdots, x_n)$ of $x_1, \cdots, x_n$ of degree not exceeding k forms a Hilbert space $L_k^{(2)}$ if the scalar product of the elements Q_1 and Q_2 is defined by:

$$(Q_1, Q_2) = E_0(Q_1 Q_2).$$

The subspace formed by the polynomials $Q \in L_k^{(2)}$ of the form

$$Q = Q(x_2 - x_1, \cdots, x_n - x_1)$$

will be denoted by Λ_k.

Consider the estimate

$$t_n^{(k)}(x_1, \cdots, x_n) = t_n^k = \bar{x} - \hat{E}_0(\bar{x} \mid \Lambda_k) \tag{S. 5.6}$$

where $\hat{E}_0(\cdot \mid \Lambda_k)$ is the operator of projection on the subspace Λ_k. Note that for the construction of the estimate (S. 5.6) we need to know only the first $2k$ moments of the distribution function, not the whole function.

Theorem S. 5.1. *For all* $\theta \in R^1$

$$E_\theta t_n^k = \theta; \;\; E_\theta (t_n^k - \theta)^2 \leq E_\theta (\bar{x} - \theta)^2, \tag{S. 5.7}$$

where the equality or the inequality in (S. 5.7) *are realized simultaneously for all* $\theta \in R^1$ *and the equality holds if and only if* $\hat{E}_0(\bar{x} \mid \Lambda_k) = 0$.

Proof. Since $Q \equiv 1 \in \Lambda_k$, we have

$$(\bar{x} - \hat{E}_0(\bar{x} \mid \Lambda_k), 1) = 0.$$

Hence

$$E_\theta(\bar{x} - \hat{E}_0(\bar{x} \mid \Lambda_k)) = E_\theta \bar{x} - E_\theta(\hat{E}_0(\bar{x} \mid \Lambda_k)) = \theta - E_0(\hat{E}_0(\bar{x} \mid \Lambda_k)) = 0.$$

Moreover,

$$E_\theta(\bar{x} - \theta)^2 = E_\theta(\bar{x} - \hat{E}_0(\bar{x} \mid \Lambda_k) - \theta + E_0(\bar{x} \mid \Lambda_k))^2$$
$$= E_\theta(t_n^k - \theta)^2 + 2E_\theta((t_n^k - \theta)\hat{E}_0(\bar{x} \mid \Lambda_k)) + E_\theta(\hat{E}_0(\bar{x} \mid \Lambda_k))^2 .$$

But

$$E_\theta((t_n^k - \theta)\hat{E}_0(\bar{x} \mid \Lambda_k)) = E_0(t_n^k \hat{E}_0(\bar{x} \mid \Lambda_k)) = 0$$

as $t_n^k = \bar{x} - \hat{E}_0(\bar{x} \mid \Lambda_k)$ is orthogonal to each function from Λ_k. Hence

$$E_\theta(\bar{x} - \theta)^2 = E_\theta(t_n^k - \theta)^2$$
$$= E_\theta(t_n^k - \theta)^2 + E_0(\hat{E}_0(\bar{x} \mid \Lambda_k))^2 \geq E_\theta(t_n^k - \theta)^2$$

and the equality sign (for all $\theta \in R^1$ simultaneously) holds under the condition

$$\hat{E}_0(\bar{x} \mid \Lambda_k) = 0.$$

In connection with Theorem S. 5.1 it is natural to raise the question: for what functions $F(x)$ is the estimate t_n^k better than $\bar{x}$? In other words, when does the knowledge of the first $2k$ moments of $F(x)$ enable us to improve upon the standard estimate $\bar{x}$.

Theorem S. 5.2. *If $F(x)$ satisfies condition* (S. 5.5), *then for $n \geq 3$ the estimate t_n^k will be better than the sample mean $\bar{x}$ as an estimate of the location parameter in all cases except when the first $(k+1)$ moments of the distribution function $F(x)$, μ_l coincide with the corresponding moments of a normal law, so that*

$$\mu_l = \begin{cases} 0, & l \text{ odd} \\ (l-1)!!\,\sigma^l, & l \text{ even} \end{cases} \qquad 1 \leq l \leq k+1$$

for a certain σ^2.

Theorem S. 5.2 is an obvious consequence of the Theorem S. 5.1 and the following lemma.

Lemma S. 5.1. *If $n \geq 3$ then $\hat{E}_0(\bar{x} \mid \Lambda_k) = 0$ if and only if the first $(k+1)$*

moments of $F(x)$ coincide with the corresponding moments of a normal law.

Proof of Lemma S. 5.1. We proceed by induction. For $k = 1$ the lemma holds trivially. Since $\Lambda_k > \Lambda_{k+1}$, it follows from $\hat{E}_0(\bar{x} \mid \Lambda_k) = 0$ that $\hat{E}_0(\bar{x} \mid \Lambda_{k+1}) = 0$. We can assume now that the first k moments of $F(x)$ coincide with corresponding normal moments and we shall prove that then the moment μ_{k+1} coincides with the $(k+1)$st normal moment.

The condition

$$\hat{E}_0(\bar{x} \mid \Lambda_k) = 0 \tag{S. 5.8}$$

is equivalent to the set of conditions

$$\hat{E}_0(\bar{x} \mid \Lambda_{k-1}) = 0 \tag{S. 5.9}$$

$$\hat{E}_0(\bar{x}(x_{j_1} - x_1) \cdots (x_{j_k} - x_1)) = 0, \tag{S. 5.10}$$

where (S. 5.10) must hold for all the sets $(j_1, \cdots, j_k)$ of integers $2, \cdots, n$. The equivalence of the conditions (S. 5.8) and (S. 5.9)–(S. 5.10) follows from the fact that Λ_{k-1} and the functions $(x_{j_1} - x_1) \cdots (x_{j_k} - x_1)$ generate the whole Λ_k.

Let

$$(x_{j_1} - x_1) \cdots (x_{j_k} - x_1) = (x_{i_1} - x_1)^{\alpha_1} \cdots (x_{i_s} - x_1)^{\alpha_s}, \tag{S.5.11}$$

where $i_1, \cdots, i_s$ are mutually distinct and $\alpha_1 + \cdots + \alpha_s = k$. We have

$$(x_{i_1} - x_1)^{\alpha_1} \cdots (x_{i_s} - x_1)^{\alpha_s}$$

$$= \sum_{l_1, \cdots, l_s = 0}^{\alpha_1, \cdots, \alpha_s} (-1)^{l_1 + \cdots + l_s} C_{\alpha_1}^{l_1} \cdots C_{\alpha_s}^{l_s} x_1^{\alpha_1} \cdots x_{i_s}^{\alpha_s} x_1^{k - \sum_{i=1}^{s} l_i}. \tag{S.5.12}$$

Put $\Sigma_1^s l_i = l$; then from (S. 5.12) we get

$$\left.\begin{aligned} &E_0(\bar{x}(x_{i_1} - x_1)^{\alpha_1} \cdots (x_{i_s} - x_1)^{\alpha_s}) \\ &= \frac{1}{n} \sum_{l_1=0}^{\alpha_1} \cdots \sum_{l_s=0}^{\alpha_s} (-1)^l C_{\alpha_1}^{l_1} \cdots C_{\alpha_s}^{l_s} \mu_{l_1} \cdots \mu_{l_s} \mu_{k+1-l} \\ &+ \frac{1}{n} \sum_{q=1}^{s} \sum_{l_1=0}^{\alpha_1} \cdots \sum_{l_s=0}^{\alpha_s} (-1)^l C_{\alpha_1}^{l_1} \cdots C_{\alpha_s}^{l_s} \mu_{l_1} \cdots \mu_{l_{q-1}} \mu_{l_{q+1}} \cdots \mu_{l_s} \mu_{k-l}. \end{aligned}\right\} \tag{S. 5.13}$$

We shall consider the cases of odd and even values of $(k-1)$ separately.

1) $k-1$ odd. By the induction assumption, all odd moments up to the order $(k-1)$ are equal to zero. Hence from (S. 5.13) we get

$$n E_0(\bar{x}(x_{i_1}-x_1)^{\alpha_1}\cdots(x_{i_s}-x_1)^{\alpha_s})$$

$$= \Sigma^*(-1)^l C_{\alpha_1}^{l_1}\cdots C_{\alpha_s}^{l_s}\mu_{l_1}\cdots\mu_{l_s}\mu_{k+1-l}$$

$$+\sum_{q=1}^{s}\Sigma_q^*(-1)^l C_{\alpha_1}^{l_1}\cdots C_{\alpha_s}^{l_s}\mu_{l_1}\cdots\mu_{l_{q-1}}\mu_{l_{q+1}}\cdots\mu_{l_s}\mu_{k-l}, \tag{S. 5.14}$$

where the summation in Σ^* is taken over all even $l_1,\cdots,l_s$ and in Σ_q^* over even $l_1,\cdots,l_{q-1},l_{q+1},\cdots,l_s$ and odd l_q, the limits being indicated by (S. 5.13). In the sum Σ^* the number l is always even and therefore $(k+1-l)$ is odd. Moreover, if $k+1-l\le k-1$, then $\mu_{k+1-l}=0$ by the induction assumption. In the sum Σ_q^*, the number l is always odd; hence $(k-l)$ is also odd; since $k-l\le k-1$ we have $\mu_{k-l}=0$. Hence the condition (S. 5.10) is equivalent to the relation

$$\mu_{k+1}=0. \tag{S. 5.15}$$

2) $(k-1)$ is even. Consider again the relation (S. 5.14). We break Σ^* and $\Sigma_{q=1}^s\Sigma_q^*$ into subsums

$$\Sigma^*=S_0+S_2+\cdots+S_{k-1},$$

where S_{2m} is the part of the sum Σ^* corresponding to all the values $l_1,\cdots$ $\cdots,l_s$, $0\le l_1\le\alpha_1,\cdots,0\le l_s\le\alpha_s$ for which $l_1+\cdots+l_s=2m$, and

$$\sum_{q=1}^{s}\Sigma_q^*=S_1+S_3+\cdots+S_k$$

where S_{2m+1} is the part of the double sum corresponding to the values $l_1,\cdots$ $\cdots,l_s$ for which $l_1+\cdots+l_s=2m+1$.

Consider first the conditions

$$0<\alpha_1<k,\cdots,\quad 0<\alpha_s<k. \tag{S. 5.16}$$

We shall show that $S_{2m}+S_{2m+1}=0$ for $0<2m\le k-1$. Note that in view of the condition (S. 5.16) in the sum $\Sigma_{q=1}^{s} \Sigma_q^*$ we have $l_q + 1 \le h$; hence by the induction assumption

$$\mu_{l_q+1} = l_q \mu_{l_q-1} \sigma^2 \tag{S. 5.17}$$

for a certain $\sigma^2 > 0$. Then

$$S_{2m+1} = -\sum_{q=1}^{s} \sum_{l_1+\cdots+l_s=2m+1}{}^{\!*}_{q} C_{a_1}^{l_1} \cdots C_{a_s}^{l_s} \mu_{l_1} \cdots \mu_{l_{q-1}} \mu_{l_q+1} \mu_{l_{q+1}} \cdots \mu_{l_s} \mu_{k-1-2m}$$

$$= - \sum_{l_1+\cdots+l_s=2m}{}^{\!*} C_{a_1}^{l_1} \cdots C_{a_s}^{l_s} \mu_{l_1} \cdots \mu_{l_s} \mu_{k-1-2m} \sigma^2$$

$$\times \left[\frac{C_{a_1}^{l_1+1}}{C_{a_1}^{l_1}} (l_1 + 1) + \cdots + \frac{C_{a_s}^{l_s+1}}{C_{a_s}^{l_s}} (l_s + 1) \right]. \tag{S. 5.18}$$

But

$$\frac{C_{a_1}^{l_1+1}}{C_{a_1}^{l_1}} (l_1 + 1) + \cdots + \frac{C_{a_s}^{l_s+1}}{C_{a_s}^{l_s}} (l_s + 1) = \sum_{q=1}^{s} (a_q - l_q) = k - 2m.$$

Since

$$\mu_{k-1-2m} \sigma^2 (k - 2m) = \mu_{k+1-2m}$$

(recall that $m > 0$), we get from (S. 5.18)

$$S_{2m} + S_{2m+1} = 0. \tag{S. 5.19}$$

In view of (S. 5.19) the condition (S. 5.10) is equivalent to

$$S_0 + S_1 = 0. \tag{S. 5.20}$$

But

$$S_0 = \mu_{k+1},$$

$$S_1 = -(a_1 + \cdots + a_s) \mu_2 \mu_{k-1} = -k \sigma^2 \mu_{k-1}.$$

Hence (S. 5.10) is equivalent to the equality

$$\mu_{k+1} = k\sigma^2 \mu_{k-1}. \tag{S. 5.21}$$

The induction assumption together with (S. 5.21) gives

$$\mu_{k+1} = k!!\,\sigma^{k+1}.$$

Now let one of the numbers $\alpha_1, \cdots, \alpha_s$ be equal to k; then all the other numbers vanish. Without loss of generality we can assume that

$$\alpha_{i_1} = k, \ \alpha_{i_2} = \cdots = \alpha_{i_s} = 0. \tag{S. 5.22}$$

In this case (S. 5.14) reduces to

$$nE_0(\bar{x}(x_{i_1} - x_1)^k) = \sum_{\substack{l=0\\ l \text{ even}}}^{k-1} C_k^l \mu_l \mu_{k+1-l} - \sum_{\substack{l=1\\ l \text{ odd}}}^{k} C_k^l \mu_{l+1} \mu_{k-l}.$$

In the second sum we put $l' = k - l$; then

$$nE_0(\bar{x}(x_{i_1} - x_1)^k) = \sum_{\substack{l=0\\ l \text{ even}}}^{k-1} C_k^l \mu_l \mu_{k+1-l} - \sum_{\substack{l'=0\\ l' \text{ even}}}^{k-1} C_k^{l'} \mu_{k+1-l'} \mu_{l'} = 0,$$

so that under condition (S. 5.22), the condition (S. 5.14) is always satisfied.

We remark that for $n \geq 3$ the subspace Λ_k always contains the function $(x_{i_1} - x_1)^{\alpha_1} \cdots (x_{i_s} - x_1)^{\alpha_s}$ under condition (S. 5.16).

Hence we have established that for $n \geq 3$ the condition (S. 5.8) is equivalent to the coincidence of the first $(k+1)$ moments of the distribution function $F(x)$ with the moments of a normal law. This completes the proof of Lemma S. 5.1 and Theorem S. 5.2.

BIBLIOGRAPHY

[1] B. L. van der Waerden, *Moderne Algebra*. Vol. 2, Springer, Berlin, 1931; 4th ed., 1959; Russian transl., GITTL, Moscow, 1947. MR 2, 120; MR 31 #1292.

[2] A. M. Kagan and V. P. Palamodov, *Conditions of optimal unbiased estimation of parametric functions for incomplete exponential families with polynomial ties*, Dokl. Akad. Nauk SSSR 1967. (Russian)

[3] ———, *Incomplete exponential families and variance unbiased minimum estimates*. I, Teor. Verojatnost. i Primenen. 12 (1967), 34–49. (Russian)

[4] A. M. Kagan, *Sample mean as an estimate of the shift parameters*, Dokl. Akad. Nauk SSSR 169 (1966), 1006–1008 = Soviet Math. Dokl. 7 (1966), 1041–1043. MR 33 # 6747.

[5] A. M. Kagan and A. L. Ruhin, *On the theory of the estimation of a scale parameter*, Teor. Verojatnost. i Primenen. 12 (1967). (Russian)

[6] V. P. Palamodov, *On verifiable functions*, Teor. Verojatnost. i Primenen. 12 (1967). (Russian)

[7] ———, *Testing multidimensional polynomial hypotheses*, Dokl. Akad. Nauk SSSR 172 (1966), 291–293 = Soviet Math. Dokl. 7 (1966), 95–97.

[8] ———, *On systems of differential equations with constant coefficients*, Dokl. Akad. Nauk SSSR 148 (1963), 523–526 = Soviet Math. Dokl. 4 (1963), 133–136. MR 29 # 1442.

[9] ———, Ph. D. Thesis, Moscow State University, Moscow, 1965. (Russian)

[10] G. M. Fihtengol'c, *A course on differential and integral calculus*. Vol. II, "Nauka", Moscow, 1966; German transl., Hochschülbucher für Mathematik, Band 62, 2nd ed., VEB Deutscher Verlag, Berlin, 1966.

[11] L. Hörmander, *Linear partial differential operators*, Die Grundlehren der math. Wissenschaften, Band 116, Academic Press, New York and Springer-Verlag, Berlin, 1963. MR 28 # 4221.

[12] A. M. Kagan, *On the estimation theory of location parameter*, Sankhya (1966).

[13] E. Lehmann, *On the non-verifiability of certain parametric functions*, Teor. Verojatnost. i Primenen. 10 (1965), 758–760. (Russian summary) MR 32 # 8445.

[14] E. J. G. Pitman, *The estimation of the location and scale parameters of a continuous population of any given form*, Biometrika 30 (1938), 391–421.

[15] J.-P. Serre, *Géométrie algébrique et géométrie analytique*, Ann. Inst. Fourier, Grenoble 6 (1955–56), 1–42. MR 18, 511.

[16] C. Stein, *The admissibility of Pitman's estimator of a single location parameter*, Ann. Math. Statist. 30 (1959), 970–979. MR 22 # 278.

[17] H. Whitney, *Elementary structure of real algebraic varieties*, Ann. of Math. (2) 66 (1957), 545–556. MR 20 # 2342.

[18] E. Lehmann, *Testing statistical hypotheses*, Wiley, New York, and Chapman & Hall, London, 1959. MR 21 # 6654.

BIBLIOGRAPHY

BAHADUR, R. R.

1. *Sufficiency and statistical decision functions*, Ann. Math. Statistics 25 (1954), 423–462. MR 16, 154.

BASU, D.

2. *On statistics independent of sufficient statistics*, Sankhyā 20 (1958), 223–226. MR 21 #4494.

BESICOVITCH, A. S.

3. *On diagonal values of probability vectors of infinitely many components*, Proc. Cambridge Philos. Soc. 57 (1961), 759–766. MR 23 #A4154.

BLACKWELL, D.

4. *Conditional expectation and unbiased sequential estimation*, Ann. Math. Statistics 18 (1947), 105–110. MR 8, 478.

BOCHNER, C. and MARTIN, W. T.

5. *Several complex variables*, Princeton Univ. Press, Princeton, N. J., 1948; Russian transl., IL, Moscow, 1951. MR 10, 366.

BRÉNY, H.

6. *L'état actuel du problème de Behrens-Fisher*, Trabajos Estadíst. 6 (1955), 111–113. MR 17, 868.

BROWN, L.

7. *Sufficient statistics,in the case of independent random variables*, Ann, Math. Statistics 35 (1964), 1456–1474.

BOURBAKI, N.

8. *Espaces vectoriels topologiques*, *Éléments de Mathématiques*, no. 1189. 1229, Hermann, Paris, 1953, 1955; Russian transl., IL, Moscow, 1959. MR 14, 880; MR 17, 1109.

WALD, A.

9. *Testing the difference between the means of two normal populations with unknown standard deviations*, Selected Papers in Probability and Statistics, McGraw-Hill, New York, 1955, pp. 669–695. MR 16, 435.

WIDDER, D.

10. *The Laplace transform*, Princeton Math. Series, vol. 6, Princeton Univ. Press, Princeton, N. J., 1946.

WIJSMAN, R. A.

11. *Incomplete sufficient statistics and similar tests*, Ann. Math. Statistics 29 (1958), 1028–1045. MR 21 #5256.

VOLODIN, I. N.

12. *On the distinction between the Poisson and Pólya distributions when a large number of small samples is available,* Teor. Verojatnost. i Primenen. 10 (1965), 364–367. (Russian) MR 31 #6299.

GLEASON, A. M.

13. *Finitely generated ideals in Banach algebras,* J. Math. Mech. 13 (1964), 125–132. MR 28 #2458.

DANTZIG, G.

14. *On the non-existence of tests of "Student's" hypothesis having power function independent of* σ, Ann. Math. Statistics 11 (1940), 186–191. MR 1, 348.

DARMOIS, G.

15. *Sur les lois de probabilité à estimation exhaustive,* C. R. Acad. Sci. Paris 260 (1935), 1265–1266.

DOETSCH, G.

16. *Handbuch der Laplace-Transformation* Vols. I, II, Birkhäuser, Basel, 1950, 1955. MR 13, 230; MR 18, 35.

DOWKER, C. H.

17. *Lectures on sheaf theory,* Tata Institute, Bombay, 1956, 1962. MR 19, 301.

DOOB, J. L.

18. *Stochastic processes,* Wiley, New York, 1953; Russian transl., IL, Moscow, 1956. MR 15, 445; MR 19, 71.

DYNKIN, È. B.

19. *Necessary and sufficient statistics for a family of probability distributions,* Uspehi Mat. Nauk 6 (1951), no. 1(41), 68–90; English transl., Selected Transl. Math. Stat. and Prob., vol. 1, Amer. Math. Soc., Providence, R.I., 1961, pp. 17–40. MR 12, 839.

ZINGER, A. A.

20. *Independence of quasi-polynomial statistics and analytical properties of distributions,* Teor. Verojatnost. i Primenen. 3 (1958), 265–284. (Russian) MR 21 #941.

21. *On a problem of A. N. Kolmogorov,* Vestnik Leningrad. Univ. 11 (1956), no. 1, 53–56. (Russian) MR 17, 863.

ZINGER, A. A. and LINNIK, Ju. V.

22. *Characterization of the normal distribution,* Teor. Verojatnost. i Primenen. 9 (1964), 692–695. (Russian) MR 30 #607.

KAGAN, A. M. and LINNIK, Ju. V.

23. *A class of families admitting similar zones,* Vestnik Leningrad. Univ. Ser. Mat. Meh. Astronom. 19 (1964), 25–36. (Russian)

KAGAN, A. M. and ŠALAEVSKIĬ, O. V.

24. *The Behrens-Fisher problem for the existence of similar regions in an algebra of sufficient statistics,* Dokl. Akad. Nauk SSSR 155 (1964), 1250–1252 = Soviet Math. Dokl. 5 (1964), 556–558. MR 28 #4627a.

CARTAN, H.

25. *Idéaux des fonctions analytiques de n variables complexes,* Ann. Sci. École Norm. Sup. (3) **61** (1944), 149–197. MR 7, 290.

26. *Sur les matrices holomorphes de n variables complexes,* J. Math. Pures Appl. 19 (1940), 1–26. MR 1, 312.

27. *Idéaux et modules de fonctions analytiques de variables complexes,* Bull. Soc. Math. France 78 (1950), 29–64. MR 12, 172.

28. *Variétés analytiques complexes et cohomologie,* Colloque sur les Fonctions de Plusieurs Variables, Bruxelles 1953, Georges Thone, Liège and Masson & Cie, Paris, 1953, pp. 41–55. MR 16, 235.

29. *Variétés analytiques réelles et variétés analytiques complexes,* Bull. Soc. Math. France 85 (1957), 77–99. MR 20 #1339.

KENDALL, M. G.

30. *The evergreen correlation coefficient,* Contributions to Probability and Statistics, Stanford Univ. Press, Stanford, Calif., 1960, pp. 274–277. MR 22 #11457.

KOLMOGOROV, A. N.

31. Izv. Akad. Nauk SSSR Ser. Mat. 6 (1942), 3–32. MR 4, 221.

32. *Unbiased estimates,* Izv. Akad. Nauk SSSR Ser. Mat. 14 (1950), 303–326; English transl., Amer. Math. Soc. Transl. (1) 11 (1962), 144–170. MR 12, 116.

CRAMÉR, H.

33. *Mathematical methods of statistics,* Princeton Math. Series, vol. 9, Princeton Univ. Press, Princeton, N. J., 1946; Russian transl., IL, Moscow, 1948. MR 8, 39.

KOOPMAN, B. O.

34. *On distributions admitting a sufficient statistic,* Trans. Amer. Math. Soc. 39 (1936), 399–409.

KULLBACK, S.

35. *Information theory and statistics*, Wiley, New York and Chapman & Hall, London, 1959. MR **21** #2325.

LEHMANN, E.

36. *Testing statistical hypotheses*, Wiley, New York and Chapman & Hall, London, 1959; Russian transl., IL, Moscow, 1963 and "Nauka", Moscow, 1964. MR **21** #6654.

LEHMANN, E. and SCHEFFÉ, H.

37. *Completeness, similar regions and unbiased estimation*. I, Sankhyā **10** (1950), 305–340. MR **12**, 511.

LINNIK, Ju. V.

38. *On polynomial statistics in connection with the analytical theory of differential equations*, Vestnik Leningrad. Univ. **11** (1956), no. 1, 35–48; English transl., Selected Transl. Math. Stat. and Prob., vol. 1, Amer. Math. Soc., Providence, R.I., 1961, pp. 171–206. MR **17**, 983.

39. *Polynomial statistics and polynomial ideals*, Calcutta Math. Soc. Golden Jubilee Commemoration Volume (1958–1959), Part I, Calcutta Math. Soc., Calcutta, 1963, pp. 95–98. MR **29** #2830.

40. *On the theory of statistically similar regions*, Dokl. Akad. Nauk SSSR **146** (1962), 300–302 = Soviet Math. Dokl. **3** (1962), 1297–1299. MR **25** #3584.

41. *Sur certaines questions de la statistique analytique*, Ann. Fac. Sci. Univ. Clermont Math. No. 8 (1962), 53–61.

42. *Complex variables in problems with nuisance parameters and finite rank sufficient statistics*, Dokl. Akad. Nauk SSSR **149** (1963), 1026–1028 = Soviet Math. Dokl. **4** (1963), 512–513. MR **27** #3041.

43. *Remarks on the Fisher-Welch-Wald test*, Dokl. Akad. Nauk SSSR **154** (1964), 514–516 = Soviet Math. Dokl. **5** (1964), 118–120.

44. *Randomized homogeneous tests for the Behrens-Fisher problem*, Izv. Akad. Nauk SSSR Ser. Math. **28** (1964), 249–260; English transl., Selected Transl. Math. Stat. and Prob., vol. 6, Amer. Math. Soc., Providence, R.I., 1966, pp. 207–217. MR **28** #5521.

45. *On the construction of optimal similar solutions of the Behrens-Fisher problem*, Trudy Mat. Inst. Steklov. **79** (1965), 40–53 = Proc. Steklov Inst. Math. no. 79 (1965), 41–56.

46. *On A. Wald's test for the comparing of two normal samples,* Teor. Verojatnost. i Primenen 9 (1964), 16–30. (Russian) MR 28 #5520.

47. *Characterization of tests of the Bartlett-Scheffé type,* Trudy Mat. Inst. Steklov. 79 (1965), 32–39 = Proc. Steklov Inst. Math. no. 79 (1965), 32–40.

48. *An application of a theorem of H. Cartan in mathematical statistics,* Dokl. Akad. Nauk SSSR **160** (1965), 1248–1249 = Soviet Math. Dokl. 6 (1965), 291–293. MR **31** #6298.

LINNIK, Ju. V., ROMANOVSKAJA, J. L. and ŠALAEVSKIĬ, O. V.

49. *Remarks on the theory of the Fisher-Welch-Wald test,* Teor. Verojatnost. i Primenen. 10 (1965), 727–730. (Russian) MR 32 #8446.

LINNIK, Ju. V., ROMANOVSKIĬ, J. V. and SUDAKOV, V. N.

50. *A non-randomized homogeneous test in the Behrens-Fisher problem,* Dokl. Akad. Nauk SSSR 155 (1964), 1262–1264 = Soviet Math. Dokl. 5 (1964), 570–572. MR 28 #4627b.

LINNIK, Ju. V. and ŠALAEVSKIĬ, O. V.

51. *On the analytic theory of tests for the Behrens-Fisher problem,* Dokl. Akad. Nauk SSSR 150 (1963), 26–27 = Soviet Math. Dokl. 4 (1963), 580–582. MR 27 #3042.

ŁOJASIEWICZ, S.

52. *Sur le problème de division,* Studia Mathematica **18** (1959), 87–136. MR 21 #5893.

LUKACZ, E.

53. *Characterization of populations by properties of suitable statistics,* Proc. Third Berkeley Sympos. Math. Stat. and Prob. 1954–1955, vol. 2, University of California Press, Berkeley, 1956, pp. 195–214. MR 18, 942.

LJAPUNOV, A. A.

54. *On completely additive vector-valued functions,* Izv. Akad. Nauk SSSR Ser. Mat. 4 (1940), 465–468. (Russian) MR 2, 315.

MALGRANGE, B.

55. *Lectures on the theory of functions of several complex variables,* Tata Institute, Bombay, 1962.

NEYMAN, J.

56. *Sur la vérification des hypothèses statistiques composées,* Bull. Soc. Math. France 63 (1935), 346–366.

57. *Un théorème d'existence*, C. R. Acad. Sci. Paris 222 (1946), 843–845. MR 7, 457.

58. *Current problems of mathematical statistics*, Proc. Internat. Congress Math. 1954, vol. 1, Noordhoff, Groningen and North-Holland, Amsterdam, 1957, pp. 349–370. MR 20 #1374.

OKA, K.

59. *Sur les fonctions analytiques de plusieurs variables.* VII: Sur quelques notions arithmétiques, Bull. Soc. Math. France 78 (1950), 1–27. MR 12, 18.

PETROV, A. A.

60. *Verification of statistical hypotheses on the type of distribution based on small samples*, Teor. Verojatnost. i Primenen. 1 (1956), 248–271. (Russian) MR 19, 76.

RAO, C. R.

61. *Information and the accuracy attainable in the estimation of statistical parameters*, Bull. Calcutta Math. Soc. 37 (1945), 81–91. MR 7, 464.

62. *Some theorems on minimum variance estimates*, Sankhyā 12 (1952), 27–42. MR 14, 1103.

ROMANOVSKAJA, J. L.

63. *On the Fisher-Welch-Wald test*, Sibirsk. Mat. Ž. 5 (1964), 1343–1359. (Russian) MR 30 #661.

ROMANOVSKIĬ, I. V. and SUDAKOV, V. N.

64. *On the existence of independent partitions*, Trudy Mat. Inst. Steklov. 79 (1965), 5–10 = Proc. Steklov Inst. Math. no. 79 (1965), 1–7.

RÜCKERT, W.

65. *Zur Eliminationsproblem der Potenzreihenideale*, Math. Ann. 107 (1932), 259–281.

SAKS, S.

66. *Théorie de l'intégrale*, Warsaw, 1933; English transl., rev. ed., Dover, New York, 1964; Russian transl., IL, Moscow, 1949.

STEIN, Ch.

67. *A two-sample test for a linear hypothesis whose power is independent of the variance*, Ann. Math. Statistics 16 (1945), 243–258. MR 7, 213.

SVERDRUP, E.

68. *Similarity, unbiasedness, minimaxibility and admissibility of statistical test procedures*, Skand. Aktuarietidskr. 36 (1953), 64–86. MR 15, 453.

HALMOS, P. R.

69. *Measure theory*, Van Nostrand, Princeton, N. J., 1950; Russian transl., IL, Moscow, 1953. MR 11, 504; MR 16, 22.

HALMOS, P. R. and SAVAGE, L. J.

70. *Application of the Radon-Nikodym theorem to the theory of sufficient statistics*, Ann. Math. Statistics 20 (1949), 225–241. MR 11, 42.

HARDY, G. H., LITTLEWOOD, J. E. and PÓLYA, G.

71. *Inequalities*, 2nd ed., Cambridge Univ. Press, New York, 1952; Russian transl., IL, Moscow, 1948. MR 13, 727; MR 18, 722.

HARDY, B.

72. *Some properties of an angular transformation of the correlation coefficient*, Biometrika 43 (1956), 219–224. MR 17, 981.

HOGG, R. and CRAIG, A.

73. *Sufficient statistics in elementary distribution theory*, Sankhyā 17 (1956), 209–216. MR 19, 188.

FEĬGEL'SON, T. S.

74. *On a simple method of establishing independence of statistics*, Vestnik Leningrad Univ. Ser. Mat. Astronom. 19 (1964), no. 3, 157–158. (Russian) MR 29 #4131.

FUKS, B. A.

75. *Introduction to the theory of analytic functions of several complex variables*, Fizmatgiz, Moscow, 1962; English transl., Transl. Math. Monographs, vol. 8, Amer. Math. Soc., Providence, R. I. 1963; reprint 1965. MR 27 #4945; MR 29 #6049.

76. *Special chapters in the theory of analytic functions of several complex variables*, Fizmatgiz, Moscow, 1963; English transl., Transl. Math. Monographs, vol. 14, Amer. Math. Soc., Providence, R. I., 1965. MR 30 #4979.

ŠALAEVSKIĬ, O. V.

77. *On the non-existence of regularly varying tests for the Behrens-Fisher problem*, Dokl. Akad. Nauk SSSR 151 (1963), 509–510 = Soviet Math. Dokl. 4 (1963), 1043–1045. MR 27 #2037.

CHATTERJEE, S. K.

78. *On an extension of Stein's two sample procedure to the multi-normal problem*, Calcutta Statist. Assoc. Bull. 8 (1959), 121–148. MR 21 #4501.

SCHEFFÉ, H.

79. *On solutions of the Behrens-Fisher problem based on the t-distribution*, Ann. Math. Statistics 14 (1943), 35–44. MR 4, 221.

SHOHAT, J. A. and TAMARKIN, J. D.

80. *The problem of moments*, Math. Surveys, vol. 1, Amer. Math. Soc., Providence, R. I., 1943; rev. ed., 1947. MR 5, 5.